Chemistry of Antibiotics and Related Drugs

Mrinal K. Bhattacharjee

Chemistry of Antibiotics and Related Drugs

Second Edition

 Springer

Mrinal K. Bhattacharjee
Department of Chemistry and Biochemistry
Long Island University
Brooklyn, NY, USA

ISBN 978-3-031-07584-1   ISBN 978-3-031-07582-7   (eBook)
https://doi.org/10.1007/978-3-031-07582-7

1st edition: © Springer International Publishing Switzerland 2016

This Springer imprint is published by the registered company Springer Nature Switzerland AG
The registered company address is: Gewerbestrasse 11, 6330 Cham, Switzerland

This book is dedicated to my wife Nayana and daughter Neha

Preface

Millions of lives have been saved by antibiotics since they were first discovered more than 90 years ago. This book discusses the various classes of antibiotics and their mechanisms of action. For the first few decades, antibiotics had a remarkable success, encouraged by which in 1969, the then-US Surgeon General, William Stewart testified before Congress that it was "time to close the book on infectious disease." Today we know how premature that statement was. Today, more than five decades later, infectious disease still remains the leading cause of death worldwide and ranks among the top ten causes of death in the USA. The main reason for all these is that bacteria are increasingly becoming resistant to antibiotics. Mechanisms by which bacteria develop resistance to antibiotics are also discussed here. Because of the increasing antimicrobial resistance, it is of utmost importance to discover new antibiotics. Alternative strategies for discovering new antibiotics are presented. The World Health Organization has recognized the threat of growing antimicrobial resistance and has adopted a Global Action Plan to combat the threat. This plan and similar plans of member countries as well as Stewardship Programs adopted by agencies at local levels are also discussed.

This book has been written keeping the student in mind. The main focus is to explain how antibiotics work in curing infectious diseases and how resistance develops to the antibiotics such that they do not work anymore. A background in Biochemistry is needed to understand the mechanism of action of the antibiotics. However, any background information that is needed is discussed in the book. So it will not be necessary for the student to consult any separate biochemistry textbook, or register for a separate course in biochemistry, in order to understand the theory of antibiotics. However, the discussion of biochemistry in this book is not meant to be complete. The only information presented is what is relevant to the understanding of antibiotics.

There are many other important aspects of antibiotics that are studied by doctors, pharmacists, and scientists but are beyond the scope of this book. Some of these aspects are briefly mentioned here but will not be discussed in much detail in the book.

1. Dosage, formulation, bioavailability, and biostability are important aspects for effective use of antibiotics. These ensure that the antibiotic will be delivered specifically to diseased site in the right amount and for the required duration. Dose depends on how much of the antibiotic is absorbed from the digestive system, how stable it is, how much enters the cells, the distribution of the drug in various tissues, and many other factors. However, discussions of these aspects are beyond the scope of this book.
2. All antibiotics do not work for all infections. Also, antibiotics have various side effects and may interact with other drugs to give unwanted effects. Information needed to decide about which antibiotic to prescribe for which infection is beyond the scope of this book.

All antibiotics will be referred to here by their common names such as penicillin, tetracycline, erythromycin, etc., even though many of these antibiotics have other brand names given by the manufacturers. There are also various derivatives with altered but similar activities for many of these antibiotics. Not all of these names are included in this book. All antibiotics will also have a systematic name based on IUPAC nomenclature of organic compounds. Those names will be mentioned for the simple molecules but not for those with complex structures.

Brooklyn, NY, USA   Mrinal K. Bhattacharjee
April 2022

Contents

Chapter 1
Introduction to Antibiotics

Abstract This chapter is an introduction to antibiotics. Topics discussed are definition of antibiotics, characteristics of an ideal antibiotic, history of antibiotics, discovery of the first antibiotics, penicillin, gramicidin, streptomycin, chloramphenicol, tetracycline, etc. Background biochemistry information needed for understanding mechanisms of action of antibiotics is briefly presented including discussions on enzymes, inhibitors, and thermodynamics of metabolic pathways.

1.1 Definition of Antibiotics

The term "antibiotic" was coined by Selman Waksman, who worked at Rutgers University and together with Albert Schatz and Elizabeth Bugie discovered the antibiotic streptomycin. According to Waksman's definition, an antibiotic is a chemical substance that is produced by microorganisms and that have the capacity, in dilute solution, to selectively inhibit the growth of and even to destroy other microorganisms. Soon scientists were able to develop synthetic compounds that had antibiotic properties. Although some scientists including Waksman strongly resisted the "misuse" of the term antibiotic, the definition was nevertheless changed to include synthetic antibiotics. In fact, synthetic antibiotics in many ways are better than the natural ones (Sect. 2.8). So antibiotics can be defined as molecules that either kill or inhibit the growth of microorganisms. Since the practical application of antibiotics is to cure infections in humans, a necessary property of an antibiotic should be selectivity. Thus, an antibiotic is now widely defined as a chemical substance that selectively inhibits the growth of microorganisms and causes minimum damage to the host cells. This definition is still very restrictive because of the term "microorganism." Most scientists would not consider viruses as microorganisms but there are antiviral drugs available and they function by mechanisms similar to that of antibiotics. Whenever, a layman talks about using an antibiotic to cure a viral infection, the learned are quick to point out that antibiotics cure only bacterial and not viral infections. However, it is to be noted that this is a matter of definition. If antibiotics are defined to include antiviral drugs, then antibiotics can be used to cure viral infections also. The terms are so confusing that in US Food and Drug

Administration (FDA) reports the terms "antimicrobial," "antibacterial," and "antibiotic" are used interchangeably while at the same time the following clarification is also made in the footnotes: "The term 'antimicrobial' refers broadly to drugs with activity against a variety of microorganisms including bacteria, viruses, fungi, and parasites [1]." Some scientists strictly adhere to the current definition of antibiotics and strongly object when anyone refers to antifungal and antiviral drugs as antibiotics. Antiprotozoal and anticancer drugs are also not called antibiotics but follow similar mechanisms as antibiotics. Since all these drugs follow similar mechanisms, most books and chapters written on antibiotics usually include a discussion on antibacterial, antifungal, antiviral as well as anticancer drugs without actually calling them antibiotics. It is therefore this author's opinion that the time has come to change the definition of antibiotics one more time to make it more inclusive rather than restrictive. In order to justify including all the chapters in this book under one title, an antibiotic is defined here as a chemical that selectively inhibits an infectious biological process but causes minimum damage to the host. Thus, the broad term "antibiotics" can include subgroups such as antibacterial antibiotics, anticancer antibiotics, antifungal antibiotics, antimalarial antibiotics, and antiviral antibiotics. The fact that there is synergy observed between these various classes of anti-infectives is another reason to not refrain from listing them all under one classification and refer to all of them as "antibiotics" (Sect. 8.1.5). There can be disagreement about whether cancer can be described as an infection. Although cancer is not an infection by a foreign species, there are many similarities between cancer and microbial infections [2]. Related to this, an interesting case of cancer has been reported [3]. The authors presented the first report of cancer caused in a human not by human cells but by cells from tapeworm. This is a novel finding of a human disease caused by parasite-derived cancer cells.

Some other commonly used terminologies are mentioned here. An alternative terminology for antibiotics that is commonly used is *anti-infectives*, which also includes the antivirals in the definition. *Antimicrobials* are defined as chemicals that kill or inhibit growth of microorganisms. Antimicrobials include three subgroups: *antibiotics*, which are used to kill or inhibit microorganisms within the body; *antiseptics*, which are applied on living tissue to prevent infection; and *disinfectants*, which are used to kill or inhibit microorganisms on non-living objects. The term *sterilization* means killing microorganisms in liquid media or on solid objects by using chemical means such as oxidizing agents or physical means such as heat or high-energy radiation. The term *sanitizing* agent is used interchangeably with disinfectant, antiseptic, and sterilizing agent. The term *antibacterial* when used in scientific literature usually means an antibiotic that kills or inhibits bacteria. However, today it is often used synonymously with disinfectant as in "antibacterial soap." The term *chemotherapy* (as coined by Paul Ehrlich in the early 1900s) meant the use of chemicals to cure diseases. The focus was mainly on the use of antibiotics to cure microbial infections. However, today the term chemotherapy usually means cancer chemotherapy in which chemicals are used to kill cancer cells (Sect. 4.3.7).

Bacterial infections can be classified based on the source of infection. Infections acquired from other infected people in the community or from the environment including air, water, or solid objects are classified as *community-acquired* infections. Infections that are acquired in the hospital after the patient is admitted for some other unrelated diseases are called *nosocomial* infections. The latter type of infections is usually caused by bacteria that are resistant to antibiotics and thus is of special concern. Of great concern in recent years are infections by methicillin-resistant *Staphylococcus aureus* (MRSA) (Sect. 3.3.2.10) which are often nosocomial.

1.2   History of Antibiotics

Ever since it was proven by Robert Koch and Louis Pasteur in the late nineteenth century that diseases can be caused by germs, scientists have been searching for ways to kill these disease-causing germs. One successful approach developed by Pasteur was to use harmless bacteria to cure diseases caused by harmful bacteria. Today we can explain those observations. Those harmless bacteria probably produced antibiotics that killed the infecting disease-causing bacteria. Another approach was to use chemicals to kill the bacteria, giving rise to the process called chemotherapy. This was started by Paul Ehrlich who understood that the first step of chemotherapy must be binding of the chemical to the bacteria. This led to the testing of dyes as antibacterial agents because they were known to stain bacteria and thus, obviously bind to them. After testing numerous dyes he found one in 1904 that could cure mice infected with trypanosomes. He named the dye Trypan Red, which was the first chemotherapeutic agent discovered (Fig. 1.1). Trypan red was later shown to

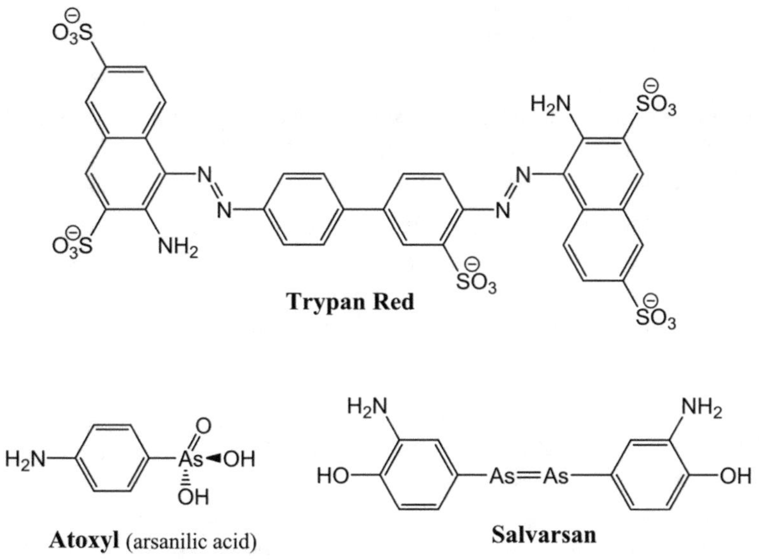

Fig. 1.1   Some early chemotherapeutic (antibiotic) agents

have antiviral activity [4]. In 1932, another antibiotic discovery was made by Gerhard Domagk, while testing other dyes (Sect. 2.8).

Arsenical compounds constituted another class of drugs used as chemotherapeutic agents. The first arsenical drug was arsanilic acid, discovered in 1863 by a French chemist, Antoine Béchamp who named it as Atoxyl [5]. It was widely used as a cure for trypanosomiasis in the early twentieth century. However, it was observed that the protective effect of Atoxyl was only temporary and in the end the parasite reappeared even if the dose was increased. Also the drug was highly toxic and eventually resulted in death of the patient [6]. In 1910, Ehrlich discovered the arsenical drug Salvarsan which was effective against trypanosomes and also against the bacteria that cause syphilis and was less toxic than Atoxyl. However, it was still significantly toxic but remained the drug of choice for the next several decades till it was replaced by penicillin in the 1940s.

Penicillin was the first scientifically studied antibiotic. However, it was not the first recorded use of antibiotic. The Greeks were known to use extracts of male fern to treat worm infestations. Extract of cinchona bark was used in Peru, Bolivia, and Ecuador to treat malaria as far back as the sixteenth century. The active component of the extract was later shown to be quinine which was the only available antimalarial drug until the 1940s when chloroquine became a more popular drug of choice. However, quinine is still recommended for the treatment of malaria [7]. In 1888, E. de Freudenreich discovered that pyocyanase, a blue pigment secreted by *P. aeruginosa* had antibiotic activity but was highly toxic for the host [8]. Ipecacuanha root was used in Brazil to cure diarrhea and dysentery. Emetine was isolated as its active component in 1817 and was shown to cure amoebic dysentery [9, 10].

1.3 The Ideal Antibiotic

There are many compounds that are able to kill microorganisms; however, they are not all called antibiotics. To be an ideal antibiotic it should have the following other properties besides being able to kill microorganisms.

1. *Selectivity*: The antibiotic must kill or inhibit the infecting microorganism but cause minimum harm to the host cells. This is discussed further in Sect. 1.4.
2. *Water solubility*: The antibiotic must be soluble in water to a sufficient extent so that it can be transported through body fluids to the infected sites. Some antibiotics are poorly soluble in water; however, some solubility is essential for effectiveness.
3. *Few side reactions*: Side reactions of the antibiotic should be minimum. These include possible allergic reactions and negative interactions with food or other drugs that the patient may be taking.
4. *Stability*: This includes shelf stability and bio-stability. The antibiotic should have a long shelf life to be economically useful. It should preferably be stable at room temperature; however, there are many antibiotics that need to be stored in a

refrigerator. Once taken by the patient, the antibiotic should remain unaltered in the body fluids for sufficient time to be able to carry out its function. Foreign molecules in the body will eventually be either degraded (usually in the liver) or excreted with urine. For the ideal antibiotic both these processes should be slow. However, rapid excretion is actually a desired property for treating urinary tract infections since a high concentration of the drug in urine can be achieved this way.

5. *Low cost*: The cost of manufacturing of the antibiotic should be low enough for patients to be able to afford it.
6. *Slow resistance development*: Microorganisms have developed resistance to most antibiotics. An ideal antibiotic will be the one to which resistance develops at a slow rate. This will depend not only on the characteristics of the antibiotic but also on its frequency of use. This is discussed further in Chap. 2.

Selectivity of antibiotics. One essential characteristic of an antibiotic is that it should selectively inhibit the infecting pathogen but not affect the host cells. Such selectivity can be achieved in two possible ways: (1) The target of the antibiotic may be present only in the infecting bacteria but not in the host and so the host will not be affected. (2) The target of the antibiotic may be present in both the infecting bacteria and the host; however, structurally or mechanistically they may be different enough such that the antibiotic inhibits only the bacterial enzyme or process but the host is not affected. One example of a target that is present only in the bacteria and not in the host is the cell wall. Human cells do not have cell walls. There are several antibiotics that inhibit the synthesis of bacterial cell wall (Chap. 3). Another target that is present in bacteria but not in humans is the folic acid biosynthetic pathway (Chap. 4). Folic acid is required in both the infecting pathogen and the host for the synthesis of the deoxynucleotide, dTMP, which is essential for the synthesis of DNA. However, humans obtain folic acid as a vitamin in the diet and so do not have the pathway for biosynthesis of folic acid. Bacteria synthesize their required folic acid and in fact, they are unable to utilize any premade folic acid provided from an external source. The folic acid biosynthetic pathway is a favorite target of pharmaceutical companies for development of antibiotics.

There are several examples of the second method in which selectivity is achieved because the targets of the antibiotics are significantly different in the bacteria and the host. Biosynthesis of DNA, RNA, and proteins are essential functions in all species. However, there are enough structural differences between the bacterial and human enzymes of these biosynthetic pathways such that there are several antibiotics known that selectively target these pathways in bacteria but not in the host (Chaps. 5 and 6).

1.4   Sources of Antibiotics

While some antibiotics are chemically synthesized, a majority of the antibiotics that are used today are produced by microorganisms. Why and when do microbes make antibiotics? Natural antibiotics made by microbes are products of their secondary metabolic pathways meaning those that are not absolutely required for their survival. The pathways for biosynthesis of antibiotics are turned off during exponential growth phase (log phase) because of the abundance of nutrients. However, in the stationary phase of growth, they face competition from other microorganisms for the limiting amount of nutrients and so they turn on the pathways for biosynthesis of antibiotics in order to win the competition by killing the neighboring bacteria. Also, the surviving bacteria utilize the nutrients that are released when the dead bacterial cells lyze. This information is important in the industrial production of antibiotics from natural sources. Note that this theory has been challenged by some scientists (Sect. 2.9).

How antibiotic producers protect themselves. Microorganisms which make anti-biotics need to protect themselves from those antibiotics. A variety of strategies are employed to do that: (1) Some antibiotics are exported into the environment imme-diately after they are synthesized so that the intracellular concentrations of the antibiotics are kept low. (2) Some antibiotic-producing microorganisms also make a resistance protein that inactivates their own antibiotic. Antibiotics in the active form are released outside to kill other bacteria; however, any antibiotic that comes back into the cell is inactivated by the resistance protein. For example, *Actinomy-cetes* that make streptomycin protect themselves by making antibiotic inactivating enzyme [11, 12]. (3) Some antibiotics such as aminoglycosides and macrolides are made and exported to the outside in an inactive form and are then converted to the active form outside the cell. (4) Some antibiotic producers alter the target of the antibiotic within themselves. For example, microorganisms that make cell wall affecting antibiotics make their own cell wall using a different enzyme that is not affected by the antibiotic. A new mechanism of self-protection has been discovered recently. *Streptomyces platensis* makes two antibiotics which inhibit enzymes in the fatty acid synthesis pathway of other bacteria, but protects itself from the two antibiotics by employing a different enzyme in its pathway for fatty acid synthesis that is not affected by the antibiotics [13].

Soil, the best place to search for antibiotics. It is known that the soil is a very complex ecosystem in which the inhabitants have developed chemical defenses against each other as a response to competition for nutrients. So it was obvious for many scientists to search for these chemicals in the soil. Also, even before the discovery of antibiotics, scientists had wondered about another aspect of the soil. In the history of mankind or animal kind, many people and animals have died because of bacterial infection. By the time someone dies because of infection, the bacteria have multiplied within the body to reach billions in number. When these dead bodies are buried or just left to decay, all those bacteria are released into the soil. So by now the soil should be full of pathogenic bacteria and so the soil should be the most dangerous thing to even touch. But that is not so. The soil does not contain any pathogenic bacteria. So where did all the pathogenic bacteria go? Now we know that there are numerous microorganisms in the soil that produce antibiotics that probably kill the pathogenic bacteria coming from the dead bodies. So if one wants to search for antibiotics, the best place to search is the soil. That is what scientists did. Of all the antibiotics known today, the majority of them have been isolated from microorganisms present in the soil. Some examples are Terramycin, Vancomycin, and Streptomycin.

Most of the antibiotics from soil were discovered within a few decades, but this was followed by half a century of unsuccessful attempts at discovering new antibiotics. One reason for this lack of success is that most of the microorganisms in soil cannot be cultured under laboratory conditions. In fact 99% of bacterial species on this planet have not yet been cultured. Kim Lewis and coworkers have developed a novel method to culture many of these organisms in soil and in the process, have discovered several new potential antibiotics [14]. One such antibiotic is teixobactin (Sect. 3.3.4).

1.5   Discovery of Modern Antibiotics

Alexander Fleming, working at St. Mary's Hospital in London is widely credited for the discovery of the first antibiotic, penicillin in 1928 for which he was awarded the Nobel Prize. However, penicillin was not really the first antibiotic to be discovered.

Discovery of lysozyme. In 1920, Fleming had discovered the antibiotic, lysozyme, a naturally occurring substance present in human tears. Lysozyme kills bacteria by lyzing (breaking) the cells walls of bacteria. This causes the bacterial cell to burst open. Fleming described the results of his experiment as, "A thick milky suspension of bacteria could be completely cleared in a few seconds by a fraction of a drop of human tears or egg white [15]." However, lysozyme is not popularly known as an antibiotic because being a protein, it could not be used for treating patients.

Discovery of penicillin. In 1928, Fleming made his second antibacterial discovery and that was penicillin. Most people have heard about this story about the discovery of penicillin. Fleming noticed that in one of the old plates (Petri dishes) left in the laboratory, colonies of the bacteria *Staphylococcus* (that causes skin diseases) had lyzed, probably because of a contaminating greenish mold growing in an adjacent area of the plate. This led to the discovery of penicillin. By extracting the substance from cultures of the mold, he was able to demonstrate its antibacterial activity not only on plates but also when given to infected mice. Fleming named this substance that killed bacteria, "penicillin" after the *Penicillium* mold from which it was obtained.

It is to be noted that this "chance" discovery in no way diminishes the credit that goes to Fleming. In his 1945 Nobel Lecture Fleming was humble in making the statement, "My only merit is that I did not neglect the observation and that I pursued the subject as a bacteriologist." All microbiologists often find contaminating molds

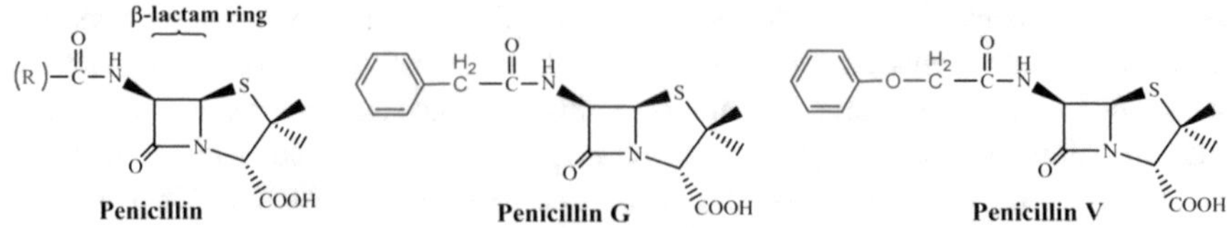

Fig. 1.2 Structures of Penicillin G and Penicillin V

in their plates even in modern day laboratories. Before Fleming, many other scientists probably had also seen this effect but ignored it and treated the contaminant problem as a nuisance. However, the discovery of penicillin is credited to Fleming because he recognized the importance of the phenomenon and pursued it further. It is not enough to just make an observation, but to follow it up and do further research on it. There are many other scientific discoveries that are attributed to similar "accidents"; however, it is not just a coincidence that such "accidents" happen only with scientists who make great discoveries from those events. Similarly, in the history of mankind, many apples had probably fallen on many people's heads before one fell on Newton's. However, all others probably just ate the apple and may have cursed or thanked the apple tree, but it took a Newton to discover the theory of gravity from this observation.

Clinical use of penicillin. Although Fleming was the first to demonstrate the antibiotic properties of penicillin, he could not purify it in sufficient quantity to be used clinically on animals or patients. This was later done by two scientists, Howard Florey and Ernst Chain, at Oxford University. In 1940, they developed the method for purification to obtain sufficient amounts of penicillin and then tested it on mice to obtain miraculous results. It was first tested on a human in 1941. In the early days the amount of penicillin available was so small that it was used exclusively for the military. Large-scale use for the general public took place in 1942 when there was a devastating fire in a nightclub in Boston that killed 492 people, making it the second deadliest fire in American history. Survivors of the fire were successfully treated with penicillin that was being purified at Merck Company, New Jersey. By 1944, penicillin could be made in sufficient quantity to make it available to the public and came to be known as the "miracle drug." Fleming, Florey, and Chain were awarded the 1945 Nobel Prize in medicine for their work on penicillin.

In 1945, the chemical structure of penicillin (Fig. 1.2) was determined at Oxford University by Dorothy Crowfoot Hodgkin. The key feature in the structure is the β-lactam ring. Note that cyclic esters and amides are called lactones and lactams, respectively, and a 4-, 5-, or 6-membered lactam ring is called a β-, γ-, or δ lactam, respectively. The first penicillin discovered was penicillin G in which the R group is a benzyl group (φ-CH$_2$). One problem with penicillin G is that it is unstable in stomach acid and so cannot be taken orally and so is better administered intravenously, which was not a very convenient process in those days. The discovery of penicillin V in which the R group is φ-O-CH$_2$, greatly improved medical treatments since it is more stable in acid and so could be given orally.

Discovery of gramicidin. In 1939, Rene Dubos, a former student of Selman Waksman, working at Rockefeller University isolated the first antibiotic-producing microorganism, *Bacillus brevis* [16, 17]. His discovery also led to the first clinically tested antibiotic, which he named Gramicidin because it killed only gram-positive bacteria. However, the drug was found to be very toxic when given intravenously. So gramicidin is used only externally for minor skin infection. For mechanism of action of gramicidin, see Sect. 7.2.2.2.

Discovery of Streptomycin. In 1943, Selman Waksman, Albert Schatz, and Elizabeth Bugie at Rutgers University screened many microorganisms from the soil and discovered an antibiotic from the bacteria *Streptomyces griseus*. They named it "Streptomycin," which was the second antibiotic obtained from soil bacteria (the first one was gramicidin). Streptomycin was effective against several diseases but its most important use was in treating tuberculosis that is caused by *Mycobacterium tuberculosis*. This was the first known antibiotic that could cure tuberculosis. Streptomycin belongs to the aminoglycoside group of antibiotics which contain amino sugars linked by glycosidic bonds (Sect. 6.2.2). One problem with streptomycin was that bacteria developed resistance to the antibiotic at a much faster rate than in case of penicillin. The subject of antibiotic resistance development is discussed in Chap. 2. The next aminoglycoside was discovered in 1949 in the laboratory of Waksman. It was named neomycin, to which bacteria did not become resistant as fast as to streptomycin. However, neomycin was found to be very toxic for the host and so is used only in topical antibiotic ointments such as Neosporin which is popularly used for skin infections associated with wounds and burns. Later several other aminoglycosides were also developed including kanamycin, amikacin, gentamicin, and tobramycin. Mechanism of action of aminoglycosides is discussed in Sects. 6.2.2 and 6.2.3.

Discovery of Broad-Spectrum Antibiotics. Penicillin functions by inhibiting the synthesis of bacterial cell wall. The first penicillin that was discovered was penicillin G (benzylpenicillin) (Fig. 1.2), which is active against only gram-positive but not gram-negative bacteria. One factor that determines the spectrum of activity of penicillins is their ability to cross the bacterial cell wall. For gram-negative bacteria they also have to cross the outer membrane, as a result, most gram-negative bacteria are intrinsically resistant to penicillin G. Based on their range of target specificities, antibiotics can be classified into two types: narrow-spectrum and broad-spectrum antibiotics. Narrow-spectrum antibiotics are effective against only some selective group of bacteria, whereas broad-spectrum antibiotics are active against a much broader group of bacteria and thus can be used to treat a larger variety of infections. These two terms were originally coined in the 1950s and were used in a comparative sense; however, the terms have never been clearly defined [18] resulting in inconsistent use of the words. Narrow-spectrum antibiotics are prescribed when the infecting microorganism has been identified. One advantage of using a narrow-spectrum antibiotic is that it will not kill as many resident nonpathogenic bacteria in the body as will the broad-spectrum antibiotics. So there is less chance of causing superinfection, which is defined as a second infection superimposed on an earlier one and happens since the normal flora of the body is suppressed by the antibiotic

[19, 20]. Also, since only a few species of bacteria are affected by narrow-spectrum antibiotics, there is lesser probability of selecting for resistant bacteria. The obvious disadvantage of using narrow-spectrum antibiotics is that they may not be able to kill all species of bacteria and so should not be used if the infecting species have not been identified. So the discovery of broad-spectrum antibiotics was a significant development in the history of antibiotics.

The first broad-spectrum antibiotic to be discovered was *chloramphenicol* (chloromycetin). Paul Burkholder at Yale University isolated chloramphenicol in 1947 from *Streptomyces Venezuela*, a soil bacterium. The antibiotic is active against a broad range of bacteria including gram-positive, gram-negative as well as anaerobic ones. It was also the first known antibiotic to be effective against typhoid. However, because of its numerous toxic side effects including bone marrow suppression, anemia, leukemia, etc., the antibiotic's use has been discontinued in most of the western world except in some special instances. In the developing world it continues to be used widely mainly because it is inexpensive and has a broad spectrum of activity.

The second broad-spectrum antibiotic, *chlorotetracycline* was discovered after a very large-scale scientific search of soil samples coordinated by Benjamin Duggar under the direction of Yellapragada Subbarao at Lederle Laboratories in New York [21, 22]. American soldiers fighting in various places in the world during World War II were asked to bring soil samples from those places for screening at Lederle Laboratories for possible antibiotic-producing fungi. This led to the discovery of a golden colored antibiotic secreted by *Streptomyces aureofaciens*. The antibiotic was named aureomycin, which was later renamed as chlorotetracycline and was the first member of the tetracycline group of antibiotics. As the name suggests, tetracycline molecules contain four hydrocarbon rings (Fig. 6.7). Other commonly used tetracycline antibiotics are tetracycline and doxycycline. Similar to chloramphenicol, tetracyclines are inexpensive and broad-spectrum antibiotics that effectively cure many infections including typhoid.

1.6   Classification of Antibiotics

Antibiotics can be classified in several different ways. For example, they can be classified based on microbial coverage: those that work against methicillin-resistant *Staphylococcus aureus* (MRSA), against vancomycin-resistant *Enterococcus* (VRE), against mycobacteria (that cause leprosy and tuberculosis), against *Pseudomonas aeruginosa*, etc. A drawback of this type of classification is that many antibiotics are effective against multiple microbial species.

Another way of classifying antibiotics is by their structures; however, that results in numerous classes of antibiotics such as aminoglycosides; β-lactams (including subclasses carbapenems, cephalosporins, monobactams, and penicillins); macrolides; oxazolidinones; polypeptides; quinolones; rifamycins; sulfonamides; streptogramins and tetracyclines. Besides these there are the following antibiotics

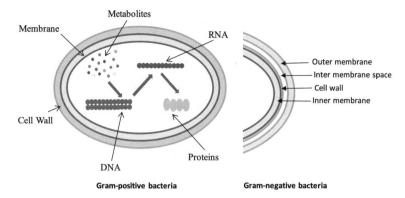

Fig. 1.3 Targets for antibiotics

that do not fit into any of the above classes and so each should be a separate class by itself: chloramphenicol; clindamycin; daptomycin; fosfomycin, lefamulin; metroni-dazole; mupirocin; nitrofurantoin; and tigecycline.

One useful way of classifying antibiotics is by their mechanism of action. The advantage of this method is that the chemistry of antibiotics can be better under-stood. One drawback of this method of classification is that some antibiotics may function by more than one related or unrelated mechanisms or the mechanism may be misunderstood or not yet known. In some cases, the knowledge of the mechanism may change with time and so classification of the antibiotic will also have to change. Some examples are nitrofurans and triclosan.

The classification of antibiotics that is followed in this book is based on the targets in the microbial cell that they interact with to cause growth inhibition. Accordingly, there are six major classes of antibiotics: (1) those that inhibit bacterial cell wall synthesis, (2) those that disrupt the cell membrane, (3) those that inhibit the synthesis of important metabolites, (4) those that inhibit DNA synthesis (replica-tion), (5) those that inhibit RNA synthesis (transcription), and (6) those that inhibit protein synthesis (translation). The six targets in the cell that are affected by the six classes of antibiotics are shown below in Fig. 1.3. Recent research has identified other new targets for development of new antibiotics. Some of these are discussed later in Chap. 7.

Bacteriostatic and bactericidal antibiotics. Another way to classify antibiotics is based on their effect on growth and survival of the bacteria. Those antibiotics that kill bacteria are called bactericidal, e.g., penicillin and those that stop the growth of bacteria but do not kill them are called bacteriostatic, e.g., chloramphenicol. Note that bacteriostatic antibiotics also can eventually kill bacterial cells if used at large concentration. Similarly, bactericidal antibiotics can also behave as bacteriostatic if the concentration used is too low. So these two terms are better defined as follows. An antibiotic is bacteriostatic if it inhibits growth of bacteria but does not kill bacterial cells at a safe and practically achievable concentration. An antibiotic is bactericidal if it kills bacterial cells at a safe and practically achievable concentration.

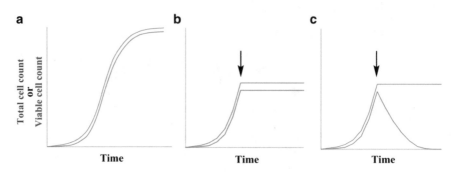

Fig. 1.4 Hypothetical growth curves for bacteria. (**a**) in the absence or (**b**) presence of bacteriostatic or (**c**) bactericidal antibiotics. The arrows indicate the time of addition of the antibiotic

If a bacteriostatic agent is removed, then the bacterial growth will resume. So does this mean that bacteriostatic antibiotics are not useful? That is not true, bacteriostatic antibiotics are often prescribed by doctors. This is because by stopping growth of the infecting bacteria it gives the body's immune system sufficient time to kill the bacteria. So they work along with the help of the immune system. In fact, even bactericidal antibiotics sometimes do not kill 100% of the bacteria. But by bringing down the total number of bacteria, they help the immune system tackle the problem.

To experimentally differentiate between bacteriostatic and bactericidal antibiotic first grow bacterial cells in broth to mid-log phase as measured by OD_{600}. Distribute the cells into three test tubes. Continue incubating one tube without any addition. To the second tube add a bacteriostatic antibiotic and to the third add a bactericidal antibiotic and continue incubating the tubes. Monitor growth of the bacteria in each tube by measuring OD_{600} at selected time intervals. Also monitor cell viability at time intervals by removing aliquots and spreading serial dilutions on agar plates. Number of viable cells can be determined after overnight incubation of the plates followed by counting of colonies. In the first tube, which functions as a control, both OD_{600} and viable cell count will continue to increase till they become constant in the stationary phase of growth. In both the second and third tubes the OD_{600} will remain constant after addition of antibiotic since the growth of the bacteria will be stopped. However, OD_{600} cannot differentiate between bacteriostatic and bactericidal antibiotics since both live cells and dead cells will scatter light to the same extent when measuring OD. The two types of antibiotics can be differentiated by their viable cell counts. If the number of viable cells remains constant after addition of the antibiotic, then the antibiotic is bacteriostatic but if the number decreases with time, then the antibiotic is considered to be bactericidal (Fig. 1.4).

Bacteriostatic antibiotics function by stopping an essential metabolic process, which can later resume when the antibiotic concentration decreases. Bactericidal antibiotics cause irreversible damage to their targets thereby killing the cells. An alternative mechanism also has been proposed for the action of bactericidal antibiotics [23, 24]. The authors have proposed a unified mechanism of killing, whereby toxic reactive oxygen species (ROS) are produced in the presence of antibiotics, leading to cell death. However, this concept has also been challenged later [25], in which the authors argued that ROS do not play any role in killing of bacterial pathogens by antibiotics.

1.7 Background Biochemistry Information

1.7.1 Enzymes

All classes of antibiotics except some of those that disrupt the membrane have some enzyme as their target. What are enzymes? They are usually proteins that act as biological catalysts. Similar to other chemical catalysts, the function of enzymes is to increase the rates of reactions. However, enzymes function better than other chemical catalysts since they are more efficient and more specific. Nearly all of the numerous biochemical reactions taking place in the body are catalyzed by enzymes. Rates of enzyme-catalyzed reactions are about 10^8–10^{12} times that of uncatalyzed reactions.

A small portion of the three-dimensional structure of an enzyme is actually involved in binding to the reactants (called substrate in biochemistry) and in catalyzing the reaction. This region of the enzyme is called the active site (Fig. 1.5). The remainder of the protein is required for the proper shape of the protein such that the active site structure is formed and maintained. Enzymes are very specific for their substrates; they catalyze reactions with one specific substrate molecule and not with other molecules even if they have similar structures. The reason for this specificity can be explained by the "*lock and key*" model. The cleft where the substrate binds is complementary in shape to the portion of the substrate

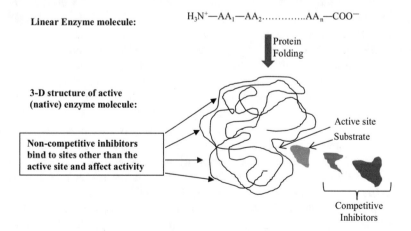

Fig. 1.5 Enzyme structure, activity, and inhibition

that binds. There has to be an exact fit of the substrate at the active site for the reaction to take place just as there has to be an exact fit of a lock with the key.

Coenzymes and metal ions. Besides the protein part, some enzymes may also require other factors for activity. These are called cofactors, and can be of two types: metal ions, and small organic molecules called *coenzymes*. The metal ion or the coenzyme can also be present as part of the active site, and participate in the catalytic reaction. For example, a Zn^{2+} ion is present at the active site of some β-lactamase enzymes and is essential for their activity (Sect. 3.3.2.6). There are many coenzymes that are needed for activity of various enzymes. Some examples include ATP, NADH, FAD, PLP, etc. In higher organisms, coenzymes can be classified into two types: metabolite coenzymes (those that can be synthesized in our body, e.g., ATP) and vitamin-derived coenzyme (those that we cannot make and so must be present in our diet, e.g., folic acid). Knowledge of one coenzyme, folic acid is essential for the understanding of antibiotics and will be discussed in Sects. 2.8 and 4.2. All species need coenzymes. Usually the lower organisms synthesize all their coenzymes while higher organisms including humans have lost their ability to synthesize some coenzymes and so must obtain them from their diet. The coenzymes are required in small amounts (microgram to milligram amounts per day) and are obtained by humans in the form of *vitamins*, which are precursors of coenzymes. Vitamins are converted to the corresponding coenzymes in the body.

1.7.2   Enzyme Inhibitors

Enzyme inhibitors are compounds that reduce the activity of the enzyme. Enzyme inhibitors can be classified in many ways; for example, they can be *reversible* or *irreversible*. It is reversible if the binding of the inhibitor to the enzyme is reversible and thus the enzyme can regain its original activity when the inhibitor is removed. It is irreversible if the inhibitor forms a permanent (covalent) bond with the enzyme and thus the enzyme loses activity permanently even if the rest of inhibitor is removed. Both these types of inhibitors can be further classified as *specific* or *nonspecific* inhibitors. Specific inhibitors target only one enzyme, whereas nonspecific ones can inhibit any enzyme. Enzyme inhibitors can function by binding to either the *coenzyme*, the small organic molecule present at the active site or *metal ion cofactor*, which is also present at the active site, or the *apoenzyme*, which is the protein part of the enzyme. The reversible apoenzyme inhibitors can be of three types: *competitive, noncompetitive*, and *uncompetitive*. For a detailed description of all types of inhibitors, please refer to biochemistry textbooks. For the purpose of understanding antibiotics, only two types of inhibitors will be described here in detail: *competitive* inhibitors, which are specific and reversible, and *suicide* or *mechanism-based* inhibitors, which are specific and irreversible.

Competitive inhibitors. Only a portion of the substrate molecule is complementary to the active site and binds to it. So if modifications are made in the portion of the molecule that is not involved in binding, the resulting molecule will be able to bind to the enzyme but the reaction will not take place because it is not the right substrate (Fig. 1.5). This results in inhibition of enzyme activity. Such molecules which inhibit an enzyme by binding to the active site are called competitive inhibitors because they compete with the normal substrate for binding to the enzyme. The higher the concentration of the competitive inhibitor, the more the inhibition. It can be said that this concept of competitive inhibition has saved more lives than anything else because a large majority of medicines function by inhibiting a certain enzyme in the body by competitive inhibition. The same is true for many antibiotics. Most antibiotics function by inhibiting a certain enzyme that is essential for survival of the infecting microorganism.

Suicide (or mechanism-based) inhibitors will be discussed later, after enzyme mechanism.

1.7.3   Enzyme Mechanisms

Enzymes can function as catalyst because they decrease the energy of activation of the reaction. Rate of a chemical reaction is given by the *Arrhenius equation*:

$$\text{Rate} = Ae^{-Ea/RT} \tag{1.1}$$

where A is the Arrhenius constant, also known as the frequency factor, e is the base of natural logarithm, 2.718, Ea is the energy of activation, R is the gas constant, 8.31 J/(K mol), and T is the absolute temperature. The value of frequency factor (A) depends on the frequency of collisions between the reactants in the proper orientation. Note that collisions between reactants in the wrong orientation will be nonproductive collisions and will not contribute towards the rate of reaction. The frequency factor is usually considered to be constant for a chemical reaction. However, it can be different if the reaction is catalyzed by an enzyme. In mammals the body temperature is constant. Thus, the only variables in enzyme-catalyzed reactions are A and Ea. Enzymes are able to increase the rate of reaction because they can either increase the value of A or decrease the value of Ea or both. Note that, because of the negative sign of Ea in the Arrhenius equation, the rate of reaction increases as Ea decreases.

The value of Ea can be lowered in two ways: either by decreasing the energy of the transition state (TS) or by increasing the energy of the reactants (Fig. 1.6). Examples for both mechanisms have been seen in enzymes. To increase the energy of the reactants, the reaction is broken into two separate steps. For any reaction that consists of several elementary steps, the slowest of those steps (one with the highest Ea), known as the rate determining step (RDS), will determine the overall rate of reaction. Enzyme-catalyzed reactions are faster because the Ea for the RDS is lower than the Ea for the uncatalyzed reaction (Fig. 1.6a, b). Another mechanism for increasing the rate of reaction is by lowering the energy of the transition state. This can happen if the enzyme binds to the TS better than to the reactant (substrate)

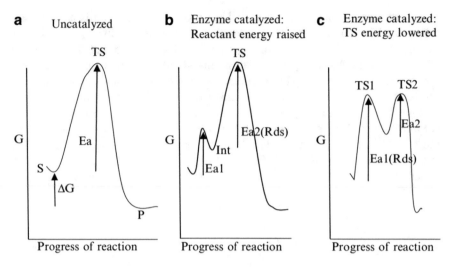

Fig. 1.6 Hypothetical energy diagrams. (**a**) for uncatalyzed and (**b** and **c**) for enzyme-catalyzed reactions. Energy of activation can be lowered by increasing the energy of reactants (**b**) or by decreasing the energy of the transition state (**c**)

and it does so because the active site is more complementary to the TS structure than to the structure of the substrate. However, the TS is a transient form that no one has ever seen. So how does one know about the TS structure? One fact for sure is that the TS structure must have some similarity to the reactant(s) and some similarity to the product(s). The extent of similarity can be predicted based on Hammond's postulate, which states that the TS looks more like the molecule it is closest in energy to. For exothermic reactions the TS is closer in energy to the reactant and thus, looks like the reactant, while in endothermic reactions the TS is closer in energy to the product and thus, looks like the product. By binding to the TS, the energy of the TS is lowered and thus the Ea of the RDS is lowered (Fig. 1.6c). Since the RDS in Fig. 1.6c is endothermic, the TS must look like the intermediate. The enzyme binds to the intermediate (and TS) and lowers the energy of the TS and thus, increases the rate of the reaction. In the past, search for synthetic antibiotics has focused on developing substrate analogs as competitive inhibitors; however, recently pharmaceutical companies have expanded their search by also developing analogs of the intermediate. Examples of both types of antibiotics are known, those that resemble the substrate and those that resemble the intermediate.

Suicide Inhibitors (Mechanism-Based Inhibitors). These are the most effective inhibitors. Similar to competitive inhibitors, they are specific for binding to the active site, but unlike competitive inhibitors they irreversibly inhibit the enzyme and so are needed only in stoichiometric amounts.

In suicide inhibition, the inhibitor closely resembles the substrate such that it not only competes with the substrate for binding to the active site but also participates as a substrate only for the first few steps of the reaction mechanism. After those first few steps it forms a stable covalent bond with the enzyme. Because of the highly stable

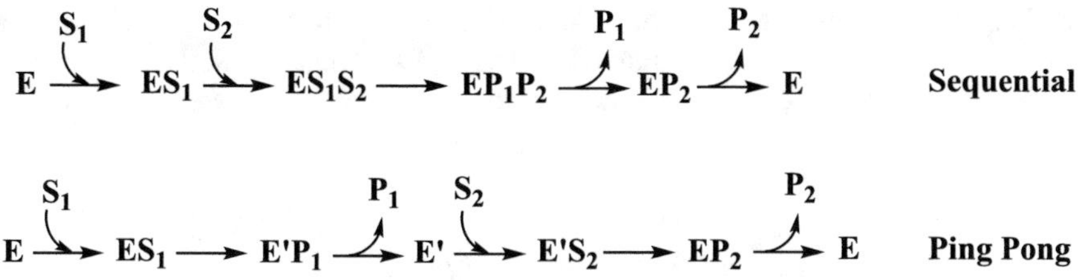

$$
\begin{array}{c}
S_1 \qquad S_2 \qquad\qquad\qquad\qquad P_1 \qquad P_2 \\
E \xrightarrow{} ES_1 \xrightarrow{} ES_1S_2 \longrightarrow EP_1P_2 \xrightarrow{} EP_2 \xrightarrow{} E
\end{array}
\qquad \textbf{Sequential}
$$

$$
\begin{array}{c}
S_1 \qquad\qquad\qquad P_1 \quad S_2 \qquad\qquad\qquad P_2 \\
E \xrightarrow{} ES_1 \longrightarrow E'P_1 \xrightarrow{} E' \xrightarrow{} E'S_2 \longrightarrow EP_2 \xrightarrow{} E
\end{array}
\qquad \textbf{Ping Pong}
$$

Fig. 1.7 Sequential and Ping Pong bi-substrate reaction mechanisms

bond with the enzyme, the reaction can neither go forward nor go in the reverse direction and thus permanently inactivates the enzyme. Sometimes this process involves a prosthetic group or coenzyme to form a reactive intermediate which then reacts with the enzyme to form a stable covalent bond. The suicide inhibitor does not necessarily have to inactivate the enzyme molecule. Equal effectiveness can be achieved by forming a stable covalent bond with a prosthetic group, which is tightly bound to the enzyme and is needed for its activity. The first demonstration of this phenomenon was with α-fluoromethylhistidine, which functions as a suicide inhibitor of the enzyme histidine decarboxylase by forming a stable covalent bond with the tightly bound coenzyme, pyridoxal phosphate [26]. A common strategy in designing a suicide inhibitor is to replace a hydrogen with a fluorine atom in the substrate molecule. Since hydrogen and fluorine atoms are of similar size the molecule is recognized as a substrate analog and effectively competes with the substrate for binding to the active site of the enzyme. The reaction is initiated by forming a covalent bond between the inhibitor and an amino acid at the active site. The hydrogen that has been replaced is normally removed as a H^+ in one of the steps in the mechanism of enzyme-catalyzed reaction. Since fluorine is highly electronegative, it cannot be removed as F^+ and so further reaction is halted permanently while the inhibitor remains covalently bound to the enzyme. Since one molecule of the inhibitor irreversibly inactivates one molecule of the enzyme (or coenzyme), a very small and stoichiometric amount of the inhibitor is needed to inactivate all the enzyme.

Several antibiotics are known that function as suicide inhibitors and so are needed in small amounts. Some examples include fosfomycin (Sect. 3.3.1.1), penicillin (Sect. 3.3.2.1), and the anticancer antibiotic (Sect. 1.1), 5-fluorouracil (Sect. 4.3.7).

Bi-substrate reactions. Many enzyme-catalyzed reactions involve two substrates, S1 and S2 and may produce one or more products, P1, P2, etc. Bi-substrate reactions can be classified into two types (Fig. 1.7):

1. Sequential bi-substrate reactions, in which both substrates must bind to the enzyme active site before reacting to form the products. If the binding of the reactants and release of the products have to be in a certain order, the reaction is said to be ordered sequential. If the order of addition and release is not important, the reaction is said to follow a random sequential bi-substrate mechanism.

2. Ping pong bi-substrate reaction, which is more relevant to the understanding of the mechanism of action of antibiotics. The first substrate binds, the first product is released; then the second substrate binds and the second product is released.

1.7.4   Metabolism and Metabolic Pathways

In order to understand biosynthesis of any molecules such as carbohydrates, proteins, lipids, or nucleic acids, it is important to know the background knowledge about metabolic pathways and the thermodynamic principles governing them. All biochemical reactions taking place in living cells are collectively called *metabolism*. Biochemical reactions never occur as single reactions but each reaction is followed by another reaction, forming a series of steps known as a *metabolic pathway*. In the pathway, the product of one reaction is a reactant of the next reaction. Thus, all these products are also intermediates in a metabolic pathway and are called *metabolites*. Some intermediates may be common for more than one pathway and thus form branches that connect one pathway with another. The entire network of reactions and pathways is called *metabolism* and can include hundreds of reactions in each cell. Each of these reactions is catalyzed by an enzyme. Metabolic pathways do not have any beginning or end although every textbook will show a beginning and an end for every pathway that are determined almost arbitrarily by scientists and authors just for our convenience. For most pathways there will be more reactions before and after the first and last steps, respectively, of the pathways shown in any publication.

All metabolic pathways consist of multiple steps. Why do metabolic pathways have so many distinct reactions? An experienced organic chemist may be able to synthesize some of these products in the laboratory in fewer steps. Degradation reactions can be performed in the laboratory in even fewer steps. For example, one can easily degrade glucose to CO_2 and H_2O in just one step by simply burning the glucose in the fire. However, in living cells the process of burning glucose to CO_2 and H_2O requires at least 28 steps. There are four main reasons for having multiple reactions in all pathways: (a) All biochemical reactions are catalyzed by enzymes and each enzyme carries out only a specific type of reaction. The overall pathway may involve many types of reactions. (b) Some compounds can be substrate or product of more than one metabolic pathway. So more steps will ensure more intermediates that can participate in other pathways. (c) Metabolic pathways in cells are highly regulated. All pathways are not on all the time. They are turned on or off as needed by the cell. This process is called regulation and is done by turning on or off one or some steps of the whole pathway. So the more steps there are, the more will be the possible points where the pathway can be regulated. (d) Finally, the reason that is most relevant for the understanding of antibiotics is the energy factor of biochemical reactions. Many of these reactions either absorb energy (endergonic, ΔG positive) or release energy (exergonic, ΔG negative). For endergonic reactions, the energy required is obtained from ATP and for exergonic reactions, the released energy is used to synthesize ATP. The equation for the reaction involving synthesis and breakdown of ATP is as follows:

$$ATP + H_2O \rightleftharpoons ADP + Pi \Delta G^{0\prime} = -30 \text{ kj/mol} \tag{1.2}$$

where Pi stands for inorganic phosphate including PO_4^{3-} ion and its protonated forms HPO_4^{2-} and $H_2PO_4^-$, the predominant form depending on the pH. The meaning of $\Delta G^{0\prime}$ is explained below (Sect. 1.7.5). The most efficient utilization or formation of ATP can happen if there are multiple steps in the metabolic pathways. For example, when glucose is oxidized to CO_2 and H_2O, it releases a total of ~2800 kJ/mol. If all this energy is released in one step, most of that energy will be wasted as heat energy since only 30 kJ/mol energy can be transferred to ADP in one step. The remaining ~2770 kJ/mol will be released as heat energy and cause a drastic increase of body temperature. By having multiple steps for the degradation of glucose, ATP can be made in many of these steps, thus accounting for a total of 32 ATP molecules for each glucose molecule oxidized. In case of synthesis reactions that require energy, only one ATP molecule can be used per step to provide energy (~30 kJ/mol). Thus, biosynthesis of glucose cannot take place in one step since one ATP cannot provide all the energy necessary (2800 kJ/mol) for the reaction. The synthesis takes place in multiple steps, with no step requiring more than one ATP.

1.7.5 Thermodynamics of Metabolic Pathways

Biochemical reactions follow the same thermodynamic principles as any other chemical reaction. For a reaction to be spontaneous, $\Delta G < 0$. ΔG, the Gibb's free energy difference is given by the equation $\Delta G = \Delta H - T \Delta S$. A reaction can take place only if ΔG is negative. If $\Delta G = 0$, the reaction is at equilibrium. If ΔG is positive, the reaction will not happen as written. Even if ΔH is positive (endothermic), the reaction can still have a negative ΔG provided ΔS is a large positive number. Such reactions that proceed because of a large positive ΔS are said to be entropy driven. Two examples of entropy driven processes that are important for understanding of antibiotics are the formation of lipid bilayers and the folding of proteins to an active conformation. Both these processes depend on what is known as the hydrophobic effect. At first sight neither of these processes appears to be entropy driven since the entropy of the lipid as well as the protein actually decreases by being more ordered (note that entropy is a measure of disorder). However, one should consider the total entropy of the whole system, including that of the surrounding water molecules. The decrease in entropy of the protein or the membrane is offset by a large increase in the entropy of the surrounding water molecules, which now has a much greater mobility because the hydrophobic regions of the protein or the membrane are now separated from the water molecules. This process of separating the hydrophobic regions from the polar solvent molecules to make the system more stable is known as the hydrophobic effect. The same explanation applies to the automatic conversion of a mixture of oil and water into two separate layers.

Chemical Equilibria. An enzyme does not change the ΔG of a reaction and cannot change the laws of thermodynamics. For example, if ΔG for a reaction is positive, an enzyme will not be able to catalyze such a reaction. An enzyme as well as all other catalysts functions by increasing the rate of reaction. It can only increase the speed at which equilibrium is reached. Gibb's free energy, ΔG for the reaction aA + bB $\rightleftharpoons$ cC + dD is given by the equation

$$\Delta G = \Delta G^0 + RT \ln \frac{[C]^c [D]^d}{[A]^a [B]^b} \tag{1.3}$$

where ΔG^0 is the free energy change of the reaction when all of its reactants and products are in their standard states. At equilibrium, $\Delta G = 0$

$$\therefore \quad \Delta G^0 + RT \ln \text{Keq} = 0$$

$$\text{or} \quad \Delta G^0 = -RT \ln \text{Keq}.$$

Thus, ΔG^0 can be calculated from the equilibrium constant. The ΔG^0 values for various reactions can be found in any Chemistry textbook. However, these numbers are different from those in Biochemistry books, which list $\Delta G^{0\prime}$ values and not ΔG^0 values. Since most biochemical reactions occur at pH 7, the concentration of H^+ (10^{-7} M) is included in the $\Delta G^{0\prime}$ values for those reactions that have H^+ ion as one of the reactants or products. Many metabolic reactions have positive $\Delta G^{0\prime}$ values but are still spontaneous (do take place) because actual ΔG is negative. The concentrations of the reactants and products in the cell are maintained at such levels that calculated ΔG based on the above equation (Eq. 1.3) is negative.

Coupled Reactions. Consider the two reactions:

$$A + B \rightarrow W + X \qquad \Delta G1 = +$$
$$W + C \rightarrow Y + Z \qquad \Delta G2 = -$$
$$A + B + C \rightarrow X + Y + Z \quad \Delta G = -$$

The first reaction cannot take place since $\Delta G1$ is positive. However, if $\Delta G2$ is sufficiently negative so that $\Delta G1 + \Delta G2$ is <0, then the two reactions can go forward. The second reaction helps the first reaction to move forward. The two reactions are said to be coupled through the common intermediate, which in this case is W. Note that, for two reactions to be coupled there has to be a common intermediate between the two reactions. In biochemistry there are numerous examples of coupled reactions. In fact, all biosynthetic reactions are coupled to breakdown of ATP which provides energy for the otherwise non-spontaneous reaction. Energy released by the oxidation of food (carbohydrates, proteins, and lipids) is stored in the form of phosphate bond in ATP (or, to a lesser extent, other nucleotide triphosphates). This ATP then provides the energy needed for all biosynthetic reactions. So ATP can be considered as the "energy currency" of the body.

1.7.6 High-Energy Compounds

Why is ATP chosen as the energy currency in cells? That is because a large amount of energy can be released when ATP is hydrolyzed to give ADP. Note that ATP contains two phospho-anhydride and one phosphoester linkages. Acid anhydrides are unstable (high energy) and will release a lot of energy when hydrolyzed. The reaction for the hydrolysis of ATP is shown in Fig. 1.8.

For ΔG of a reaction to be highly negative, either the energy of the reactants must be high or the energy of the products must be low or both. In other words, unstable reactant(s) and stable product(s) can make the ΔG highly negative. The following factors explain the highly negative ΔG value for the hydrolysis reaction of ATP. (a) In ATP, the repulsion between the four negative charges makes it more unstable (higher energy) compared to ADP which has three negative charges. (b) The products of hydrolysis, ADP and Pi or AMP and PPi are better hydrated since they have greater charge densities (charge/surface area). This makes the products more stable (less energy). (c) The products HPO_4^{2-} and ADP^{3-} are stabilized more by resonance than the reactant ATP^{4-}. Resonance structures make a compound stable. Each terminal phosphate has three while each internal phosphate has two resonance forms as shown in Fig. 1.9. Hydrolysis of ATP replaces one terminal phosphate of ATP with two terminal phosphates of ADP and HPO_4^{2-}. This makes the products more stable (lower energy) than the reactants.

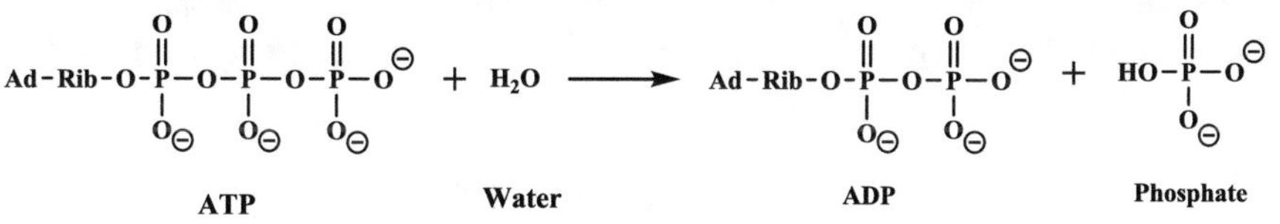

Fig. 1.8 Reaction for hydrolysis of ATP. Ad = adenosine, Rib = ribose

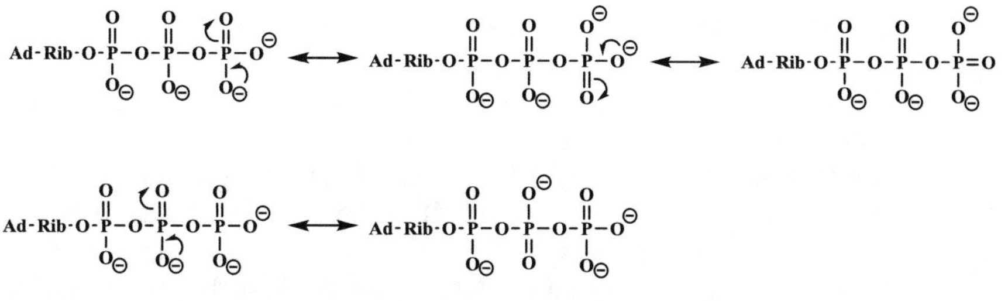

Fig. 1.9 Resonance structures of ATP. Resonance for only one internal phosphate is shown

Because of the three reasons mentioned above, hydrolysis of ATP is accompanied by the release of large amount of energy. Compounds such as ATP and all other nucleotide triphosphates (NTPs) that release large amount of energy when hydrolyzed are referred to as high-energy (or energy rich) compounds. The phosphate bonds in ATP are sometimes mistakenly called high-energy bonds. The term "high energy compound" is often used in Biochemistry and is different from and unrelated to the term "high energy bond" that is often used in Chemistry and is defined as the energy needed to break (not hydrolyze) a bond. The reaction for breaking a bond has a highly positive ΔG as opposed to highly negative ΔG for hydrolysis of high-energy compounds. Note that the hydrolysis reaction in Fig. 1.8 is shown to explain the concept of "high energy compounds." The energy that can be released if the compound is hydrolyzed represents the amount of potential energy present in the compounds. Actually, such hydrolysis reactions never take place in the cells. If these compounds were to be hydrolyzed, the energy released would be "wasted" as heat energy. Instead, nature has developed methods for utilizing the energy for useful purposes such as synthesizing biochemicals, doing mechanical work, or transporting various molecules across membranes.

Phosphoryl group transfer potential. Although hydrolysis of ATP has a highly negative ΔG, this reaction actually never takes place in the cell because of two reasons: (1) as explained above, if hydrolyzed, the energy of hydrolysis will not be utilized for any useful purpose other than generation of heat and (2) even though the reaction is thermodynamically feasible, it is not kinetically feasible because the energy of activation for the reaction is very high making the reaction extremely slow. Moreover, there is no enzyme available that can catalyze the reaction to make it faster since it is not a desirable reaction for the cell. The enzyme pyrophosphatase is an exception as will be explained below. Hydrolysis of ATP can be also viewed as transfer of phosphate group to the OH group of water. However, this reaction does not happen; instead, the phosphate is transferred to OH group of certain metabolites. For example ATP can phosphorylate glucose to form glucose-6-phosphate. Henceforth, a phosphate group (PO_4^{2-} or HPO_4^{-}) linked to organic compounds will be written as P.

$$ATP + Glc \rightarrow Glc\text{-}6\text{-}P + ADP$$

The glucose-6-P formed is also a high-energy compound and is capable of transferring the phosphate to other compounds. Since ATP can transfer a phosphate to glucose, ATP is said to have a higher *phosphoryl group transfer potential* than Glc-6-P. A list of high-energy compounds and their $\Delta G^{0\prime}$ values for hydrolysis reaction can be found in any biochemistry textbook. Some high-energy compounds in the list and their $\Delta G^{0\prime}$ for hydrolysis are as follows:

Phosphoenolpyruvate (PEP) (-62 kJ/mol), 1,3-bis-phosphoglycerate (-49 kJ/mol)), ATP (-31 kJ/mol), Glu-6-P (-14 kJ/mol). One higher in the list can transfer a phosphate to make the one lower in the list when catalyzed by an enzyme. Thus, ATP can transfer a phosphate to glucose to make glucose-6-P. In the same way, PEP and 1,3-BPG can transfer a phosphate to ADP to make ATP. Note that $\Delta G^{0\prime}$

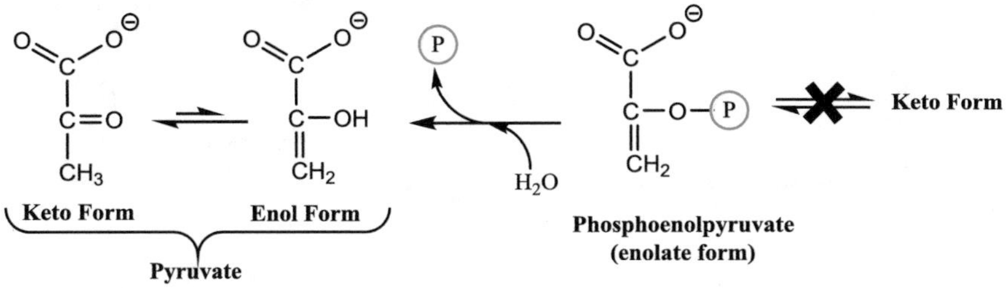

Fig. 1.10 Phosphoenolpyruvate (PEP) is a high-energy compound. Hydrolysis of PEP gives the enol form of pyruvate which quickly equilibrates to the keto form and releases energy

represents the standard free energy changes. The actual ΔG depends not only on the $\Delta G^{0'}$ values but also on the actual concentrations of the reactants and products in the cell as shown in Eq. (1.2).

Note that 1,3-bis-phosphoglycerate, similar to ATP, also has a phosphoanhydride linkage and so is a high-energy compound because hydrolysis of the anhydride bond can release a lot of energy (Fig. 1.8). Phosphoenolpyruvate is a high-energy compound for a different reason. As the name suggests, it is an enolphosphate. In aqueous solution, any ketone exists in equilibrium with the enol form and the process is called keto-enol tautomerism. Of the two, the enol form is much less stable and so a majority of the molecules exist in the keto form. Phosphoenolpyruvate contains an ester bond between the enol (an alcohol) and a phosphate (an acid). Thus, the compound is locked in the unstable enol form due to the phosphoester bond. If the enolphosphate bond is hydrolyzed, it immediately equilibrates to the more stable keto form releasing a large amount of energy, making phosphoenolpyruvate one of the highest energy compounds (Fig. 1.10).

Thioesters are high-energy compounds. Another type of high-energy compound that is frequently used in cells is a thioester. Examples include acetyl CoA (coenzyme A) and acetyl ACP (acyl carrier protein). Oxo-esters are stabilized by two resonance structures in each of which a π-bond is formed between carbon and oxygen by overlap of p orbitals of the two atoms. Thioesters cannot have one of these resonance forms in which the π-bond is formed between carbon and sulfur because of inefficient overlap between two unequal size p orbitals (Fig. 1.11). This makes thioesters more unstable compared to oxo-esters. So more energy is released when thioesters are hydrolyzed. This classifies thioesters as high-energy compounds.

1.7.7 Metabolically Irreversible and Near-Equilibrium Reactions

A metabolic pathway consists of multiple steps. By definition, the overall ΔG of any pathway that is turned on should be negative. Each step of a pathway can either be a near-equilibrium reaction (ΔG very close to zero but negative) or a metabolically

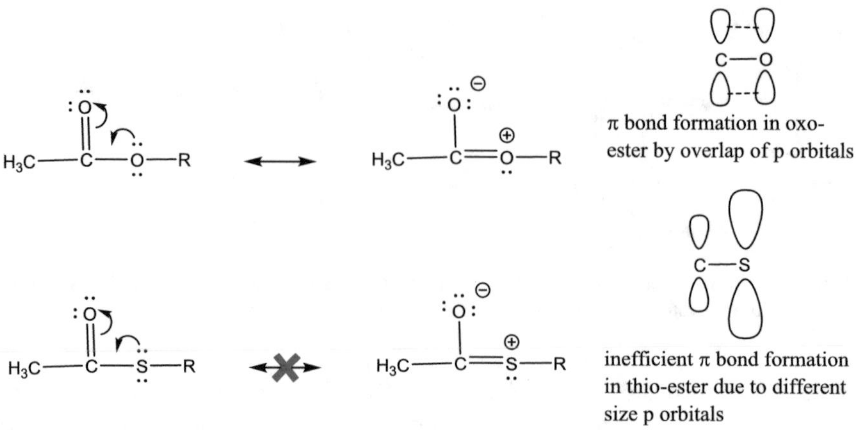

π bond formation in oxo-ester by overlap of p orbitals

inefficient π bond formation in thio-ester due to different size p orbitals

Fig. 1.11 Greater resonance stabilization of oxo-esters compared to thioesters

Fig. 1.12 A hypothetical metabolic pathway showing reversible and irreversible steps and branch points

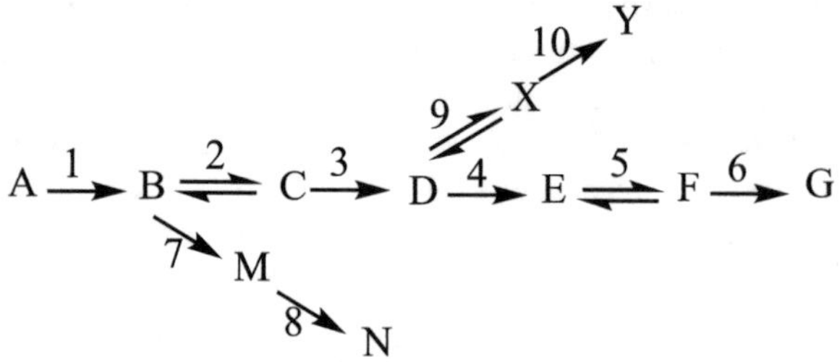

irreversible reaction (ΔG highly negative). The concentrations of the reactants and products in the cell are always adjusted to maintain these negative ΔG values. Each pathway must have at least one irreversible step for the overall pathway to move forward. Even biosynthetic pathways, which require energy must contain at least one irreversible step. This is often achieved by "sacrificing" a high-energy molecule in order to make ΔG highly negative as explained below. The irreversible steps are of special importance since pathways are regulated (turned on or off) by regulating the activities of the enzymes of these steps.

First committed step. Some intermediates of a pathway may be part of more than one pathway. These are called the branch points. A pathway consists of near-equilibrium and metabolically irreversible steps and will also contain a first committed step. This is the first irreversible step after the last branch point in the pathway. After this step takes place, the metabolites are committed to continue to the final product of the pathway. The enzyme catalyzing this step is an attractive target for development of antibiotics to inhibit the pathway. Inhibition of this step will ensure that the final product of the pathway will not be made and no other pathway will be affected since this step is after the last branch point in the pathway. In the following hypothetical scheme of reactions, steps 4, 7, and 10 are the first committed steps for biosynthesis of the compounds G, N, and Y, respectively (Fig. 1.12).

Fig. 1.13 Two equivalents of energy are spent to make a product that contains one equivalent of energy higher than the substrate

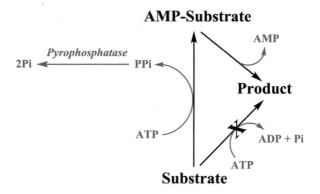

Pyrophosphatase makes reactions irreversible. As mentioned above, all pathways, including biosynthetic ones must have at least one metabolically irreversible step. Synthetic reactions such as those joining two monomers to form a dimer involve formation of new bonds. The energy required for the new bond is provided by a high-energy compound such as ATP. Because of the new bond formed, the product will be of higher energy than reactant, which is the opposite of what is needed for the reaction to be irreversible. So, to make this an irreversible reaction the reactant energy must be raised to a higher level than the energy of the product. If energy for the reaction is provided by one molecule of a high-energy compound, the reaction can become at best an equilibrium reaction but not an irreversible reaction, for which the energy of the reactant must be raised to a much higher level than that of the product. This is achieved by using two equivalents of high-energy molecules and not just one (Fig. 1.13). However, as mentioned before (Sect. 1.7.4), a biochemical reaction can utilize only one high-energy molecule at a time. So the reaction is broken into two steps, each step using one high-energy compound. Of these, the second step is usually a hydrolysis reaction of the ion pyrophosphate ($P_2O_7^{4-}$) to give two phosphate ions (HPO_4^{2-}) and is catalyzed by the enzyme pyrophosphatase. This can also be understood based on Le Chatelier's principle which states that a reaction will move forward if one of the products (in this case, pyrophosphate) is removed. For such biosynthetic reactions it is a common strategy to first "activate" the reactant to a high-energy level and in the next step use the extra energy to form the new bond in the product (Fig. 1.13).

An example in which this strategy is seen is the following reaction for the extension of a glucose polymer by adding one more glucose residue.

$$G\text{-}1\text{-}P + (G)_n \rightarrow (G)_{n+1}$$

The energy for the formation of G-G bond in the product is provided by the energy in the glucose-phosphate bond in the reactant. However, that cannot make the reaction irreversible (highly negative ΔG). To make the reaction feasible and also metabolically irreversible, the G-1-P is further activated using ATP (or other NTPs).

$$G\text{-}I\text{-}P + P\text{-}P\text{-}P\text{-}Rib\text{-}Ad \rightarrow G\text{-}I\text{-}P\text{-}P\text{-}Rib\text{-}Ad + PPi$$
$$\quad\text{(ATP)}\qquad\qquad\qquad\qquad\text{(ADP-glucose)}$$

$$G\text{-}I\text{-}P\text{-}P\text{-}Rib\text{-}Ad + (G)_n \rightarrow (G)_{n+1} + ADP$$

A similar example can also be seen in bacterial cell wall biosynthesis pathway as will be discussed later (Sect. 3.2.1).

1.7.8   Membrane Transport

Biochemistry of membrane composition and structure will be discussed later (Sect. 7.1.2). One of the functions of the cell membrane is to act as a barrier for entry and exit of molecules into and out of the cells. Knowledge of this process is necessary for understanding the mechanisms of action of antibiotics and resistance development to the antibiotics. Transport through both bacterial and the human membranes is relevant to the function of antibiotics but the focus in this book is on the transport through bacterial plasma membrane. Transport of molecules across the membrane can be either by simple diffusion or mediated by protein(s) present in the membrane. Because of the hydrophobic nature of the membrane, small hydrophobic molecules can easily cross the membrane by simple diffusion without the assistance of any mediator. Polar molecules are unable to cross the hydrophobic membrane and so are transported with the aid of proteins in membrane. One way the protein can function is as a carrier that binds to the molecule and transports it from one side of the membrane to the other. In another method the proteins form transmembrane pores through which the molecules can be transported from one side to the other. The carriers and pores are usually specific for certain molecules or ions. The specificity is due to the various sizes of the pores. Specificity for certain charge of the ions (positive or negative) is due to charged amino acids present in the proteins forming the pores. The ion to be transported interacts with the oppositely charged amino acids in order to be transported.

Energetics of membrane transport. In both simple diffusion and protein mediated diffusion molecules always move from high concentration to low concentration. No extra energy from ATP or other high-energy compounds is needed for such transport since a difference of concentration, also known as a concentration gradient itself is equivalent to energy. Based on the energy needs of the process, membrane transport can be classified into two types: passive and active transport. In passive transport, molecules move from high concentration level to low concentration level without requiring any extra energy. In active transport, molecules are transported against a concentration gradient (from low to high concentration level) and this process requires extra energy. Depending on the source of this extra energy, active transport can be classified into two types. It is called primary active transport if the source of energy is a high-energy compound such as ATP and is called secondary active transport if the energy comes from an existing concentration gradient of some other molecule or ion. The existing gradient may be that of Na^+ or K^+, as in transmission of nerve signals or a proton gradient that is used in respiration (oxidative phosphorylation). Note that these gradients are first formed by using energy from ATP, so the ultimate source of energy for both types of active transports is a high-energy compound. These existing gradients in the cell can then be used to transport other molecules against their concentration gradients. Active transport is mediated by proteins in the membrane which form pores through which the molecules are transported. The proteins involved in the transport are transmembrane proteins. These protein chains usually go through the membrane several times to form loops on both sides of the membrane. The amino acids in the membrane bilayer are hydrophobic since they form hydrophobic interaction with the interior of the membrane while those forming the loops are usually hydrophilic since they have to interact with the aqueous environment on the two sides of the membrane and also have to interact with the polar molecules to be transported. Note that any existing gradient of any molecule or ion is a potential source of energy irrespective of the direction of the gradient. So based on the direction of the existing gradient, secondary active transport can be of two types: symport and antiport. In symport, the molecule or ion that is transported and the molecule or ion that forms the existing gradient are both moved through the transmembrane protein machinery in the same direction, whereas in antiport they move in opposite directions.

References

1. FDA (2012) Guidance for industry #209. The judicious use of medically important antimicrobial drugs in food-producing animals. http://www.fda.gov/downloads/AnimalVeterinary/GuidanceComplianceEnforcement/GuidanceforIndustry/UCM216936.pdf
2. Benharroch D, Osyntsov L (2012) Infectious diseases are analogous with cancer. Hypothesis and implications. J Cancer Educ 3:117–121
3. Muehlenbachs A, Bhatnagar J, Agudelo CA, Hidron A, Eberhard ML, Mathison BA, Frace MA, Ito A, Metcalfe MG, Rollin DC, Visvesvara GS, Pham CD, Jones TL, Greer PW, Vélez Hoyos A, Olson PD, Diazgranados LR, Zaki SR (2015) Malignant transformation of *Hymenolepis nana* in a human host. N Engl J Med 373:1845–1852
4. Wood HG, Rusoff II (1945) The protective action of Trypan Red against infection by a neurotropic virus. J Exp Med 82:297–309
5. Burke ET (1925) The arseno-therapy of syphilis; stovarsol, and tryparsamide. Br J Vener Dis 1: 321–338
6. Breinl A, Todd JL (1907) Atoxyl in the treatment of trypanosomiasis. Br Med J 1:132–135
7. World Health Organization (2010) Guidelines for the treatment of malaria, 2nd edn
8. von Freudenreich E (1888) De l'antagonisme et de l'immunit6, qu'il conf6re au milieu de culture. Ann Inst Pasteur 2:200
9. Vedder EB (1914) Origin and present status of the emetin treatment of amebic dysentery. JAMA LXII(7):501–506
10. Franklin TJ, Snow GA (1989) Biochemistry of antimicrobial action, 4th edn. Chapman and Hall, London

11. Benveniste R, Davies J (1973) Aminoglycoside antibiotic-inactivating enzymes in actinomy-cetes similar to those present in clinical isolates of antibiotic-resistant bacteria. Proc Natl Acad Sci U S A 70:2276–2280

12. Hopwood DA (2007) How do antibiotic-producing bacteria ensure their self-resistance before antibiotic biosynthesis incapacitates them? Mol Microbiol 63:937–940

13. Peterson RM, Huang T, Rudolf JD, Smanski MJ, Shen B (2014) Mechanisms of self-resistance in the platensimycin and platencin producing Streptomyces platensis MA7327 and MA7339 strains. Chem Biol 21:389–397

14. Ling LL, Schneider T, Peoples AJ, Spoering AL, Engels I, Conlon BP, Mueller A, Till F, Schäberle TF, Hughes DE, Epstein S, Jones M, Lazarides L, Steadman VA, Cohen DR, Felix CR, Fetterman KA, Millett WP, Nitti AG, Zullo AM, Chen C, Lewis K (2015) A new antibiotic kills pathogens without detectable resistance. Nature. https://doi.org/10.1038/nature14098

15. Fleming A (1945) Penicillin. Nobel Lecture. www.nobelprize.org

16. Dubos RJ (1939) Studies on a bactericidal agent extracted from a soil Bacillus. I. Preparation of the agent Its activity in vitro. J Exp Med 70:1–10

17. Dubos RJ (1939) Studies on a bactericidal agent extracted from a soil Bacillus. II. Protective effect of the bactericidal agent against experimental Pneumococcus infections in mice. J Exp Med 70:11–17

18. Acar J (1997) Broad and narrow-spectrum antibiotics: an unhelpful categorization. Clin Microbiol Infect 3:395–396

19. Appelbaum E, Leff WA (1948) Occurrence of superinfections during antibiotic therapy. JAMA 138:119–121

20. Dagan R, Leibovitz E, Cheletz G, Leiberman A, Porat N (2001) Antibiotic treatment in acute otitis media promotes superinfection with resistant Streptococcus pneumoniae carried before initiation of treatment. J Infect Dis 183:880–886

21. Duggar BM (1948) Aureomycin: a product of the continuing search for new antibiotics. Ann N Y Acad Sci 51:177–181

22. Sengupta D, Sangu K, Chattopadhyay MK (2012) Unsung heroes in the history of science. Sci Rep 49:31–34

23. Kohanski MA, Dwyer DJ, Hayete B, Lawrence CA, Collins JJ (2007) A common mechanism of cellular death induced by bactericidal antibiotics. Cell 130:797–810

24. Kohanski MA, Dwyer DJ, Wierzbowski J, Cottarel G, Collins JJ (2008) Mistranslation of membrane proteins and two-component system activation trigger antibiotic-mediated cell death. Cell 135:679–690

25. Karen I, Wu Y, Inocencio J, Mulcahy LR, Lewis K (2013) Killing by bactericidal antibiotics does not depend on reactive oxygen species. Science 339:1213–1216

26. Bhattacharjee MK, Snell EE (1990) Pyridoxal 5'-phosphate-dependent histidine decarboxylase: mechanism of inactivation by α-fluoromethylhistidine. J Biol Chem 265:6664–6668

Chapter 2
Development of Resistance to Antibiotics

Abstract Widespread use and misuse of penicillin and other antibiotics have resulted in development of resistance to most antibiotics. The mechanisms by which microorganisms develop resistance to antibiotics are discussed. Topics covered include the acquisition of point mutations and antibiotic resistance genes, methods of transfer of resistance genes between bacteria, and the advantages of synthetic antibiotics. Contribution of subtherapeutic use of antibiotics to resistance development and the response of the governments and other regulatory agencies to address the problem are also discussed.

2.1 Antibiotics Are No Longer Considered to Be Miracle Drugs

Antibiotics, which were hailed as the miracle drugs that cured most infected people before, do not work in many cases today. This is because bacteria are increasingly becoming resistant to antibiotics at an alarming rate and the resistance is spreading throughout the world among all species of bacteria. The main reason for this resistance development is the excessive use of the antibiotics. After the discovery and introduction of penicillin people were so excited about its miracle properties in curing infections that the drug was not only available as over-the-counter medicine (one that does not require a prescription) but was also added to a large variety of household items such as ointments and cosmetics. Later the practice was banned and penicillin was made a prescription drug, but by then there was already widespread resistance to the antibiotic.

It is not surprising that the development of antibiotic-resistant bacteria is a major concern in the scientific and medical community. This is evident from the amount of research that is being done on the subject. When significant amount of research has been done on an important scientific topic, journals usually publish review articles on the topic. So, one way to determine the importance of a research topic is to count how many review articles are being published on that subject. To do that, one can search in the "PubMed" website of NCBI for all review articles published on "antibiotic resistance" and "antimicrobial resistance." Results of such a search are

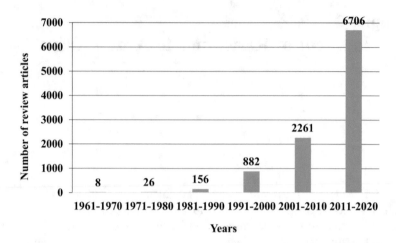

Fig. 2.1 Number of review articles published per decade on antibiotic resistance and antimicrobial resistance

shown as a bar graph in Fig. 2.1. As can be seen from the graph, very little was known about antibiotic resistance in the 1960s but today it is an extremely important subject of study. Interest in the topic has increased dramatically in the last decade.

2.2 Detection of Antibiotic Resistance

How effective an antibiotic is, can be determined by its minimum inhibitory concentration (MIC), which is the minimum concentration of the antibiotic that can stop growth of a particular microorganism. The lower the MIC, the stronger is the antibiotic. The MIC is usually determined by either the "broth dilution" or "agar dilution" method. In broth dilution method, increasing amounts of the antibiotic are added to liquid growth medium (broth) in test tubes, which are then inoculated with the same number of cells. The minimum concentration of the antibiotic that prevents growth as seen visually by the lack of turbidity is designated as the MIC for the antibiotic for that microorganism. For the agar dilution method, varying concentrations of the antibiotic are added to the molten agar immediately prior to pouring the plates. Serial dilutions of the cells are then spread on the plates and MIC is determined as the minimum concentration of the antibiotic that prevents growth of colonies on the plates.

Another method that is easier and less expensive than the broth or agar dilution method is the Kirby–Bauer method, also known as the disk diffusion method [1]. In this method, cells are first spread on a plate and a filter paper disk containing a known amount of the antibiotic is placed in the center of the plate. As the antibiotic diffuses from the paper to the agar medium in the plate, a gradually decreasing concentration gradient of the antibiotic is created. After overnight incubation, an absence of growth observed around the disk indicates sensitivity to the antibiotic. The visibility of the zone of inhibition can be dramatically improved by staining the cells with cationic dyes such as crystal violet or methylene blue [2]. The diameter of the zone of inhibition indicates the strength of the antibiotic but is only a qualitative indicator of the strength since the zone also depends on other factors such as the depth of the agar as well as size and water solubility of the antibiotic molecules. The MIC of the antibiotic is the concentration at the boundary of the zone of inhibition; however, it is not feasible to measure that concentration. An easier method for determining the MIC is the E-test in which a commercially available plastic strip containing a gradually decreasing and known concentration of an antibiotic is placed on an agar plate. The minimum concentration in the E-strip that shows no cell growth is the MIC. For a diagrammatic representation of the results of zone of inhibition and E-test, see Fig. 3.24.

2.3 Classification of Antibiotic Resistance

Antibiotic resistance in bacteria can be of two types: intrinsic and acquired.

Intrinsic resistance: Intrinsic resistance is when by virtue of their structural or functional features, some bacteria are naturally resistant to some antibiotics without having prior exposure to the antibiotics. For example, gram-negative bacteria are intrinsically resistant to vancomycin, which is too large a molecule to cross the outer membrane (Sect. 3.3.3.4). Aerobic bacteria are intrinsically resistant to metronidazole, which requires an anaerobic environment to be reduced to its active form (Sect. 5.4). These and other examples will be discussed separately for each antibiotic in later chapters.

Acquired resistance: In a population of antibiotic sensitive bacteria, some cells may acquire the ability to be resistant to the antibiotic. Thus, unlike intrinsic resistance, which is effective in all cells of a certain species, acquired resistance can be observed in only a subpopulation of any bacterial species. Acquired resistance development can take place by two different mechanisms: (1) By point mutations and (2) by resistance gene acquisition.

2.4 Resistance Development by Point Mutations

Methods for the development of point mutations can be of two types: natural methods and induced methods.

Natural methods: Replication errors: Many bacteria usually have a generation time of about 20 min, which means that the number of bacteria will double every 20 min. So one bacterium in 10 h will double 30 times to give 2^{30} or approximately one billion bacteria. This high number facilitates the development of mutants as explained below. Before every cell division the chromosomal DNA has to be duplicated by a process called DNA replication. This copying of DNA is catalyzed by the enzyme DNA polymerase III, which binds to the template DNA, brings in the correct complementary base from the surrounding medium, and ligates (joins) it to the previous base in the chain. It does all this at the incredibly fast rate of about 1000 bases per second, which is the fastest known enzymatic polymerization reaction. Because of this very fast rate the enzyme makes some mistakes during replication. The error rate is about 10^{-5}, i.e., 1 in 10^5. About 99% of these errors are corrected by the DNA polymerase itself. This is because the polymerase enzyme has a proof-reading activity. Every time it incorporates a wrong base, it removes that and brings in the right base in its place. Since only 99% errors are corrected, 1% still remains. So the overall error rate in replication is 1% of $10^{-5} = 10^{-7}$. Each of these errors can give rise to a mutation. Besides the proofreading function of the DNA polymerase enzyme there are other repair enzymes that correct most of these errors after the replication process is completed giving an overall error rate of 10^{-9}. That means, out of every billion bases copied, one will be a mistake. Bacteria usually have about 3×10^6 base pairs of DNA in their genome. So every time the bacterial cell multiplies, there is a $3 \times 10^6 \times 10^{-9} = 3 \times 10^{-3}$ probability of containing a mutation. This may not appear to be a high rate of mutation. However, because of the high number of bacteria present in an infected patient, the probability of a mutant being present in the population of bacteria becomes very significant. In a typical infection there can be about 100,000 bacterial cells per gram of tissue or per ml of urine. Assuming there are about 10^8 bacterial cells in an infected person, the number of mutants in this population will be about $3 \times 10^{-3} \times 10^8 = 3 \times 10^5$. Since these are random mutations, also known as spontaneous mutations, they can be expected to be equally distributed over the 3000 genes in the bacteria. So for every gene there will be about 100 bacteria that will have a mutation somewhere in the gene. It should be clarified here that all 100 mutations are not in the same bacterial cell but rather, there are 100 bacteria in the population having one mutation each in a particular gene. Actually the mutations are not uniformly distributed in all genes because mutations are more tolerated in some genes than in others. Mutations in housekeeping genes (those that are expressed in all cells all the time) are less tolerated because they code for proteins that perform indispensable functions.

Induced methods: Point mutations can also happen due to harsh environmental conditions such as strong ionizing radiations or oxidizing or alkylating chemicals. These methods are not relevant for resistance development in infecting bacteria because such harsh conditions do not exist in the host. However, recently it has been discovered that use of antibiotics by the host can induce generation of point mutations in the infecting bacteria. Lethal concentration of bactericidal antibiotics will, of course, kill the bacteria but if sublethal concentration of antibiotic is used, it triggers the formation of reactive oxygen species, which can cause mutations in the DNA [3]. Since this has relevance to the use of antibiotics in farm animals, this phenomenon will be discussed further under "Subtherapeutic use of antibiotics" (Sect. 2.10.2).

Effect of point mutations: Most mutations (but not all) will result in a change in protein sequence (Sect. 6.1). Some of these changes (but not all) can affect the activity of the protein. For example, changes in the sequence at the active site of an enzyme can have an effect on its activity while mutations in the structural part of the enzyme will have less or no effect. Mutations do not always result in a loss of function; some mutations can cause a gain of function. For example, a mutation can give the bacterium a selective advantage for survival in the presence of an antibiotic. Such a mutant is said to be antibiotic resistant.

2.5 Selection for Resistance

In the absence of antibiotic in the growth medium, the mutants and wild-type bacteria will all grow at the same rate as they compete with each other for nutrients. However, if grown in the presence of antibiotic, only the mutants that are resistant to the antibiotic will grow fast while all others will either die or grow slowly. Thus, at the beginning there may have been only one mutant cell out of several million but at the end the culture will consist of 100% antibiotic-resistant cells. This process is called "selection." Note that this selection process could take place because the mutant was resistant to the antibiotic at the concentration that was used. If a higher concentration was used, it is possible that all bacteria including the mutant would have died and so there would have been no selection for the mutant. The minimum concentration of an antibiotic that can stop the growth of bacteria is called its minimum inhibitory concentration (MIC). In the above example, the wild-type cells have a much lower MIC than the mutant. If the antibiotic concentration prescribed to the patient is not high enough to kill the mutant, then a false sense of curing will result since the wild-type bacteria which constitutes the majority of the infecting bacterial population will be killed. However, this good feeling is temporary since the few resistant bacteria that survive will then multiply rapidly without any competition from wild-type bacteria. Within a few days the mutants will be sufficiently high in number to continue the disease as before. The only difference now will be that the antibiotic will not work anymore against this infection. So it is important for the doctor to prescribe the right dose of the antibiotic that can kill all infecting bacterial cells.

Even if the doctor prescribed the right dose of antibiotic, selection for antibiotic-resistant mutant can still take place if the patient failed to complete the full course of the prescription. After the antibiotic enters the bloodstream it is cleared from the body within a certain time either through urine or by degradation. So the concentration of the antibiotic decreases with time. In order to maintain the concentration at a level higher than the MIC, more doses need to be taken as prescribed. Within the first one or two doses most of the wild-type bacteria will be killed giving the patient a false impression that the disease has been cured. However, the few resistant mutants will survive and unless more doses are taken, they will gradually multiply to bring back the symptoms of the disease. The resistant bacteria can then be transmitted to other people and thus increase the pool of antibiotic-resistant bacteria in the environment. To stop the selection of antibiotic-resistant bacteria it is the doctors' responsibility to prescribe the right dose of the antibiotic and it is the patients' responsibility to complete the prescribed dose of the antibiotic and not stop taking the remaining doses just because they start to feel better. Because of such misuse of antibiotics, resistance has developed to most antibiotics available. The more prescriptions a doctor writes for a certain antibiotic the more probable is the development of resistance to that antibiotic.

Misuse of antibiotics selects for gradually increasing level of antibiotic resistance: Point mutations usually confer low-level resistance to the bacteria. However, the mutants that have low-level resistance can develop new mutations that provide increased level of resistance. Probability of developing chromosomal mutations in any species is very low, about 10^{-9} replication error per cell per cell division (Sect. 2.4). Probability of developing two mutations in the same cell in a population of wild-type bacteria is even lower, about 10^{-18} per cell. So direct acquisition of high-level resistance by point mutations is very unlikely. However, the probability of developing the second mutation will be higher if improper use of the antibiotic has already resulted in selection for the first mutation which confers low-level resistance since 100% of the bacteria would then have the first mutation. The two mutations (the old and the new one) can act by the same or different mechanism. For example, the first mutation may decrease the binding of the antibiotic to the active site of the enzyme while the second mutation may decrease it even further. Another possibility is that the first mutation decreases the permeability of the antibiotic through the cell membrane while the second mutation may decrease the binding of the antibiotic to the active site. It is easy to treat low-level antibiotic resistance since the bacteria can be killed by usual therapeutic doses of antibiotics. However, an incomplete dose of antibiotic will select for the low-level resistant bacteria, which can later acquire new mutations giving rise to gradual selection of high-level antibiotic resistance which cannot be treated by the usual therapeutic dose of the same antibiotic which would have been sufficient before.

The development of antibiotic resistance is not a recent concept. Way back in 1945 in his Nobel Lecture Alexander Fleming had warned, "It is not difficult to make microbes resistant to penicillin in the laboratory by exposing them to concentrations not sufficient to kill them" [4]. He had also warned that misuse of penicillin could lead to the selection of mutant bacteria that are resistant to the drug. This was not just a prediction; he actually produced these penicillin-resistant mutants in his lab to demonstrate this. When penicillin V became available for oral use, there was a dramatic increase in the use of the drug. In those days, it was available without a prescription and was advertised to the public as a miracle drug. Easy access resulted in rampant misuse of the antibiotic. People were taking it even for diseases that were not caused by bacteria. Excessive use rapidly increased the development of resistance to the antibiotic.

2.6   Resistance Development by Resistance Gene Acquisition

Alexander Fleming's prediction that bacteria can become resistant to antibiotics by point mutations has been proven by him as well as numerous others later. However, if point mutations were the only mechanism for resistance development, then this would have been only a nuisance and not a serious health problem since the resistance level will be low and the bacteria can be killed by the usual dose of the antibiotic. Actually the situation has become more serious than what Fleming had predicted. Today penicillin does not work for most infections. This has happened because of another mechanism of resistance development that is more important than point mutations and this was not known during Fleming's time. This mechanism of resistance development is by acquisition of resistance genes that already preexist in nature. Humans have been using antibiotics for only about 90 years since their first discovery but these natural antibiotics have been present in the environment for a billion years since the existence of bacteria, which produce them. This is a long time during which other bacteria have evolved by developing multiple mutations that have resulted in the formation of new genes coding for enzymes which degrade various antibiotics or make them ineffective. So, long before humans started to use penicillin, there already existed bacterial strains that can make penicillin degrading enzymes. These enzymes are called *β-lactamases* since they can break the β-lactam ring in the structure of penicillin. We will study about the enzyme in more detail later (Sect. 3.3.2.5). Enzymes that degrade antibiotics are much more effective in providing resistance than point mutations in proteins. So if such a highly resistant bacterium exists in the infecting bacterial population, it will be selected for and the infection will not be cured. Under such circumstances a full course of a different antibiotic should then be taken. Until all the resistant bacteria are killed there will be a finite possibility of spreading the resistant bacteria to other people.

Plasmids, Transposons, and Integrons: Antibiotic resistance genes usually reside in either the chromosome or plasmids or transposons or integrons. Plasmids are small (up to a thousand fold smaller than the chromosome) pieces of extrachromosomal DNA, usually circular. They can be present in multiple copies and use the cellular proteins for their replication. Replication is initiated at a specific site called the origin of replication (oriV). Of the various genes that are present in plasmids, one important gene that is usually present is a resistance gene for some antibiotic. Many plasmids contain resistance genes for more than one antibiotic. Plasmids do not carry out any useful function for the cell and so they may be lost if one daughter does not receive a copy of the plasmid when the cell divides. If the particular antibiotic is present in the environment, then the plasmid will be maintained in all cells since those daughter cells that do not receive a copy of the plasmid will be killed by the antibiotic. The process is called selection. Some plasmids have specific genes which ensure that each daughter cell always receives a copy of the plasmid during cell division even in the absence of antibiotic in the environment. These genes together form the "plasmid maintenance system."

Transposons (also known as insertion sequence (IS) elements) are small pieces of DNA that can insert into the chromosome mostly at random sites or in some cases, specific sites, and are commonly known as "jumping genes." They can also be excised from the chromosome and then inserted at a different location in the chromosome. Transposons were first discovered by Barbara McClintock, for which she was awarded the Nobel Prize in 1983. There are two main requirements for a transposon to function. It contains a direct or inverted repeat sequence at the two ends and the transposon sequence is preceded or followed by a gene sequence for the enzyme Transposase. The enzyme cleaves the DNA at the two repeat sequences and inserts it, along with any DNA sequence in between, into the target DNA. Various genes including antibiotic resistance genes can be present in between the two ends of the transposon. There are two ways by which transposon insertions into the genome can confer antibiotic resistance to the bacteria: (1) The transposon may have an antibiotic resistance gene in it. So when it inserts at any region in the chromosome, the antibiotic resistance gene will be expressed. (2) If a transposon does not contain any resistance gene but inserts into a gene that is essential for proper functioning of an antibiotic, then the cell will become resistant to the antibiotic. One common example of the latter type is by insertion into genes for porins, which are needed for transport of the carbapenem antibiotics from the outside to inside the cell (Sect. 3.3.2.12). Note that most natural plasmids as well as some transposons contain resistance genes for more than one antibiotic. So improper use of any one of these antibiotics will select for the whole plasmid or transposon and thus, will automatically select for resistance to more than one antibiotic.

Similar to transposons, another type of mobile genetic elements is called integrons, which have the added ability to capture various genes such as antibiotic resistance genes from the DNA that they are inserted into [5]. So integrons can acquire resistance genes for multiple antibiotics. Unlike transposons, integrons do not have any direct or indirect repeats on two sides of the resistance gene. They contain an integrase gene that is needed for the insertion process.

Transfer of resistance genes between bacteria: If some bacteria in the infecting population contain an antibiotic resistance gene, they can transfer the gene to other bacteria in the population that are sensitive to the antibiotic. This way, bacteria that acquire the resistance genes also become resistant. Transfer of genes between bacteria can take place in three ways: (1) *Bacterial Conjugation*. In 1946, Joshua Lederberg, who was awarded the Nobel Prize in 1958, discovered that bacteria can mate with each other and transfer DNA between each other by a process called conjugation. The chromosomal DNA is usually not transferred (with some exceptions). Plasmids that are capable of being transferred between bacteria are called conjugative plasmids. Those that contain all the genes necessary for the transfer process are called self-transmissible plasmids. Another type of plasmids called mobilizable plasmids contain some but not all the genes necessary for conjugative transfer. So these plasmids can be transferred only in the presence of a self-transmissible plasmid, which acts as a helper plasmid. In order to be transferred, one strand of the plasmid DNA is cut at a site called the origin of transfer (oriT). The cut single strand is transferred and then joined to make it circular [6]. The second strand is synthesized in both the donor and recipient strain. Thus, by the method of bacterial conjugation, more and more antibiotic sensitive bacteria can acquire antibiotic resistance genes and become resistant. (2) *Bacterial Transformation*. This is the process by which bacteria take in DNA from outside which is usually released from dead bacteria. Some bacteria can be artificially made transformable in the presence of added chemicals such as calcium chloride. Some, but not all bacterial species are capable of natural transformation. This takes place by an active process in which competence genes present in the bacteria are expressed and the proteins facilitate the process of transformation. Natural transformation can be of two types: nonspecific, in which case any DNA can be taken in or specific, in which case only DNA from the same species can be taken in. The bacteria recognize DNA from the same species by the presence of an "Uptake Signal Sequence" (USS) that is repeated numerous times throughout the genome [7]. (3) *Transduction/Transfection*. This is the process by which DNA is transferred between bacteria using *bacteriophages* as intermediaries. Bacteriophages are viruses that infect bacteria. After infecting the cell one bacteriophage can multiply to give more bacteriophages which are then released from the bacteria and can then infect other bacteria. In the process of multiplication the bacteriophage can incorporate some of the bacterial DNA into their own DNA and can then transfer the DNA to other bacteria that they infect. This way, antibiotic resistance genes can be transferred between bacteria thus contributing to spread of resistance.

Note that transposons that are present in the plasmid or chromosome can also be transferred to other bacteria along with the plasmid or chromosomal DNA during transformation, conjugation, or transduction. There are some transposons called conjugative transposons that are capable of transferring by the process of conjugation [8].

Antibiotic Resistance Pool: Taking insufficient dose, not completing the full course, taking the wrong antibiotic, and taking antibiotics for viral infections such as common cold constitute misuse of antibiotics. Note that antibiotics, which, by the current definition are antibacterials, will not be able to cure viral infections. Misuse of antibiotics increases the population of antibiotic-resistant bacteria, which is also known as "antibiotic resistance pool." The more an antibiotic is used, the greater will be the resistance pool for that antibiotic. With time, more and more bacterial strains will become resistant to the antibiotic. Historically, resistance to an antibiotic has been observed within a few years after its introduction into the market. The Center for Disease Control (CDC) estimates that, each year, nearly two million people in the USA acquire an infection while in a hospital (nosocomial infection, Sect. 1.1), resulting in 90,000 deaths. More than 70% of the bacteria that cause these infections are resistant to at least one of the antibiotics commonly used to treat them. Table 2.1 shows the timeline for introduction and resistance development for some antibiotics. A more complete list of antibiotics and resistance development to them can be found in the 2013 CDC report [9]. As can be seen in Table 2.1, resistance developed to most antibiotics within a few years. Although penicillin was discovered in 1928, it was not until 1943 that it was marketed. Resistance development to the antibiotic had already been reported in 1940, 3 years before it was marketed (Sect. 2.5). Resistance

Table 2.1 Timeline for antibiotic resistance development

Name	Year introduced	Year resistance first reported	Years taken for resistance development
Penicillin	1943	1940	−3
Tetracycline	1950	1959	9
Erythromycin	1953	1968	15
Methicillin	1960	1962	2
Gentamicin	1967	1979	12
Vancomycin	1972	1988	16
Imipenem	1985	1998	13
Levofloxacin	1996	1996	0
Linezolid	2000	2001	1

development against vancomycin was one of the slowest. The reason for this will be explained later in Sect 3.3.3.7.

2.7 Mechanism of Antimicrobial Resistance

Mechanisms of resistance development to antibiotics can be classified into two types: (1) Altering the target of the antibiotic such that it is no longer affected by the antibiotic [10]. (2) Decrease the concentration of the antibiotic to a level that is lower than the MIC such that it will not have a significant inhibitory effect on the bacteria. The low concentration can be achieved in three different ways: (a) Preventing entry of the antibiotic into the cell, (b) Pumping out the antibiotic after it enters the cell, and (c) Degrading or inactivating the antibiotic by enzyme-catalyzed chemical modification before it can bind to its target. The actual mechanism used depends on the type of bacteria as well as the type of antibiotic. For example, the mechanism of resistance development in gram-positive and gram-negative bacteria may be different because of the differences in their structures. Cell membranes act as selective barriers for various molecules including antibiotics. Gram-positive bacteria have one membrane (cytoplasmic) whereas gram-negatives have two membranes (one outer and one inner or cytoplasmic). Because of the double membrane, gram-negative bacteria have intrinsic resistance to many antibiotics. Entry of the antibiotic may be prevented by either the outer or the inner or both membranes. Entry of various molecules through membranes takes place through specific pores present in the membranes. Pores in some bacteria may be specific for transporting only positively charged molecules while those in other bacteria may be specific for only negatively charged molecules. If this does not match with the charge of the antibiotic, then the bacteria will be resistant to that antibiotic. For details, see the effect of penicillins on various gram-negative bacteria (Sect. 3.3.2.4). Resistance to tetracyclines is by pumping out the antibiotic after it enters the cell (Sect. 6.2.4). Resistance to β-lactam antibiotics can be by degradation of the antibiotics by enzymes called β-lactamases (Sect. 3.3.2.5). Resistance to aminoglycosides is also by modification of the drugs (Sect. 6.2.3). Resistance development to quinolones is by target modification (Sect. 5.3). More than one mechanism may be applicable for some antibiotics.

2.8 Synthetic Antibiotics

Semisynthetic antibiotics were made in the laboratory by chemically modifying natural antibiotics with the purpose of improving their properties. Some desired properties are broader spectrum of activity, less side effects, lower cost, greater shelf life or bio-stability, and lower frequency of resistance development. The first example of semisynthetic antibiotic was tetracycline which was made by catalytic hydrogenation of chlorotetracycline (the Cl was replaced with H) at Pfizer laboratory.

Various modifications were made synthetically to the penicillin structure to obtain broader spectrum of activity as well as other improved properties. Some examples include amoxicillin, ampicillin, carbenicillin, and ticarcillin (Fig. 3.14). The modified drugs ampicillin (α-aminobenzylpenicillin) and amoxicillin, both semisynthetic penicillins, are important because they are effective against both gram-negative and gram-positive bacteria (broad spectrum), while penicillin G works only for gram-positive bacteria. One important semisynthetic penicillin is methicillin, which was introduced in the early 1960s. Methicillin is not degraded by β-lactamase (Sect. 3.3.2.10) and so it is effective against bacteria that are resistant to penicillin. However, resistance to methicillin has also increased over the years, especially in the case of infection by MRSA (methicillin-resistant *Staphylococcus aureus*, Sect. 3.3.2.10) which has become a major concern today.

Advantage of Synthetic Antibiotics: Although most antibiotics are natural products or their semisynthetic derivatives, there are some antibiotics that are entirely man-made. The first chemically synthesized antibiotics were the sulfonamides. In later years, several more synthetic antibiotics were developed including trimethoprim, nalidixic acid and its derivatives. Mechanisms of action of all these antibiotics are discussed in later chapters. Except for a few synthetic antibiotics, most other antibiotics are natural products made by other microorganisms, namely bacteria and fungi, which have existed on this earth for millions of years. As discussed before (Sect. 2.6) mutations can develop within a day of bacterial growth. So in the millions of years, many mutations have developed including those that have resulted in formation of genes for antibiotic resistance. Thus, for all natural antibiotics there already exists resistance genes that can be transferred from one microorganism to another. Synthetic antibiotics, on the other hand, have been in existence for not more than 90 years since the first one was made in the 1930s. In this short period of time some point mutations may have developed but no gene is expected to be present in nature for resistance to these antibiotics. As discussed before, point mutations confer much weaker resistance than antibiotic resistance genes. So for pharmaceutical

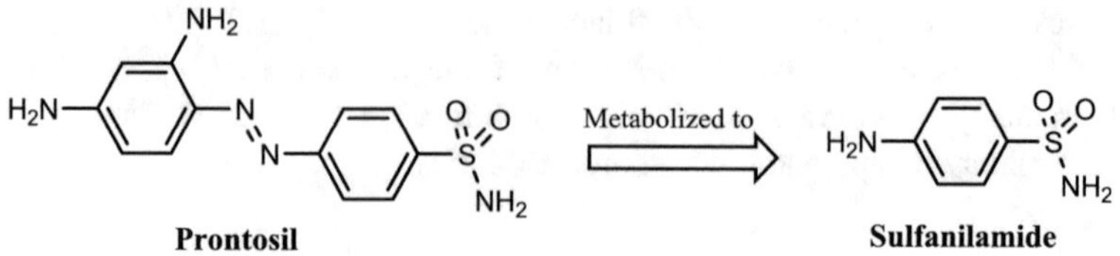

Fig. 2.2 The prodrug prontosil metabolized to the antibiotic sulfanilamide

companies it is always more desirable to develop new synthetic antibiotics than to discover new natural antibiotics. However, unfortunately very few synthetic antibiotics have been developed or are in the pipeline.

Discovery of sulfonamides: In 1932, Gerhard Domagk, in Germany, was examining various dyes for their antibacterial activities. There was precedence for testing dyes as potential chemotherapeutic agents. That is how Ehrlich had discovered Trypan Red (Sect. 1.2). The reasoning for testing dyes is that they stain bacteria. So it was thought that dyes may interfere with their growth. Domagk tested thousands of dyes for antibiotic property and discovered the dye Prontosil, which had such property. He received the 1939 Nobel Prize for his discovery of Prontosil. Normally when scientists test potential new antibiotics, they will first test them on bacteria growing in a test tube (in vitro). If it works then they will test it in animals that have been infected with the bacteria (in vivo). However, when Domagk did the test, he simultaneously used the dyes against bacteria growing in test tubes and also injected them in mice that were infected with bacteria. He made the strange discovery that one of the dyes, named Prontosil was effective against bacteria present in mice. However, in test tube assay the dye had no effect on the bacteria. So, if he had relied only on test tube assay, his discovery would never have been made. Usually when new drugs are tested, many of them work in vitro but not in vivo and so those drugs will not be useful clinically. In this case it was just the opposite. The reason it worked only when administered to mice but not in a test tube is because during metabolism in the mice the prontosil molecule was broken down to its smaller part called sulfanilamide (para-aminobenzenesulfonamide) which had the antibiotic property (Fig. 2.2). The dyeing property of the molecule was unrelated to its antibiotic property. Thus, prontosil can be classified as a "prodrug," which is defined as a medication that is administered in inactive form and is converted to the active form by metabolic reaction in the body. Once the active part of the prontosil dye was determined, scientists synthesized various derivatives of sulfonamides and tested them for their antibiotic properties. Thus, the first class of synthetic antibiotics, called "sulfa drugs," was born and continues to be used even today.

Other synthetic antimicrobials: The second synthetic antibiotic to be made was trimethoprim. In the 1960s, Hitchings and Elion explored the idea of synthesizing new antibiotics by targeting bacterial or viral DNA synthesis. Their work resulted in the synthesis of trimethoprim and pyrimethamine which are used to treat malaria, meningitis, septicemia, and a variety of other infectious diseases. They were the awarded the 1988 Nobel Prize for their work [11]. Similar to the sulfonamides, trimethoprim inhibits the biosynthesis of folic acid in bacteria. In fact, for a long time the two drugs were administered together as combination drugs under the brand names Septra, Bactrim, etc. Trimethoprim is able to penetrate deep into tissues, which made it a drug of choice for diseases such as typhoid. The mechanism of action of trimethoprim is discussed in Sect. 4.3.6. Another group of synthetic antibiotics are nalidixic acid and its derivatives called fluoroquinolones which can be taken by mouth and still achieve high concentration in the blood. These are discussed in Sect. 5.3.

Multidrug-resistant microorganism: Those microorganisms that are resistant to at least three out of the four antibiotic classes (those that affect the cell membrane, the cell wall, nucleic acid synthesis, and protein synthesis) are said to be multidrug resistant. Instances of infections by multidrug-resistant microorganisms have been rapidly increasing and are of great concern because most of the available antibiotics do not work against them. A group of bacteria known as ESKAPE are of particular concern. This group comprises *Enterococcus faecium* (vancomycin resistant), *Staphylococcus aureus* (methicillin or vancomycin resistant), *Klebsiella pneumoniae, Acinetobacter baumannii, Pseudomonas aeruginosa,* and *Enterobacter* spp., the latter four being carbapenem resistant. Some of these are discussed in later chapters.

New Antibiotics: We see that most of the antibiotics in use today were discovered in the first four decades since the discovery of penicillin. Bacteria have been found to develop resistance to all of them. The antibiotic crisis is made even worse by the fact that pharmaceutical companies are not developing many new antibiotics. Only three classes of synthetic or semisynthetic antibiotics were developed in the past five decades: fluoroquinolones in1986 (Sect. 5.3), linezolid, an oxazolidinone in 2000 (Sect. 6.2.7.1), and lefamulin, a pleuromutilin in 2019 (Sect. 6.2.7.2). The biggest advantage of totally new antibiotic is that there will be no resistance gene that neutralizes the effect of the antibiotic and point mutations conferring resistance to the antibiotic have not yet been selected for. Very few "truly new" antibiotics have been developed in the last three decades. Most "new" antibiotics developed are modified versions of already existing ones. Pharmaceutical companies are not investing much for research that can lead to developing new antibiotics because it is more profitable to develop drugs for fighting chronic problems such as high cholesterol or heart problems rather than antibiotics which are used by the patient for a short time only. Moreover, it is likely that the new antibiotic will lose effectiveness soon because of excessive use. The ideal use of a new antibiotic is to use it very little so that resistance development is delayed. However, that goes against any reasonable business model. Companies would want to make the maximum profit possible before the term of the patent expires. Solutions to these problems will be to significantly increase government funding and subsidizing of antibiotic research and manufacturing and revising patent validity period for new antibiotics.

2.9   Alternative Approaches for Studying Antibiotics

Fitness cost of antibiotic resistance. It is generally believed that antibiotic resistance makes bacteria weaker because the gain of antibiotic resistance will result in a loss of some function. For example, if a hypothetical antibiotic enters the bacterial cell through a certain pore in the cell membrane, resistance can develop by mutating the protein that makes the pore. However, that will also affect the transport of nutrients that normally enter through that pore, thus decreasing the fitness of the bacteria. Similarly, if resistance develops by decreasing the binding of the corresponding enzyme to the antibiotic, that may also decrease the binding to the natural substrate. If the resistance is due to the presence of an antibiotic resistance gene inserted into the chromosome, it will be an extra burden on the cell to express that gene. If the antibiotic resistance gene is on a plasmid, the cell will have to spend even more resources for not only expression of the resistance gene but also other genes that are on the plasmid. Also more energy and resources will be needed to replicate and maintain the plasmid.

Fitness cost of antibiotic resistance has been well documented. For example, resistance to fluoroquinolones has been shown to cause impaired motility in pseudomonads [12] and resistance to aminoglycosides is known to affect the structure of the ribosomes [13]. This fitness cost suffered by the bacteria is of benefit for us because resistant bacteria grow slower than nonresistant ones even in the absence of added antibiotic. Widespread use of antibiotics has resulted in a large increase in antibiotic-resistant strains. The best way to control the spread of antibiotic resistance is by stopping or limiting the use of the antibiotics. Since the fitness of the resistant strains is less than that of the nonresistant strains, with time, the proportion of resistant strain in the population will gradually decrease. A careful analysis of many mutations causing antibiotic resistance showed that most of these resistance mutations confer a fitness cost on the bacteria since most antibiotics target important cellular processes and resistance to them disrupts those processes [14].

Some antibiotic resistances can confer enhanced fitness on the bacteria: Contrary to the popular belief that antibiotic-resistant strains are less fit than wild-type strains and so can be easily eliminated by stopping the use of the antibiotic, alarm bell has been sounded by reporting that this is not universally true for all antibiotic resistance. So better methods such as developing vaccines against the bacteria are needed to tackle the problem of antibiotic resistance [15]. The authors reported that they found enhanced fitness in *Pseudomonas aeruginosa* because of antibiotic resistance. A transposon containing resistance gene for carbapenem was inserted in the *oprD* gene of the bacteria. Inactivation of the outer membrane porin (oprD) confers carbapenem resistance to the bacteria since the antibiotic normally enters the cell through this porin (Sect. 3.3.2.12). However, it was observed that in addition to gaining the carbapenem resistance property, the bacteria at the same time acquired enhanced resistance to killing at acidic pH. The reason proposed for this enhanced fitness is that the inactivation of the *oprD* gene led to changes in transcription pattern of numerous genes, some of which may be the cause of the new beneficial property.

This result is of immense concern because if antibiotic resistance develops during antibiotic therapy (due to inadequate concentration of antibiotic used), it may lead to increased fitness and virulence of the bacteria and may be more difficult to cure even if the particular antibiotic is no longer used by the patient.

Fitness may not always be directly correlated with increased rate of growth. For example, mutants of the periodontal pathogen *Aggregatibacter actinomycetemcomitans* that grow slower than the wild type may actually be more viable. This is because the bacteria are extremely sensitive to even mildly acidic pH resulting from metabolic acids secreted during fast growth [16]. It was demonstrated that inactivation of a sugar transport gene actually resulted in greater cell viability as a result of slow growth of the mutant bacteria [17].

Bacterial Persistence: Although antibiotic resistance development is the major reason why some bacteria are not killed by an antibiotic, another reason can be that some bacteria enter a slow-growing physiological state in which antibiotics cannot kill them, a phenomenon first reported for *Staphylococcus aureus* [18]. However, it is believed that this applies to all bacteria. This phenomenon by which a population of antibiotic sensitive cells produces some transiently resistant cells is known as "persistence" [19]. Later the persister cells can switch back to the growing state after surviving the antibiotic treatment. Since persisters are only transiently resistant, when the cells multiply, the daughter cells will not be antibiotic resistant. This phenomenon of forming persisters is not dependent on the use of the antibiotic but can happen anytime and the transient resistance developed is not just to the antibiotic being used but to all antibiotics. Thus, persister cells are transiently multidrug-resistant cells. This is because antibiotics usually inhibit active targets and thus in slow-growing or dormant bacteria the antibiotics cannot inhibit the targets.

Although this phenomenon applies to all bacteria, it does not create an alarming problem because the multidrug resistance is only transient and secondly, only a very small percentage of the bacterial population become persisters, so the body's immune system can effectively cope with this small number of resistant bacteria. However, the persisters are of serious concern in tuberculosis. The molecular mechanism of bacterial persistence has been reviewed [20] in which the authors summarize the role of the bacterial stress alarmone, $5'$-diphosphate-$3'$-diphosphate guanosine (ppGpp) as a central regulator of persistence. An alarmone is an intracellular signal molecule that is produced in response to harsh environmental conditions.

Alternative views regarding the function of antibiotics and antibiotic resistance genes: True scientists should always be open to new theories that are proposed with proper evidence. So it is worthwhile to mention some alternative views about the function of antibiotics and antibiotic resistance genes. It is widely accepted that microorganisms secrete antibiotics to inhibit or kill other microorganisms in response to competition for limited resources. However, that concept has been challenged by some scientists [21, 22] because the natural concentrations of antibiotics made are much lower than the therapeutic concentrations needed for killing other neighboring microorganisms. It has been proposed that the principal roles of these so-called antibiotics are cell–cell communication and not antibiosis.

It is also widely accepted that the function of antibiotic resistance genes in microorganisms is to protect them from the effect of antibiotics. However, that theory has also been challenged by some scientists [23] because the expression levels of most antibiotic resistance genes are much higher than what is required to effectively combat the therapeutic doses of antibiotics. In fact, as explained above, the actual concentrations of antibiotics present in nature is even less than therapeutic doses. This suggests that antibiotic resistance genes can possibly have an alternative physiological role.

2.10   Antibiotic Use in Animals

Antibiotic use in animals is a major cause of resistance development. There are two types of use of antibiotics in animals.

2.10.1   Therapeutic Use

Similar to infections in humans, infections in animals are also cured with antibiotics. There are many times more farm animals than there are humans. According to the US Department of Agriculture report of 2010, more than ten billion animals (excluding fish) are raised and killed in the USA every year. Of these, 91% are chickens raised for meat, 4.5% are chickens raised for eggs, 2.5% are turkeys, and 2% are cows, pigs, and other animals. This corresponds to an average of 28 land animals per person per year plus, according to another estimate, about 175 aquatic animals per person per year. Total number of land animals killed every year for food worldwide is about 65 billion. Use of antibiotics to cure infectious diseases in these animals is understandable but one should be aware that all use of antibiotics will contribute to increase in antibiotic resistance pool (Sect. 2.6).

Similar to animals, plants can also suffer from infections, which can be cured with antibiotics. These antibiotics are also sprayed on the plants, a process by which most of the antibiotics end up in the soil thereby increasing the resistance pool. Subtherapeutic use (see below, Sect. 2.10.2) of antibiotics in fish for growth promotion has been discontinued in Europe and North America; however, therapeutic use in fish is still a common practice. Since the antibiotics are added to fish food, the whole body of water gets contaminated with the antibiotic. This increases the antibiotic resistance pool and selects for antibiotic-resistant bacteria in the water.

Antibiotics are also used for pets. According to estimates by the US Humane Society, about 62% of US households have at least one pet. They are also given antibiotics for treatment of bacterial infections. One can buy antibiotics from a pet store without a prescription. As a result there is a misuse of these antibiotics. Since the antibiotics given for pets are the same as those given for humans, people often buy antibiotics for themselves from pet stores. This way they save money because antibiotics in pet stores are cheaper and moreover, they avoid seeing a doctor. This kind of self-medication is potentially dangerous and also increases the antibiotic resistance pool.

Misleading safety standards. Fruits and vegetables are declared to be safe for consumption if antibiotic residue is below a certain limit. However, it is not the antibiotic residue that is the main concern but the process by which the produce is obtained. If antibiotic was used during its growth, it will contribute to the antibiotic resistance pool even though all residues are later washed away from the produce. The resistant bacteria that are selected for due to the antibiotics can then transfer the resistance to other bacteria including those that infect humans.

According to rules, fish that have previously been treated with antibiotics can be harvested and sold only after waiting for a certain period of time, known as the withdrawal period. The withdrawal time, which depends on the fish as well as the antibiotic used ensures that there will be no antibiotic residue left in the fish. However, once again it should be noted that the antibiotic residue should be only a minor concern, the greater concern should be the history of the fish and how much the farming has contributed to the antibiotic resistance pool.

2.10.2   Subtherapeutic Use

Contribution of therapeutic use of antibiotics to the antibiotic resistance problem actually appears to be insignificant when compared to another type of use of antibiotics in animals. This is known as subtherapeutic use and is not related to any infection. According to FDA reports, only about 20% of the approximately 18,000 tons of antimicrobials sold in the USA are used by humans while the rest 80% are used in animals [24,25]. Of course, our farm animals are not so sick that they will need this amount of therapeutic antibiotics. Most of this antibiotic is added to animal feed to increase their body weight, which means more profit for the farmers. Antibiotics used this way are also called antimicrobial growth promoters (AGPs). This phenomenon of growth promotion was an accidental discovery when scientists were testing random food additives to discover new vitamins. In 1948, Robert Stokstad and Thomas Jukes added cellular debris of *Streptomyces aureofaciens* to chicken feed, after the antibiotic chlorotetracycline was extracted from the bacterial culture and observed faster growth of the chicken. Initially they thought that it was due to vitamin B12 present in the additive but later it was understood that the growth promotion was due to small amounts of chlorotetracycline still remaining in the bacterial cell debris. Further studies confirmed the surprise discovery that addition of a small amount (much less than therapeutic dose) of chlorotetracycline increased the growth rate of farm animals. Soon it was found that the same effect was observed with many other antibiotics. Since only a small amount of antibiotic was sufficient to get this effect, the practice was allowed by the government and antibiotics for farm animals was allowed to be sold without a prescription. The simple logic was not considered that therapeutic use is only for a short time (about 10 days per infection) whereas subtherapeutic use is for lifetime of the animal.

The mechanism of growth promotion by subtherapeutic use of antibiotics is not clearly understood. It is possible that antibiotics kill bacteria that compete with beneficial bacteria in the intestines of the animals, thereby promoting growth of the animals. Although it is not clearly understood how antibiotics promote growth, one observation made is that the antibiotic does not have to enter the bloodstream to show the growth-promoting effect because its site of action is in the intestines. Bacitracins, which are not absorbed through the intestinal walls are not used internally in humans but are used only as ointments for skin infections. However, it is one of the most commonly used growth-promoting antibiotic in animals. Of all the bacitracin manufactured, 90% is used for growth promotion in farm animals.

Negative aspects of subtherapeutic use: With time it was realized there is a big negative effect of the subtherapeutic use of antibiotics that far outweighs the minor monetary benefit of growth promotion. The constant exposure of the bacteria present in the animals to the antibiotics selects for antibiotic-resistant bacteria as explained in Sect. 2.5. If these bacteria are opportunistic pathogens, they may later infect the animals under conditions of weakened immune system. Or it may be possible that antibiotic-resistant beneficial bacteria in the animals may transfer (Sect. 2.6) the antibiotic-resistant genes to any infecting bacteria making it difficult to cure the disease.

Antibiotic-resistant bacteria can also be transferred from animals to humans. This can take place by any of the following ways: (a) by eating contaminated meat that is not cooked properly, (b) by every-day direct contact of farm workers with the animals, (c) by transfer of resistant bacteria from animal manure to soil then to plants and then to humans through the food chain, and (d) by transfer of bacteria from dead non-farm animals or farm animals who died because of disease, to the soil and then to plants and then to humans.

Call to ban subtherapeutic use of antibiotics: Because of the negative effects of subtherapeutic use, there has been calls to ban the practice particularly in the developed world where the practice was more prevalent. In 1969, the UK Government asked the Swann Committee to report on the use of antibiotics in both humans and animals. The committee concluded that AGPs contribute significantly to the development of antibiotic-resistant infections. The committee recommended that growth promotion in animals with antibiotics that are also used for human therapy should be banned. The use of tetracycline and later penicillin as growth promoters was gradually phased out by the European Common Market in the 1970s [26]. Later the European Union banned the subtherapeutic use of avoparcin in 1997 and bacitracin, spiramycin, tylosin, and virginiamycin in 1999. In 2000, Denmark and in 2006, the European Union banned all AGP's. India ranks second in aquaculture production globally and fourth in the use of AGP's [27] but, has not yet banned the subtherapeutic use of antibiotics. In 2019, India banned the subtherapeutic use of colistin. This is a significant step since colistin is an important antibiotic for humans and is usually prescribed as a last resort when other antibiotics have failed. China, which is the largest user of antibiotics for growth promotion, had also banned the use of colistin in animals in 2017 before they finally banned the subtherapeutic use of all antibiotics in 2020.

Following the ban by the European Union, numerous studies have shown that there has been a decline in the cases of antibiotic resistance [28]. It should be pointed out that although most studies show that banning subtherapeutic use of antibiotics has a positive effect, there is not a 100% agreement on that conclusion. It is believed by some that subtherapeutic use has a prophylactic effect and is needed for proper health of the animals. Banning the practice will increase diseases in the animals which can then be passed on to humans [29].

In the USA in 1977, based on the recommendations of a 1970 FDA Task Force Report, the FDA proposed to withdraw drug approvals for subtherapeutic uses of penicillin and tetracyclines in animal feed [30]. These two drugs were chosen because of their importance in human medicine. However, the proposal was criticized for lack of adequate evidence and US Congress directed FDA to hold the proposed withdrawal until further studies are conducted. In the meantime there have been numerous reports of antibiotic resistance (including multidrug resistance) development related to subtherapeutic use of antibiotics.

In 1999, in an open letter to the Commissioner of FDA, 53 eminent scientists from universities and research institutions throughout the USA urged that swift action be taken to protect the effectiveness of antibiotics by limiting their subtherapeutic use in agriculture [31]. In the letter they pointed out that although the FDA initiated proceedings to ban the subtherapeutic use of antibiotics in animal feed in 1977, that work was never completed while new research continued to demonstrate that subtherapeutic use increases antibiotic resistance in pathogens. Those resistant bacteria can be transferred to humans via contaminated food products or through direct or indirect contact with animals.

Transfer of resistance genes between bacteria puts humans at risk: The fact that subtherapeutic use of antibiotics increases the development of antibiotic resistance in bacteria has been demonstrated repeatedly in numerous studies. Some of these bacterial species can then infect humans who come in contact with the animals. In another scenario, the resistance genes can be transferred to other bacteria which can then infect humans. It was shown that when tylosin (a macrolide antibiotic, Sect. 6.2. 6.1) was given to pigs for growth promotion it resulted in the appearance of erythromycin (also a macrolide) resistant enterococci in the pigs' guts and at the same time erythromycin-resistant staphylococci was detected in the pigs' skin. These results demonstrate that use of one antibiotic can promote resistance to a different antibiotic and then the resistance can be transferred to another species of bacteria [32]. Bacteria that cause diseases in animals and plants may not infect humans. However, these bacteria may belong to the same family as those that infect humans. One example is the bacterial species *Erwinia* that causes fruit disease but does not infect humans. However, it is in the same family of bacteria (*Enterobacteriaceae*) as *E. coli*, *Salmonella*, and *Shigella* which infect humans. Transfer of resistance genes between these species of bacteria has been well documented [33, 34].

It was shown in 1975 that adding low-dose oxytetracycline in chicken feed resulted in the appearance of tetracycline-resistant *E. coli* in the intestinal flora of not only the chickens but also the farm workers who routinely handle the chickens [35]. Thus, this demonstrated the selection of antibiotic-resistant bacteria in the chicken and then transmission of the bacteria from chicken to humans. Another example is the antibiotic enrofloxacin, which is currently approved by the FDA for the treatment of individual pets and domestic animals in the USA. Both therapeutic and subtherapeutic use of the antibiotic in chicken feed can result in the development of resistance in *E. coli*, which can then transfer the resistance to *Campylobacter*, another resident bacterial species present in chicken. The *Campylobacter* is harmless to the chickens but can infect humans. It is estimated that more than 80% of the chicken meat in the USA is contaminated with *Campylobacter*, which is the most common cause of food-borne bacterial infection in the USA. If the bacteria have acquired resistance to enrofloxacin that was given to the chicken, the same resistance will also be effective against ciprofloxacin (Sect. 5.3), which is very similar to enrofloxacin and is the most widely used antibiotic for food-borne illnesses in humans. One encouraging news came in 2005 when the FDA withdrew approval of Baytril (brand name for enrofloxacin) for use in water to treat flocks of poultry. The reason cited was that this practice was known to promote the evolution of fluoroquinolone-resistant strains of the bacterium *Campylobacter*, a human pathogen [36].

There is another well-known example of subtherapeutic use of one antibiotic resulting in resistance to other related antibiotics. A class of antibiotics called streptogramins is often used as an antibiotic of last resort when other antibiotics, including vancomycin have failed because of infection by multidrug-resistant bacteria. One such antibiotic combination, Synercid was approved by FDA for human use in 1999. However, before its first use in humans, the effectiveness of Synercid had already been compromised because another streptogramin antibiotic, virginiamycin was already approved for use in animals and had been extensively used not just for curing infections but to a much greater extent for growth promotion in animals. As a result, turkeys that were fed with virginiamycin were found to harbor bacteria that had developed resistance to Synercid even though they had not been previously exposed to Synercid [37]. Bacteria resistant to Synercid were detected in humans even before the antibiotic was first used in humans in Germany [38]. Thus, virginiamycin-resistant bacteria arising due to subtherapeutic use in animals had been transmitted to humans either through food or through people who handle the animals.

Sweden was the first country to ban subtherapeutic use of antibiotics in as far back as 1986. At that time this was done mainly out of concern for residues of antibiotics that remain in food due to their subtherapeutic use. However, antibiotic residue in food is actually a very minor concern because of the small amount of residual antibiotic. Also most of the residual antibiotic can be easily removed by washing the food and are also destroyed during cooking the meat. Of much greater concern is the fact that subtherapeutic use increases the antibiotic resistance pool, creates resistant mutants and, what is even worse, results in resistance to antibiotics that are used in humans as described above. In 1998, the World Health Organization (WHO) called for a ban on the subtherapeutic use of those antibiotics that either

(1) are used therapeutically in humans or (2) are known to select for cross-resistance to antibiotics used in humans. As mentioned above, by 1999, the European Union banned the use of some antibiotics used in animal feed because of concerns about cross-resistance to antibiotics used in humans. In 2006, it banned subtherapeutic use of all antibiotics.

The USA and Canada stand out: By now most developed nations except the USA and Canada have banned the subtherapeutic use of some or all antibiotics. In the USA, a slight progress was made in 2013 when the FDA issued not a ban but a nonbinding recommendation for voluntary withdrawal of medically important antibiotics from growth promotion [39]. This slight progress is too little too late while the problem of antibiotic resistance keeps increasing. The situation in Canada is the same if not worse. In Canada also there is denial by people in authority regarding any link between subtherapeutic use of antibiotic and development of antibiotic resistance. In some parts of Canada, farmers do not need prescriptions to buy antibiotics even the ones that are important for human medicine. Like in the USA, in Canada also there is expected to be self-regulation by farmers and drug manufacturers to use antibiotics sensibly. Leaving such important decisions to agencies that benefit financially from it is not expected to give the desired results.

Multidrug resistance development caused by subtherapeutic use: While some people continue to deny that subtherapeutic use of antibiotics causes development of antibiotic resistance, a recent publication has provided evidence for a direct link between the two. There are two methods of antibiotic resistance development that have been well established: (1) by selection of naturally occurring point mutations and (2) by resistance gene acquisition (Sect. 2.6). In the recent publication a new method of antibiotic resistance development has been described. It was shown that the presence of subtherapeutic level of an antibiotic induces generation of point mutations in the bacterial genome. Some of these mutations can confer resistance to other antibiotics that may not be related to the antibiotic that the cells have been subjected to [3]. The authors have demonstrated that low levels of bactericidal antibiotics stimulate the production of reactive oxygen species, which are known to cause mutations which result in emergence of resistance to various other antibiotics including multidrug resistance. Formation of reactive oxygen species in response to low concentrations of antibiotics has previously been shown to take place for quinolones, β-lactams, and aminoglycosides [40, 41, 42].

Whatever is the mechanism of development of multidrug-resistant bacteria in animals, there is a definite threat of transfer to humans even if they are not in direct contact with the animals. The most vulnerable are those people who are already taking an antibiotic for some other unrelated infection. This is because the antibiotic kills all bacteria in the body including the beneficial ones. So the infecting multidrug-resistant bacteria can more easily cause disease because (1) they are not killed by the antibiotic and (2) they face no competition from any resident bacteria in the body.

Growth-promoting effect is less than what was previously thought: Subtherapeutic use of antibiotics in farm animals was banned by the European Union in 1999 and since then there has been a decline in the prevalence of antibiotic-resistant bacteria [28]. This is encouraging news for proponents of a ban on subtherapeutic use of antibiotics. Productivity of Danish swine farms was monitored for 8 years before and 8 years after the ban. Stoppage of subtherapeutic use of antibiotics had no negative effect on pig productivity, in fact there was an increase in the number of pigs, and mortality rate of pigs remained constant [43]. This result challenges the "(mis)conception" that antibiotics have any significant growth-promoting effect.

Recent reports show that the amount of growth promotion is about 1–2% as opposed to 10% that was originally reported in the 1950s [44]. Considering the diminishing returns, and the certainty of increasing antibiotic resistance, it needs to be decided whether it is worth the risk to obtain about 1–2% increase in profit. Today the amount of antibiotic needed to obtain the level of growth increase supposedly obtained in the 1950s is gradually increasing to that of therapeutic doses. Thus, today it makes even more sense to ban all subtherapeutic use of antibiotics.

2.11 Prevention of Antibiotic Resistance Development

Every misuse of antibiotics contributes to the development of antibiotic resistance. Antibiotics are misused by many people in most countries. People often request antibiotics from doctors for any disease because of the misconception that antibiotics, which are antibacterials, can cure viral infections such as common cold. Many times doctors agree to the patients' demands just to appear nice to their customers. Oftentimes antibacterial antibiotics are prescribed for viral infection in order to prevent secondary bacterial infection. There is also the misconception that antibiotics can do no harm. In many countries antibiotics and most other drugs are available without prescription, so people themselves decide that they need an antibiotic. In the developed countries this is theoretically not possible, but still many people manage to get antibiotics without a prescription. One common source of antibiotics is half used antibiotic dose from a previous infection. This creates double the problem. There is a chance of resistance development after the first infection since the complete dose was not taken and then the second time, because of self-medication the patient may be unnecessarily taking the antibiotics for a viral infection. Even if the antibiotics is the right one for the second infection, the patient will get only half the required dose, again increasing the chance of resistance development. Thus, everyone, including doctors, patients, as well as everyone involved in the subtherapeutic use of antibiotics in animals shares the responsibility for resistance development to antibiotics.

What is the solution to the problem? There is no simple solution. Doctors should stop overprescribing antibiotics. Certainly, the sale of antibiotics needs to be regulated. In the poor countries this is not an enforceable solution because of the scarcity of doctors. Even if patients have access to a doctor, they may not be able to afford their fees. This will encourage the creation of a black market for antibiotics. A long-term but more effective solution is to educate the people about the antibiotic resistance development problem. Subtherapeutic use is a major contributor and the farming industry needs to stop the practice. The government has a big role to play in the prevention of antibiotic resistance development. One important step that has already been taken by many governments is to ban subtherapeutic use of all antibiotics. The next important step is to provide financial incentive for developing new antibiotics. It costs millions of dollars for discovery, development, and clinical trials of new drugs. Once approved by the FDA, the companies have only a limited amount of time before the patent expires and they want to make as much money as possible in that short time. The patent expiry date is probably justified for most other products but there should be a special consideration for antibiotics because the best use of an antibiotic is to use it very little. The government needs to recognize this dilemma and change patent laws and tax laws so that the pharmaceutical companies will be willing to invest their time and money in developing new antibiotics. The government should also provide more grant money to scientists to develop new antibiotics. If enough action is not taken now, very soon the problem will reach a crisis situation.

The National Action Plan: According to Centers for Disease Control and Prevention (CDC), drug-resistant bacteria cause 23,000 deaths and two million illnesses each year in the USA. The threat of increasing antibiotic resistance has been taken seriously by the government. In a White House Press Release on March 27, 2015, President Barack Obama's office released a comprehensive plan to combat the rise of antibiotic-resistant bacteria [45]. The National Action Plan for Combating Antibiotic Resistant Bacteria describes five goals, one of which is to slow the emergence of and prevent the spread of antibiotic resistance. According to the CDC, about half of all human antibiotic use is unnecessary. By the year 2020, it is expected that inappropriate use of antibiotics will be cut by 50% in outpatient settings and the use of medically important antibiotics for growth promotion in food-producing animals will be completely eliminated. Another goal of the Action Plan is to accelerate basic and applied research and develop new antibiotics. The federal funding for research on antibiotics has been nearly doubled in the President's FY 2016 Budget.

References

1. Bauer AW, Kirby WM, Sherris JC, Turck M (1966) Antibiotic susceptibility testing by a standardized single disc method. Am J Clin Pathol 45:493–496
2. Bhattacharjee MK (2015) Better visualization and photodocumentation of zone of inhibition by staining cells and background agar differently. J Antibiot 68:657–659
3. Kohanski MA, DePristo MA, Collins JJ (2010) Sublethal antibiotic treatment leads to multi-drug resistance via radical-induced mutagenesis. Mol Cell 37:311–320
4. Fleming A (1945) Penicillin. Nobel Lecture. www.nobelprize.org
5. Mazel D (2006) Integrons: agents of bacterial evolution. Nat Rev Microbiol 4:608–620
6. Bhattacharjee MK, Meyer RJ (1991) A segment of a plasmid gene required for conjugal transfer encodes a site-specific, single-strand DNA endonuclease and ligase. Nucleic Acids Res 19: 1129–1137

7. Smith HO, Tomb J-F, Dougherty B, Fleischman RD, Craig VJ (1995) Frequency and distribution of DNA uptake signal sequences in the *Haemophilus influenzae* Rd genome. Science 269: 538–540

8. Salyers AA, Shoemaker NB, Stevens AM, Li LY (1995) Conjugative transposons: an unusual and diverse set of integrated gene transfer elements. Microbiol Rev 59:579–590

9. CDC Report (2013) Timeline of key antibiotic resistance events. http://www.cdc.gov/drugresistance/threat-report-2013/pdf/ar-threats-2013-508.pdf#page=28

10. Spratt BG (1994) Resistance to antibiotics mediated by target alterations. Science 264:388–393

11. Hitchings GA (1988) Selective inhibitors of dihydrofolate reductase. Nobel Lecture. www.nobelprize.org

12. Stickland HG, Davenport PW, Lilley KS, Griffin JL, Welch M (2010) Mutation of nfxB causes global changes in the physiology and metabolism of Pseudomonas aeruginosa. J Proteome Res 9:2957–2967

13. Springer B, Kidan YG, Prammananan T, Ellrott K, Bottger EC, Sander P (2001) Mechanisms of streptomycin resistance: selection of mutations in the 16S rRNA gene conferring resistance. Antimicrob Agents Chemother 45:2877–2884

14. Melnyk AH, Wong A, Kassen R (2014) The fitness costs of antibiotic resistance mutations. Evol Appl. ISSN 1752-4571. https://doi.org/10.1111/eva.12196

15. Skurnik D, Roux D, Cattoir V, Danilchanka O, Lu X, Yoder-Himes DR, Han K, Guillard T, Jiang D, Gaultier C, Guerin F, Aschard H, Leclercq R, Mekalanos JJ, Lory S, Pier GB (2013) Enhanced in vivo fitness of carbapenem-resistant oprD mutants of Pseudomonas aeruginosa revealed through high-throughput sequencing. Proc Natl Acad Sci U S A 110:20747–20752

16. Bhattacharjee MK, Childs CB, Ali E (2011) Sensitivity of the Periodontal Pathogen, *Aggregatibacter actinomycetemcomitans* at Mildly acidic pH. J Periodontol 82:917–925

17. Bhattacharjee MK, Anees M, Patel A (2018) Increased viability of sugar transport-deficient mutant of the periodontal pathogen *Aggregatibacter actinomycetemcomitans*. *Curr Microbiol* 75:1460–1467

18. Bigger JW (1944) Treatment of staphylococcal infections with penicillin by intermittent sterilisation. Lancet 244:497–500

19. Lewis K (2010) Persister cells. Annu Rev Microbiol 64:357–372

20. Maisonneuve E, Gerdes K (2014) Molecular mechanisms underlying bacterial persisters. Cell 157:539–548

21. Davies J (2006) Are antibiotics naturally antibiotics? J Ind Microbiol Biotechnol 33:496–499

22. Goh EB, Yim G, Tsui W, McClure J, Surette MG, Davies J (2002) Transcriptional modulation of bacterial gene expression by subinhibitory concentrations of antibiotics. Proc Natl Acad Sci U S A 99:17025–17030

23. Krulwich TA, Lewinson O, Padam E, Bibi E (2005) Do physiological roles foster persistence of drug/multidrug-efflux transporters? A case study. Nat Rev Microbiol 3:566–572

24. FDA (2012) Drug use review. http://www.fda.gov/downloads/Drugs/DrugSafety/InformationbyDrugClass/UCM319435.pdf

25. FDA (2015) FDA annual summary report on antimicrobials sold or distributed in 2013 for use in food-producing animals. http://www.fda.gov/AnimalVeterinary/NewsEvents/CVMUpdates/ucm440585.htm

26. Castanon JIR (2007) History of the use of antibiotic as growth promoters in European poultry feeds. Poult Sci 86:2466–2471

27. Walia K, Sharma M, Vijay S, Shome BR (2019) Understanding policy dilemmas around antibiotic use in food animals & offering potential solutions. Indian J Med Res 149:107–118. https://doi.org/10.4103/ijmr.IJMR_2_18

28. Marshall BM, Levy SB (2011) Food animals and antimicrobials: impacts on human health. Clin Microbiol Rev 24:718–733

29. Casewell M, Friis C, Marco E, McMullin P, Phillips I (2003) The European ban on growth promoting antibiotics and emerging consequences for human and animal health. J Antimicrob Chemother 52:159–161

30. FDA (1977) Federal registers: 42 FR 43772 (August 30, 1977) and 42 FR 56264 (October 21, 1977)
31. CSPI Report (1999) Letter to Dr. Henney of the FDA. http://www.cspinet.org/reports/letterhenney.htm
32. Aarestrup FM, Carstensen B (1998) Effect of tylosin used as a growth promoter on the occurrence of macrolide-resistant enterococci and staphylococci in pigs. Microb Drug Resist 4:307–312
33. Chatterjee AK, Starr MP (1972) Transfer among Erwinia spp. and other enterobacteria of antibiotic resistance carried on R factors. J Bacteriol 112:576–584
34. Sato M, Wei W, Watanabe K (2003) Multidrug-resistance plasmid of *Enterobacter cloacae*: transfer to *Erwinia herbicola* on the phylloplane of mulberry and weeds. J Gen Plant Pathol 69: 391–396
35. Levy SB, FitzGerald GB, Macone AB (1976) Changes in intestinal flora of farm personnel after introduction of a tetracycline-supplemented feed on a farm. N Engl J Med 295:583–588
36. FDA (2005) Withdrawal of Enrofloxacin for poultry. http://www.fda.gov/animalveterinary/safetyhealth/recallswithdrawals/ucm042004.htm
37. Welton LA, Thal LA, Perri MB, Donabedian S, McMahon J, Chow JW, Zervos MJ (1998) Antimicrobial resistance in enterococci isolated from turkey flocks fed virginiamycin. Antimicrob Agents Chemother 42:705–708
38. Witte W (1998) Medical consequences of antibiotic use in agriculture. Science 279:996–997
39. FDA (2013) Guidance for industry #213. New animal drugs and new animal drug combination products administered in or on medicated feed or drinking water of food-producing animals: recommendations for drug sponsors for voluntarily aligning product use conditions with GFI #209. http://www.fda.gov/downloads/AnimalVeterinary/GuidanceCompliance Enforcement/GuidanceforIndustry/UCM299624.pdf
40. Kohanski MA, Dwyer DJ, Hayete B, Lawrence CA, Collins JJ (2007) A common mechanism of cellular death induced by bactericidal antibiotics. Cell 130:797–810
41. Kohanski MA, Dwyer DJ, Wierzbowski J, Cottarel G, Collins JJ (2008) Mistranslation of membrane proteins and two-component system activation trigger antibiotic-mediated cell death. Cell 135:679–690
42. Dwyer D, Kohanski M, Hayete B, Collins J (2007) Gyrase inhibitors induce an oxidative damage cellular death pathway in *Escherichia coli*. Mol Syst Biol 3:91
43. Aarestrup FM, Jensen VF, Emborg HD, Jacobsen E, Wegener HC (2010) Changes in the use of antimicrobials and the effects on productivity of swine farms in Denmark. Am J Vet Res 71: 726–733
44. Key N, McBride W (2014) Antibiotics used for growth promotion have a small positive effect on hog farm productivity. United States Department of Agriculture. http://www.ers.usda.gov/amber-waves/2014-july/antibiotics-used-for-growth-promotion-have-a-small-positive effect-on-hog-farm-productivity.aspx#.VdvcVmrwuGI
45. White House Press Release (2015) The National Action Plan for combating antibiotic resistant bacteria. https://www.whitehouse.gov/the-press-office/2015/03/27/fact-sheetobama-administration-releases-national-action-plan-combat-ant

Chapter 3
Antibiotics That Inhibit Cell Wall Synthesis

Abstract Structure of the bacterial cell wall, the metabolic pathway for the biosynthesis of the cell wall, and various antibiotics affecting the different stages of cell wall synthesis are presented. Mechanisms of action of the antibiotics and the mechanisms of resistance development to the antibiotics are discussed. The antibiotics include fosfomycin, cycloserine, β-lactams, carbapenems, bacitracin, moenomycin, mersacidin, vancomycin, and teixobactin. β-lactamase and β-lactamase inhibitors are discussed in the context of antibiotic resistance.

3.1 Background Biochemistry Information

3.1.1 Carbohydrates

An introduction to sugars and polysaccharides is necessary in order to understand the structure of the bacterial cell wall.

Monosaccharides Carbohydrates or saccharides, which have the general formula $(C.H_2O)_n$ where $n > 3$, can be considered as "carbon hydrates." The basic unit of a carbohydrate is called a monosaccharide, which usually has 3, 4, 5, or 6 carbons, the most common being C6 monosaccharides (Fig. 3.1). Polysaccharides have many covalently linked monosaccharides and can have molecular weights of millions of daltons.

Monosaccharides can be aldehydes or ketones and are called aldoses or ketoses, respectively. Ketoses have one chiral center less than aldoses. For example, glucose and fructose are both hexoses. Glucose is an aldose with four chiral centers and so has $2^4 = 16$ stereoisomers. Fructose is a ketose with three chiral centers and so has $2^3 = 8$ stereoisomers. Only half of these sugars are natural: those that have the OH at C5 on the right in Fischer projection formula and are called D-sugars. Sugars also exist in ring form, which is formed by the reaction of C5-OH with the carbonyl carbon to form a hemiacetal or hemiketal. By this reaction the carbonyl carbon is changed to a new chiral center (called anomeric carbon) and so two stereoisomers (called anomers) are formed for each sugar. These are called α (anomeric OH

M. K. Bhattacharjee, *Chemistry of Antibiotics and Related Drugs*,
https://doi.org/10.1007/978-3-031-07582-7_3

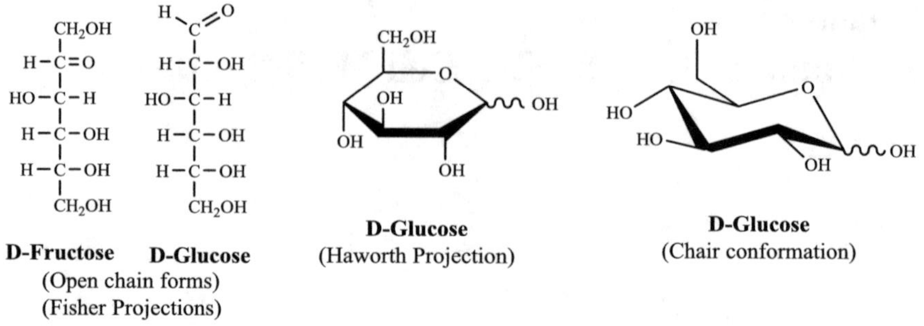

Fig. 3.1 Structures of fructose and glucose. The wavy bond indicates that the bond can be pointing up (α-D-glucose) or down (β-D-glucose)

pointing down) and β (anomeric OH pointing up) anomers. One way of drawing the ring structure is the Haworth projection formula in which all six atoms (5 carbons and an oxygen) of the ring are shown in one plane. However, this is not the true structure of the rings. Each carbon is sp^3 hybridized and so is tetrahedral. In order to maintain ~108° angle (for sp^3) with least torsional strain, the sugars exist in chair like structures in which four atoms of the ring are in one plane while C1 and C4 are alternately above and below the plane of the ring. Substituents can be either equatorial (pointing along the plane of the ring) or axial (perpendicular to the ring). Molecules are more stable if larger substituents are in equatorial position. If they are in axial positions, they will experience steric hindrance (known as 1,3-diaxial interaction) from axial hydrogens at the third carbons from the carbon in question. β-D-glucose is the only D-aldohexose that has all substituents in equatorial positions and so is the most abundant naturally occurring monosaccharide.

There are several biologically important sugar derivatives. Sugars in which an OH is replaced by H are known as deoxy sugars. 2-Deoxy ribose, an aldopentose, is a biologically important sugar present in DNA. Amino sugars are components of many polysaccharides. They usually have an NH$_2$ group at C2; for example, α-D-glucosamine (full name: 2-amino-2-deoxy-α-D-glucopyranose). In many natural carbohydrates an acetyl group is linked to the nitrogen as an amide bond forming N-acetylglucosamine (Fig. 3.2). An example of this is found in the cell wall of bacteria.

Disaccharides and Polysaccharides Disaccharides are formed by linking two monosaccharides by a covalent bond. The anomeric OH condenses with any OH of another sugar and releases a molecule of water resulting in the formation of α and β-glycosides. The new bond is called a glycosidic bond (Fig. 3.3) and is written as α(1 → 4) or β(1 → 4) or α(1 → 2), etc. Polysaccharides, also known as glycans, are formed when many monosaccharides are glycosidically linked to one another to form long chains. Since glycosidic linkage can be formed to any one of the OH groups of a sugar, it is possible for one sugar to have more than one glycosidic bond, thus resulting in branch points in the chains. Unlike proteins and nucleic acids (DNA

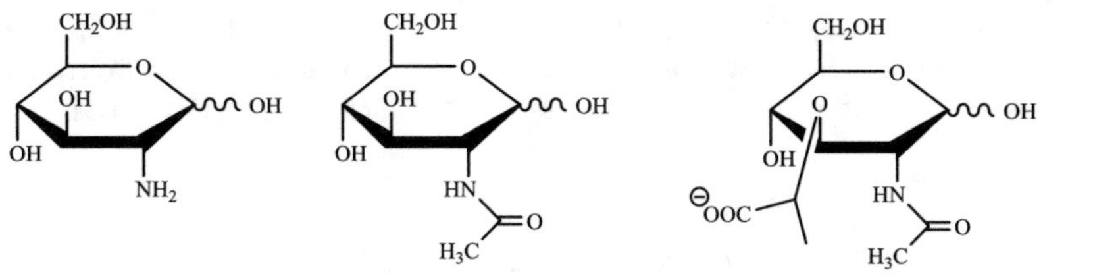

2-D-Glucosamine   N-acetyl-2-D-Glucosamine   N-acetyl-2-D-Glucosamine-3-lactate
                                                (N-acetyl muramic acid)

Fig. 3.2 Modified sugars: Glucosamine, *N*-acetylglucosamine, and *N*-acetyl muramic acid

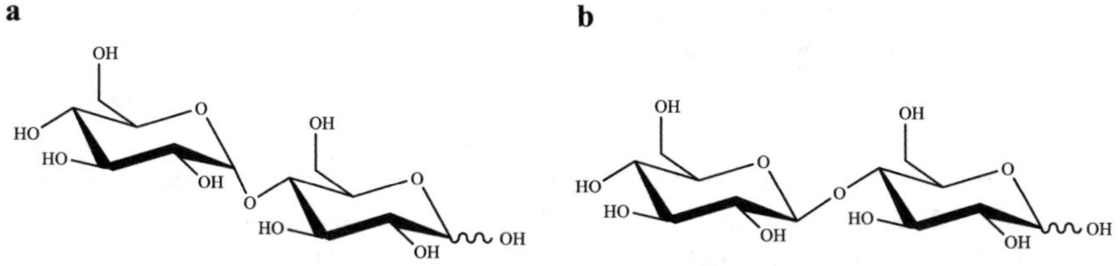

Fig. 3.3 (**a**) $\alpha(1 \rightarrow 4)$ and (**b**) $\beta(1 \rightarrow 4)$ glycosidic bonds between two glucose units

and RNA), polysaccharides can form branched as well as linear chains. Functions of polysaccharides are (a) storage (e.g., starch, glycogen) for future use as source of energy and (b) structural (e.g., cellulose in plants and peptidoglycan in bacteria).

Structural Polysaccharides Plant cell walls are made of cellulose, a linear polymer of up to 15,000 D-glucose residues linked by $\beta(1 \rightarrow 4)$ glycosidic bonds. Plant cell walls will not be discussed further since antibiotics do not affect plant cell walls. However, there is a class of antibiotics that functions by inhibiting cell wall synthesis in bacteria. Cell walls of bacteria have a more complex structure than plant cell wall and the enzymes involved in the two processes are different and thus the antibiotics are specific for bacterial cell wall. Structures of gram-positive and gram-negative bacterial cells are shown in Fig. 1.3. Gram-positive bacteria have a plasma membrane that is surrounding by a thick (20–50 nm) cell wall, whereas gram-negative bacteria have two membranes and the inter membrane space contains a thin (10–15 nm) cell wall. Human cells do not have cell walls because the cells are always present in an isotonic environment (salt concentrations inside and outside the cells are the same). Bacterial cells encounter variable environments. If they are present in a hypotonic environment (lower concentration outside), in the absence of a cell wall, water would enter the cell and create high osmotic pressure. Function of the cell wall is to maintain cell shape and withstand the high osmotic pressure of the cytoplasm. Total concentrations of all solutes in the cytoplasm are higher in gram-positive than in gram-negative bacteria. So the cell walls of gram-positive bacteria are thicker in order to withstand a higher osmotic pressure. Antibiotics such as penicillin inhibit the synthesis of cell wall which causes the cell to swell and lyze because of the osmotic pressure of the cytoplasm. However, only growing bacteria are affected this way and so penicillin is bactericidal only for growing cells. Similarly, the enzyme lysozyme, which was the first antibiotic discovered by Fleming, hydrolyzes the peptidoglycan layer and also causes the cells to lyze.

3.1.2   Molecular Structure of Bacterial Cell Wall

The major component of the cell wall is a peptidoglycan (also called mucopeptide or murein). As the name suggests, a peptidoglycan molecule consists of mostly a carbohydrate polymer (glycan) that contains some peptides linked to it. The glycan chains are arranged parallel to each other and are cross-linked to each other by peptide bonds involving a few amino acids. The glycan chains consist of alternating monomer units, N-acetylglucosamine (NAG) and N-acetylmuramic acid (NAM) (Fig. 3.2) linked by $\beta(1 \rightarrow 4)$ glycosidic bonds. The NAM monomer is the same as the NAG except that carbon 3 of the sugar is bonded to a lactic acid. The lactic acid of NAM is further linked by peptide bonds to four amino acids in series: L-Alanine, γ-D-Glutamate, L-Lysine, and D-Alanine. The third amino acid is L-lysine in most gram-positive bacteria and a *meso*-diaminopimelic acid in most gram-negative and some gram-positive bacteria. None of the peptide bonds are normal peptide bonds, which are usually bonds between the α-COOH of one L-amino acid and the α-NH$_2$ of the next L-amino acid. The first peptide bond in NAM is to a lactic acid, which is not an amino acid, the second peptide bond is to an uncommon D-glutamate. The third peptide bond is also unusual since the peptide bond is to the γ-COOH (not the usual α-COOH) of the D-glutamate. The fourth peptide bond is to the unusual D-amino acid, D-Ala. It is believed that these unusual peptide bonds in the cell wall help the bacteria to evade destruction by proteases that are used by the host as defense against foreign proteins.

The tetrapeptide chain on each NAM unit is cross-linked to the tetrapeptide chain of another NAM unit of an adjacent glycan chain through a peptide chain consisting of five glycines. The pentaglycine chains connect the ε-NH$_2$ group of lysine (or a diaminopimelic acid) to the COOH of the terminal D-Ala. The glycan chain forms a helical structure and thus the peptide chains protrude in all directions and are able to form cross-links with adjacent peptidoglycan chains in all directions [1]. This helps to make the cell wall strong. Depending on the bacterial species, there can be considerable variation from the structure described here (Fig. 3.4).

3.2   Biosynthesis of Peptidoglycan of the Cell Wall

Biosynthesis of cell wall takes place in three stages. The process starts in the cytoplasm (Stage 1) and is followed by reactions in the membrane (Stage 2). The product then crosses the membrane and the final reactions take place in the cell wall (Stage 3). There are several enzymes that are needed to catalyze these reactions.

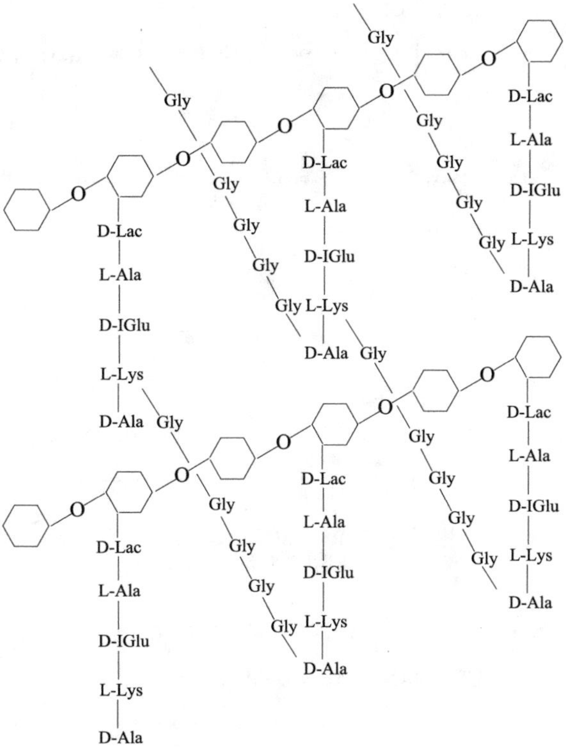

Fig. 3.4 Structure of
bacterial cell wall showing
cross-linking of adjacent
peptidoglycan strands

3.2.1   Stage 1: The Cytosolic Phase of Synthesis

Reaction scheme for the cytosolic phase of cell wall synthesis is shown in Fig. 3.5. As discussed before (Sect. 1.7.4), metabolic pathways do not have any beginning or end. Scientists and authors almost arbitrarily (but with some reasoning) decide about which steps should be considered as the first and last steps of a pathway. The enzymes catalyzing reactions 5–9 and 12 in Fig. 3.5 have been named as MurA-F [2] where mur stands for "murein," which is another name for the peptidoglycan of the cell wall.

The following are the names of the enzymes catalyzing the 12 steps in the scheme.

1. Glucosamine-6-phosphate synthase (or, L-glutamine:D-fructose-6-phosphate amidotransferase)
2. Glucosamine mutase
3. Glucosamine-1-phosphate acetyltransferase
4. UDP-NAG Synthase (or, UDP-GlcNAc pyrophosphorylase)
5. Phosphoenolpyruvate transferase (UDP-GlcNAc enolpyruvyl transferase, or MurA)
6. UDP-NAG-enolpyruvate reductase (MurB)

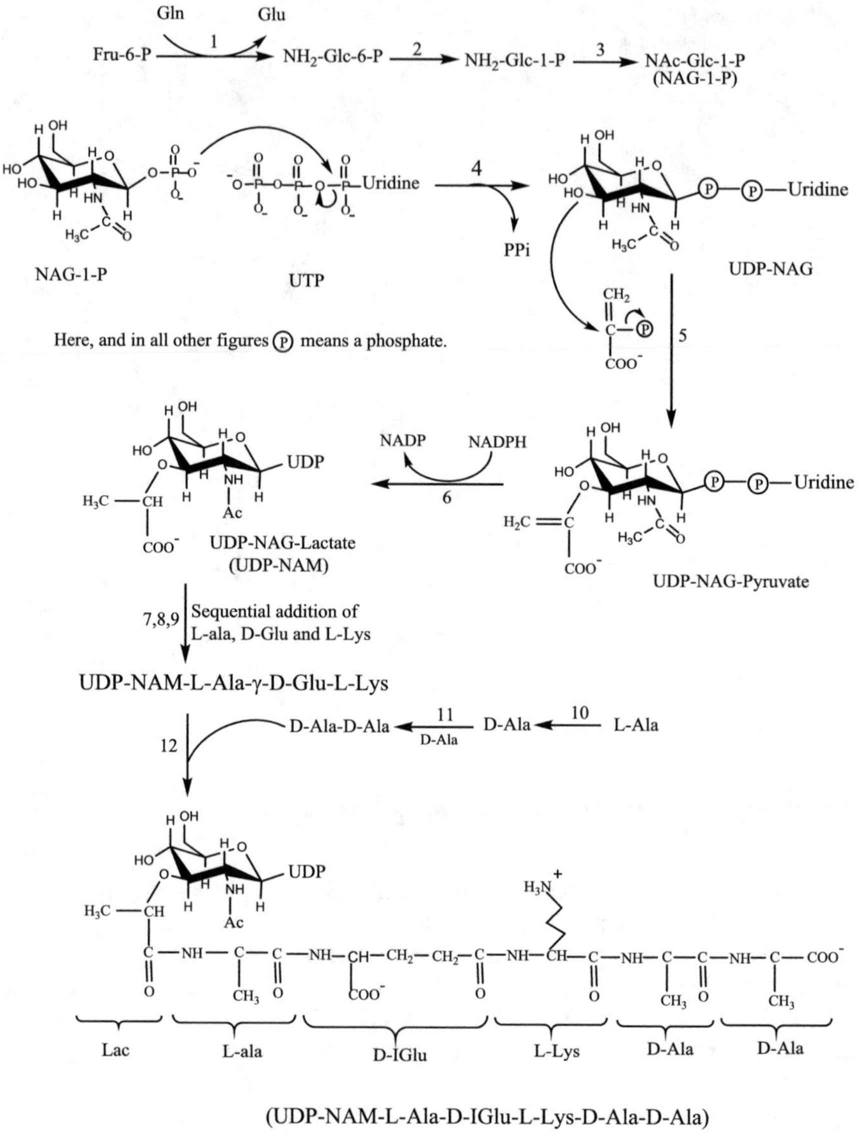
Gln
Glu
Fru-6-P
1
NH_2-Glc-6-P
2
NH_2-Glc-1-P
3
NAc-Glc-1-P
(NAG-1-P)
NAG-1-P
UTP
4
PPi
UDP-NAG
Here, and in all other figures (P) means a phosphate.
5
NADP
NADPH
6
UDP-NAG-Lactate
(UDP-NAM)
UDP-NAG-Pyruvate
7,8,9
Sequential addition of
L-ala, D-Glu and L-Lys
UDP-NAM-L-Ala-γ-D-Glu-L-Lys
12
D-Ala-D-Ala
11
D-Ala
D-Ala
10
L-Ala
Lac
L-ala
D-IGlu
L-Lys
D-Ala
D-Ala
(UDP-NAM-L-Ala-D-IGlu-L-Lys-D-Ala-D-Ala)

Note: Steps 7, 8, 9, 11 and 12 require a molecule of ATP each to provide energy for the new bonds formed

Fig. 3.5 Reactions in the cytosolic phase of cell wall synthesis

7. MurC
8. MurD
9. MurE

10. Alanine racemase
11. D-Ala-D-Ala synthetase (D-Ala-D-Ala ligase)
12. MurF

The cytosolic phase (Fig. 3.5) may be considered to start from fructose-6-phosphate, which receives an amino group from the usual amino group donor, glutamine to form 2-glucosamine-6-phosphate (Step 1), which is then isomerized to 2-glucosamine-1-phosphate (Step 2) and then acetylated to form N-acetyl glucosamine-phosphate (Step 3) [2–4].

The NAG-1-P is then activated by reacting with UTP to form UDP-NAG as shown in Eq. (3.1) and in Step 4, Fig. 3.5. For meaning and significance of activation of the reactant, see Sect. 1.7.7.

$$\text{NAG-1-}\textcircled{P} \;+\; \textcircled{P}\text{--}\textcircled{P}\text{--}\textcircled{P}\text{--U} \xrightarrow{\quad\text{(Pyrophosphate)}\;\text{PPi}\quad} \text{NAG-1-}\textcircled{P}\text{--}\textcircled{P}\text{--U}$$

(N-acetylglucosamine-   **(UTP)**   **(UDP-NAG)**
1-phosphate)

$$(3.1)$$

It is a common strategy in cells to form PPi (pyrophosphate, $P_2O_7^{4-}$) as a product if the reaction is thermodynamically unfavorable (ΔG not highly negative). The net ΔG is made negative because PPi is immediately hydrolyzed by the enzyme pyrophosphatase, which is a constitutive enzyme (that is always present in the cell). PPi $\rightarrow$ 2 Pi, ΔG for this reaction is highly negative. So by coupling the hydrolysis reaction, the previous reaction is made thermodynamically favorable (see Sect. 1.7.7).

A second precursor of peptidoglycan, phosphoenolpyruvate (PEP) is formed from glucose in the ninth step of glycolysis, which is a sequence of 10 steps in the metabolism of glucose. As discussed before in Sect. 1.7.6, PEP is a high-energy compound because it is locked by the phosphate in the unstable enol form (Fig. 1.10). In Step 5, PEP transfers a pyruvate to the 3-OH of the glucose unit of UDP-NAG. The energy in PEP is used to form a C-O-C ether linkage between pyruvate and the hydroxyl group at C3 of UDP-NAG. This step (Fig. 3.5 step 5) is the target of the antibiotic fosfomycin (Sect. 3.3.1.1). The pyruvate residue is then reduced (Step 6) to lactate by NADPH, which is a common reducing agent in biochemical reactions. Note that NAG-Lactate is also known as *N*-acetyl muramic acid or NAM.

Three amino acids are then added in sequence (Steps 7, 8, and 9) to form UDP-NAG-L-Lac-L-Ala-γ-D-Glu-L-Lys (aka UDP-NAM-tripeptide). Note that this enzyme-catalyzed peptide bond formation does not involve ribosomes and t-RNA and thus is different from the usual protein synthesis process. Another difference between the two processes is that ribosome-catalyzed protein synthesis takes place from N-terminal to C-terminal direction while in bacterial peptidoglycan biosynthesis, amino acids are added from the C-terminal to N-terminal direction.

The MurE enzyme links the L-Lys to γ-carboxyl group of D-Glu instead of the α-carboxyl group, which is normally seen in proteins. Although ATP is required to provide energy for peptide bond formation, no ribosomes or t-RNAs are involved.

To form the UDP-NAM-pentapeptide, first a L-Ala is isomerized to D-Ala (Step 10) by the enzyme alanine racemase and then two D-Ala amino acids are joined to form a peptide bond (Step 11), a reaction catalyzed by the enzyme D-Ala-D-Ala synthetase. Both these enzymes are the targets of the antibiotic cycloserine (Sect. 3.3.1.2). The D-Ala-D-Ala dipeptide is then added to the UDP-NAM-tripeptide to form UDP-NAM-pentapeptide (Step 12).

$$L\text{-ala} \longrightarrow D\text{-ala} \longrightarrow D\text{-ala-D-ala}$$

UDP-NAM-Tripeptide $\longrightarrow$ UDP-NAM-Pentapeptide

(3.2)

3.2.2 Stage 2: The Membrane Phase of Synthesis

The reactions described above take place in the cytosol while the cell wall is located on the other side of the membrane. The products of these reactions are polar and need to be transported through the membrane, which has a hydrophobic interior. Polar compounds cannot enter or cross the lipid bilayer of the membrane because of the hydrophobic environment. A membrane carrier is needed to bring a polar compound into the membrane where the next phase of cell wall synthesis takes place. The membrane carrier used is undecaprenyl phosphate (C55-P), a 55-carbon lipid which consists of the five carbon unit, isoprene repeated 11 (undeca) times. Because of the long 55 carbon hydrophobic chain, it is soluble in the membrane and is able to bring the polar products through the membrane. It forms phosphoanhydride linkage to P-NAM-pentapeptide (Fig. 3.6, Step 1). The reaction takes place at the interphase of the membrane and the cytoplasm.

$$U\text{-P-P-NAM-pentapeptide} \; + \; C55\text{-P}$$

$$\downarrow\!\!\!\searrow UMP$$

(3.3)

$$C55\text{-P-P-NAM-pentapeptide}$$

Since a phosphoanhydride bond in the reactant (U-P-P-NAM-pentapeptide) is replaced by another phosphoanhydride bond in the product (C55-P-P-NAM-penta-peptide), the reaction is energetically favorable. This is then followed by a reaction with a molecule of UDP-NAG to form a β(1 → 4) bond between the NAG and NAM (Fig. 3.6, Step 2). The energy of the phosphoester bond between NAG and phosphate of UDP is used to form the new glycosidic bond between NAG and NAM.

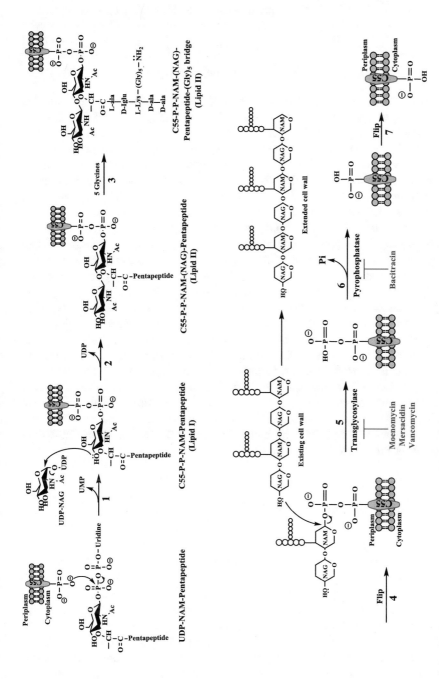

Fig. 3.6 Membrane and cell wall phases of cell wall synthesis

$$C55\text{-}P\text{-}P\text{-}NAM\text{-pentapeptide} \ + \ U\text{-}P\text{-}P\text{-}NAG$$

$$\searrow \ UDP \tag{3.4}$$

$$C55\text{-}P\text{-}P\text{-}NAM\text{-pentapeptide-}NAG$$

Five glycine residues are then added sequentially to form peptide bonds, the first bond being to the lysine ε-amino group of the pentapeptide (Step 3). The donor is glycyl-tRNA, which is different from the usual glycyl-tRNA that is involved in protein biosynthesis. The product is C55-P-P-NAM-(NAG)-pentapeptide-(Gly)$_5$ bridge, also known as "Lipid II" and is the monomer for the formation of the peptidoglycan polymer. The C55 then transports the monomer through the membrane from the cytoplasmic side to the periplasm side (Step 4). The next step takes place at the junction of the membrane and the cell wall since the monomer is transferred to the existing peptidoglycan polymer of the cell wall while the C55-P-P remains in the membrane (Step 5). The reaction is catalyzed by transglycosylase enzyme, which is a target for several antibiotics including moenomycin, mersacidin, and vancomycin (Sects. 3.3.3.2–3.3.3.4). The released C55-P-P again is important for thermodynamics of the reaction. The pyrophosphate is hydrolyzed by a specific pyrophosphatase to give C55-P. This serves two purposes. The C55-P is recycled again for another round of reaction, which cannot happen unless the C55-pyrophosphate is hydrolyzed. The reaction also lowers the concentration of the C55-P-P, which is the product of the previous transfer reaction. Lowering the concentration of one of the products drives the reaction forward (Le Chatelier's principle, Sect. 1.7.4). Another way of stating this is that the ΔG of the reaction becomes more negative due to coupling of the reaction with the highly exergonic (ΔG negative) hydrolysis reaction of pyrophosphate, thus making the reaction thermodynamically favorable and irreversible (Eq. 1.3). Hydrolysis of C55-P-P is a target for the antibiotic bacitracin (Sect. 3.3.3.1).

3.2.3   Stage 3: The Cell Wall Phase of Synthesis

The final stage of the biosynthesis takes place in the cell wall. As mentioned before, the transglycosylation reaction involves both the membrane and the cell wall. From the membrane, the peptidoglycan monomer is added to the existing peptidoglycan polymer in the cell wall. However, a small cell already has a complete cell wall. So when the cell is growing, where are the new monomer units added? In order to add new monomer units some covalent bonds need to be broken first in the existing cell wall. There are various lytic enzymes present which break (hydrolyze) different covalent bonds in the existing cell wall. New monomer units are added at these places to form new covalent bonds. The main reaction taking place is a transglycosylation (formation of a β(1 → 4) glycosidic bond) reaction between

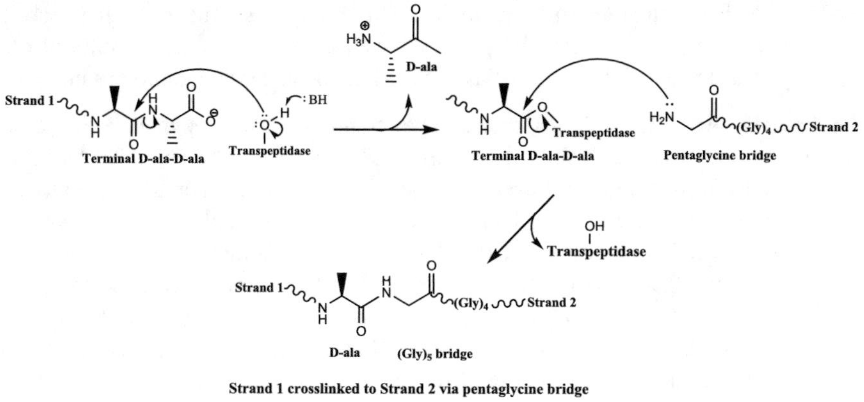

Fig. 3.7 Cell wall phase of cell wall synthesis: peptidoglycan cross-linking by transpeptidase

two new monomer units or between a new monomer and the existing peptidoglycan (Fig. 3.6, step 5). The transglycosylation reaction is the target of three different antibiotics: Moenomycin (Sect. 3.3.3.2), Mercacidin, a lantibiotic, (Sect. 3.3.3.3) and Vancomycin (Sect. 3.3.3.4).

The final step in cell wall biosynthesis is the cross-linking of the peptidoglycan strands by transpeptidation. The cross-linked structure of the cell wall is shown in Fig. 3.4. The cross-linking is important because it makes the cell wall strong and rigid. Note that the peptidoglycan monomer has five amino acids linked to NAM, whereas the final mature cell wall has only four amino acids (Fig. 3.4). The last amino acid, D-alanine is released during transpeptidation reaction (Fig. 3.7). The NH_2 of the fifth glycine of the pentaglycine of one glycan chain forms a peptide bond to the second last D-ala of the pentapeptide of an adjacent glycan chain. Nature's design should be appreciated here since a bond is formed between adjacent glycan chains but no ATP is used. In fact, there is no ATP or any other NTP available in the periplasm where the cross-linking reaction takes place. The transpeptidation reaction for cross-linking does not require any ATP because the energy required for the new bond formation is obtained from the energy of the terminal D-ala-D-ala peptide bond.

Note that all monomer units are not cross-linked because that will make the cell wall too rigid. Depending on the strain and growth conditions the percentage of monomers cross-linked can vary from 45 to 90%. If a monomer unit is not cross-linked, its terminal D-Ala residue is usually removed by hydrolysis reaction catalyzed by a carboxypeptidase enzyme [1]. Both transpeptidase and carboxypeptidase have similar reaction mechanisms in the initial part. During transpeptidation reaction mechanism, the enzyme first reacts with the fourth D-ala of the pentapeptide to form a covalent bond with it in order to preserve the energy of the peptide bond while the fifth D-ala is released. Then a new peptide bond is formed with the NH_2 group of Gly of the pentaglycine bridge of another strand, thus cross-linking the two strands. In case of carboxypeptidase reaction, instead of transferring the D-ala from the enzyme-D-ala covalent intermediate to the amino group of glycine to form a peptide bond, it is transferred to the OH group of water. This effectively results in the hydrolysis of the terminal D-ala-D-ala bond of the pentapeptide monomer units.

Hydrolases Are Also Transferases The transpeptidase enzyme belongs to the category of enzymes called transferases. The name derives from the fact that the D-ala residue is transferred from the D-ala-D-ala part of the peptidoglycan to the glycine of the pentaglycine unit. Transferases always follow a ping pong mechanism (Fig. 1.7) in which an intermediate covalent complex is formed with the enzyme as explained above. The second substrate then displaces the enzyme from the complex. Both these steps are nucleophilic substitution (SN2) reactions. In the first step the OH group of the serine residue is the nucleophile and in the next step, the second substrate is the nucleophile. In case of transpeptidation reaction, the second nucleophile is the NH_2 group of the glycine. For all transferase reactions, one question always arises: why does the transfer take place to the second substrate acting as the nucleophile but not to the more abundant nucleophile, water? Since all biochemical reactions take place in aqueous solution, the concentration of water is extremely high (55 M) compared to the concentration of the second substrate (usually in the mM range). Also, water being a small molecule can easily get into the active site compared to the much larger second substrate (in this case, the pentaglycine unit of the peptidoglycan chain). Moreover, reaction with the second substrate, for example, NH_2 group of glycine in case of transpeptidation reaction, has the same energy requirement as reaction with H_2O. In spite of the greater concentration of water, all transferases transfer the intermediate to the second substrate and not to water. The reason, which applies to all transferases is that, after the second substrate binds to the active site, the enzyme undergoes a conformational change that closes the active site to make its shape more complementary to the shape of the substrate so that the enzyme binds tightly to the substrate and entry of water molecules into the active site is prevented. This is known as Koshland's "induced fit" model. If water does act as a nucleophile, the reaction would be called a hydrolysis reaction, in which the intermediate group is transferred to OH group of water. There are many enzymes whose functions are to catalyze hydrolysis reactions and are known as hydrolases. Examples include all the digestive enzymes which hydrolyze peptide bonds of proteins or glycosidic bonds of carbohydrates or acyl bonds of lipids. Thus, hydrolases are a special category of transferases in which water acts a nucleophile. The difference between a transferase and a hydrolase is that a transferase does not allow water to enter the active site after the second substrate binds, whereas hydrolases let water enter easily. So in case of transpeptidation reaction, after a covalent bond is formed between a serine at the active site of the transpeptidase enzyme and the penultimate D-ala of the monomer unit, the enzyme undergoes a conformational change in order to bind more tightly to the intermediate and to prevent water from entering the active site.

$$\text{Enz-Ser-O}^- + \text{Tetrapeptide-D-ala} \longrightarrow \text{Enz-Ser-O-tetrapeptide} + \text{D-ala}$$

$$\text{NH}_2\text{-(Gly)}_5 \diagup \qquad \diagdown \text{H}_2\text{O}$$

$$\begin{array}{cc} \text{Tetrapeptide-NH}_2\text{-(Gly)}_5 & \text{Tetrapeptide} \\ + & + \\ \text{Enz-Ser-O}^- & \text{Enz-Ser-O}^- \\ \text{(transpeptidase)} & \text{(hydrolase)} \end{array}$$

$$(3.5)$$

Carboxypeptidases Are Hydrolases Not all polypeptide chains of the cell wall are cross-linked. This is because besides transpeptidase, there is another type of enzyme involved in cell wall synthesis that also binds to the terminal D-ala-D-ala of the peptidoglycan and catalyzes the release of the terminal D-ala. However, instead of forming a cross-link with an adjacent peptidoglycan chain, it reacts with water that is allowed to enter the active site. The net result is the hydrolysis of the D-ala-D-ala peptide bond (Eq. 3.5). Note that in Eq. 3.5, the Ser is shown with a negative charge because it is an active serine (Sect. 3.3.2.6). Since the hydrolysis reaction takes place at the carboxy end of the polypeptide, the enzyme is called a carboxypeptidase. The energy of the peptide bond is released as heat energy and so no cross-linking takes place with lysine of the next strand. So the function of carboxypeptidase in cell wall biosynthesis is to compete with transpeptidase in order to limit the extent of cross-linking and thus prevent excessive rigidity of the cell wall. The extent of cross-linking varies in different bacteria and is determined by the extent of carboxy-peptidase activity.

3.3 Antibiotics That Inhibit Cell Wall Biosynthesis

As discussed before, the cell wall synthesis takes place in three stages: the cytosolic stage, the membrane stage, and the cell wall stage. There are antibiotics that target each of these stages of the cell wall synthesis.

3.3.1 Antibiotics Targeting the Cytosolic Phase of Synthesis

3.3.1.1 Fosfomycin

The antibiotic fosfomycin (aka phosphomycin) is a broad-spectrum antibiotic that is produced by some *Streptomyces*. The antibiotic inhibits the enzyme, pyruvyl trans-ferase (aka UDP-*N*-acetylglucosamine enolpyruvyl transferase, or MurA), which

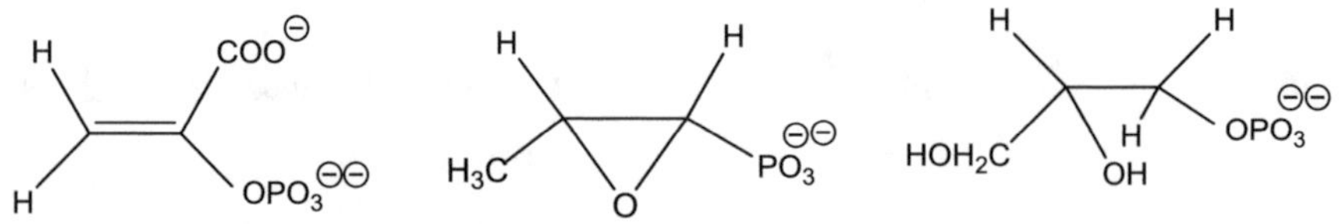

Phosphoenolpyruvate   Fosfomycin   Glycerol-3-phosphate

Fig. 3.8 Fosfomycin mimics phosphoenolpyruvate and glycerol-3-phosphate

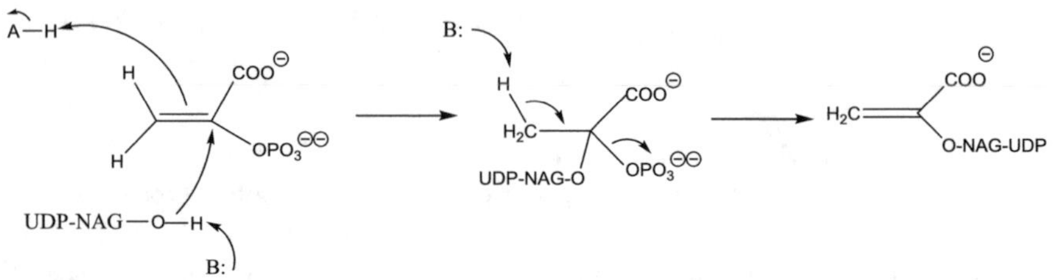

Fig. 3.9 Reaction between UDP-NAG and PEP catalyzed by pyruvyl transferase. The B: and A-H are basic and acidic amino acids, respectively, at the active site

catalyzes the transfer of pyruvate from PEP to UDP-NAG (Fig. 3.5 step 5). This is the first committed step (Sect. 1.7.7) of cell wall biosynthesis. Deletion of the *murA* gene is lethal in both gram-positive [5] and gram-negative [6] bacteria. There is no homolog of this enzyme in mammalian cells and so is an ideal target for development of antibiotics.

Mechanism of Action The structure of fosfomycin resembles PEP to some extent; however, the resemblance is not much (Fig. 3.8). In a way this lack of resemblance is good because PEP is a common metabolite in several other pathways in both the pathogen and the host. If fosfomycin resembled PEP more closely, it would act as a competitive inhibitor in all these pathways and so would adversely affect the host. However, fosfomycin is highly specific for this enzyme and because of this, it has very low toxicity.

Fosfomycin is an effective antibiotic because it functions as a suicide inhibitor (Sect. 1.7.3) of MurA enzyme. The mechanisms of the reaction catalyzed by the enzyme with the natural substrates UDP-NAG and PEP and the reaction of the enzyme with the antibiotic fosfomycin are shown in Figs. 3.9 and 3.10, respectively. Fosfomycin forms a covalent bond to a cysteine amino acid present at the active site of the enzyme [7]. The mechanism of this reaction is similar to the first part of the mechanism of the reaction between UDP-NAG and the PEP catalyzed by the enzyme (Fig. 3.9). The thiol (-SH) group of the cysteine reacts with the unstable epoxide ring of fosfomycin and forms an irreversible covalent bond with it, thereby permanently inactivating the enzyme. This mechanism is supported by the fact that mutation of this Cys115 in *E. coli* MurA enzyme to an Asp makes the bacteria resistant to Fosfomycin [8]. It is to be noted that in PEP there is an oxygen bonded to

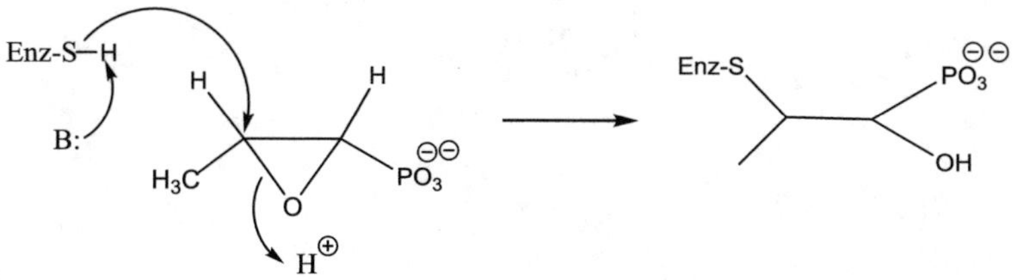

Fig. 3.10 Mechanism-based (suicide) inactivation of pyruvyl transferase by Fosfomycin. The SH group belongs to a cysteine residue at the active site of the enzyme

a carbon (C-OPO_3^{2-} bond), whereas in fosfomycin there is a phosphorous bonded to carbon (C-PO_3^{2-}) bond. In the normal reaction with PEP the C-O bond breaks because phosphate ion (OPO_3^{3-}) is a good leaving group but the C-P bond in fosfomycin does not break because PO_3^{3-} is not a good leaving group. Instead, reaction takes place at the adjacent carbon, breaking the epoxide ring.

Resistance Development Even though fosfomycin is highly specific and is not toxic for the host, it is not considered a good antibiotic because bacteria can develop resistance to the drug quickly. One mechanism of resistance development is by prevention of uptake into the bacterial cells. Fosfomycin is normally transported into the cell through the transporter for glycerol-3-phosphate to which it has some structural similarity (Fig. 3.8). The transporter is not absolutely essential for the bacteria. Mutations in the transporter protein can prevent entry of fosfomycin, thus making the cells resistant. Entry of glycerol-3-phosphate will also be prevented but the compound is not essential for the bacteria.

Another method of resistance to fosfomycin is by modification of the drug. Three enzymes, FosA, FosB, and FosX, present in various microorganisms, catalyze the modification of fosfomycin that make it inactive [9]. The thiol transferase enzymes FosA and FosB catalyze the inactivation of fosfomycin by the addition of glutathione and cysteine to C1 of the oxirane, respectively. FosX catalyzes the inactivation by the transfer of a water molecule to the C1 of the oxirane. The resistance gene may also be present on small plasmids in some strains of bacteria.

3.3.1.2 D-Cycloserine

D-cycloserine (oxamycin) is an antibiotic that inhibits two enzymes in the cytoplasmic phase of bacterial cell wall synthesis. The two enzymes are alanine racemase, which converts L-ala to D-ala (Fig. 3.5 step 10) and D-ala-D-ala synthetase, which joins two D-ala molecules to form D-ala-D-ala dipeptide (Fig. 3.5 Step 11). Cycloserine (Fig. 3.11) functions as an antibiotic since it resembles D-alanine and thus acts as a competitive inhibitor of both enzymes. The growth inhibition can be antagonized by D-alanine and to a lesser extent by L-alanine. As can be seen from the structure of the antibiotic, the name cycloserine is a misnomer since the formula has an oxygen atom of the carboxyl group of serine replaced by a nitrogen atom.

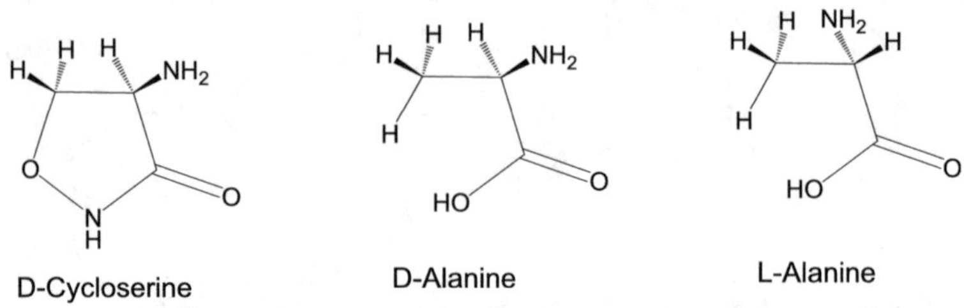

Fig. 3.11 Cycloserine as analog of D-Alanine

Cycloserine has not been used much as an antibiotic because of its side effects on the host. It is able to penetrate the central nervous system and cause neurological disorders. As an antibiotic it is effective against *Mycobacterium tuberculosis*. Cycloserine is used clinically as a second-line drug for the treatment of tuberculosis if the first-line drugs fail.

Resistance Development Against D-Cycloserine One method of resistance development is by increasing the expression of the enzyme alanine racemase or the enzyme D-ala-D-ala synthetase or both [10]. Another method is by mutations that inhibit the entry of D-cycloserine into the bacteria, which takes place by utilizing the transport system that normally transports D-alanine and glycine into the cell. Mutations in this transport system can decrease the entry of the antibiotic. Transport of L-alanine, which uses a different transport system, is not affected [11].

3.3.2 Antibiotics Targeting the Cell Wall Phase of Synthesis

Based on the sequence of reaction in the biosynthesis of the cell wall (Sect. 3.6), the membrane phase of the synthesis takes place before the cell wall phase. However, antibiotics targeting the membrane phase will be discussed (Sect. 3.3.3) after the antibiotics targeting the cell wall phase. This is because of two reasons: (1) Chronologically, the β-lactam antibiotics acting in the cell wall phase of biosynthesis were discovered first and (2) Discussion of the antibiotics affecting the membrane phase of synthesis will have frequent references to the β-lactam antibiotics and so prior knowledge of the latter will be necessary. For the same reason, antibiotics targeting the transglycosylase reaction will also be discussed after the β-lactam antibiotics even though the transglycosylation reaction actually takes place before the cross-linking step in the sequence of reactions for the synthesis of the cell wall.

β-Lactam Antibiotics The β-lactams include penicillins, cephalosporins, and carbapenems and is clinically the most important class of antibiotics. About 55% of all antibiotics used globally belong to this class. Figure 3.12 shows the nomenclature of some common ring systems related to β-lactam antibiotics. Cyclic esters are called lactones while cyclic amides are called lactams and "β" indicates that the

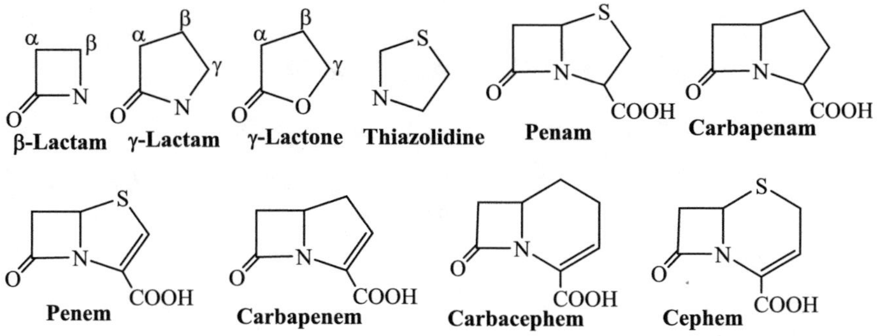

Fig. 3.12 Nomenclature of some relevant ring systems

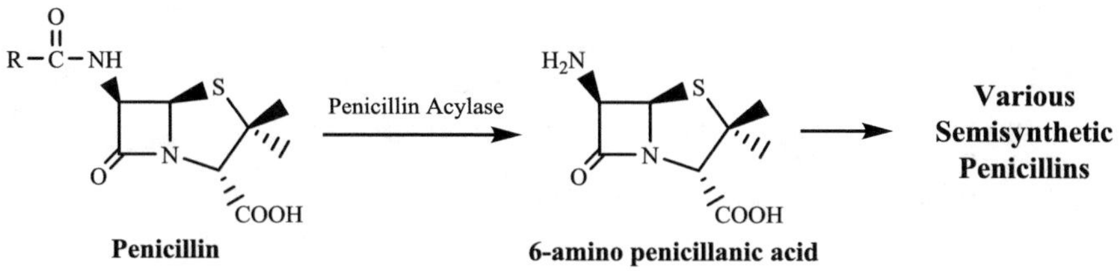

Fig. 3.13 Synthesis of penicillin derivatives

amino group is at β position from the carboxyl carbon. Thus, β-lactams are four-membered rings, γ-lactams are five-membered rings, etc.

3.3.2.1   Penicillin

Penicillin structure is shown in Fig. 3.13. The five-membered ring in penicillin contains a sulfur atom and is called the thiazolidine ring. The two rings together is called the penam ring and along with the methyl and carboxyl substituents is called penicillanic acid and is biosynthetically formed from the amino acids, cysteine and valine. Penicillins are acyl derivatives of 6-amino penicillanic acid. Other semisynthetic derivatives of penicillin are made from 6-amino penicillanic acid which is obtained by deacylation of penicillin (Fig. 3.13). The various penicillin derivatives, penicillin G, penicillin V, methicillin, ampicillin, amoxicillin, carbenicillin, and ticarcillin differ from each other in the R group as shown in Fig. 3.14.

3.3.2.2   Cephalosporin

Cephalosporin antibiotics have structures similar to penicillin. They also have a β-lactam ring but it is fused to a six-membered ring instead of a five-membered ring that is present in penicillins (Fig. 3.15). Also cephalosporins have more variable

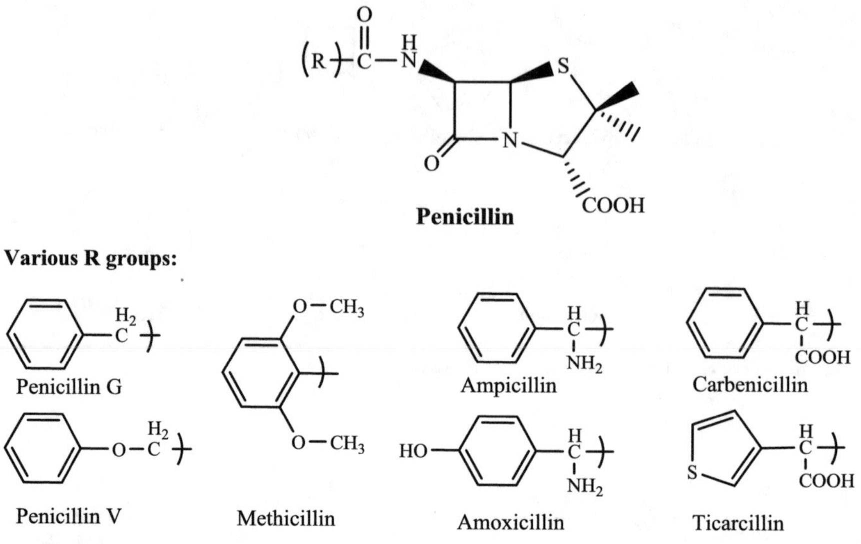

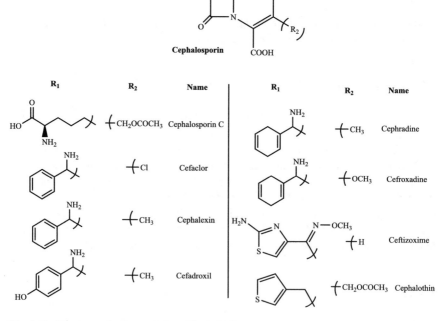

Fig. 3.14 Structures of some penicillin derivatives

Fig. 3.15 Structures of some cephalosporin derivatives

substituents (R1 and R2) in the rings. Cephalosporins and penicillins are produced by different microorganisms but the pathways for their synthesis are similar. Their mechanisms of action are also similar (Sect. 3.3.2.3). However, cephalosporins have a broader spectrum of activity than penicillins and are effective against both gram-negative and gram-positive bacteria.

Although cephalosporin C, the first discovered natural cephalosporin, has not been used much as an antibiotic, there are many semisynthetic derivatives of the drug that have found significant use as antibiotics. Cephalosporins are the largest and most diverse family of antibiotics. There can be many variations in the R1 and R2 groups shown in Fig. 3.15 resulting in numerous natural and semisynthetic cephalosporins. Only a few representative ones are shown in Fig. 3.15.

3.3.2.3   Mechanism of Action of Penicillin

A lot of research is involved in determining the mechanism of action of any drug. In this book, the mechanism of action of all classes of antibiotics is presented but the extensive research done to establish those mechanisms will not be discussed for most antibiotics. Since historically, penicillin was the first antibiotic to be discovered and studied in detail, some of the initial research that leads to the mechanism have also been discussed. Penicillin functions as an antibiotic by inhibiting the transpeptidase enzyme, which catalyzes the cross-linking of the peptidoglycan strands in the cell wall phase of the cell wall biosynthesis (Sect. 3.2.3). Penicillin is able to bind to the active site of transpeptidase enzyme since its structure resembles that of the substrate, which is the terminal D-ala-D-ala dipeptide of the pentapeptide of each monomer unit (Fig. 3.16). Note that D-ala-D-ala dipeptide of the substrate can

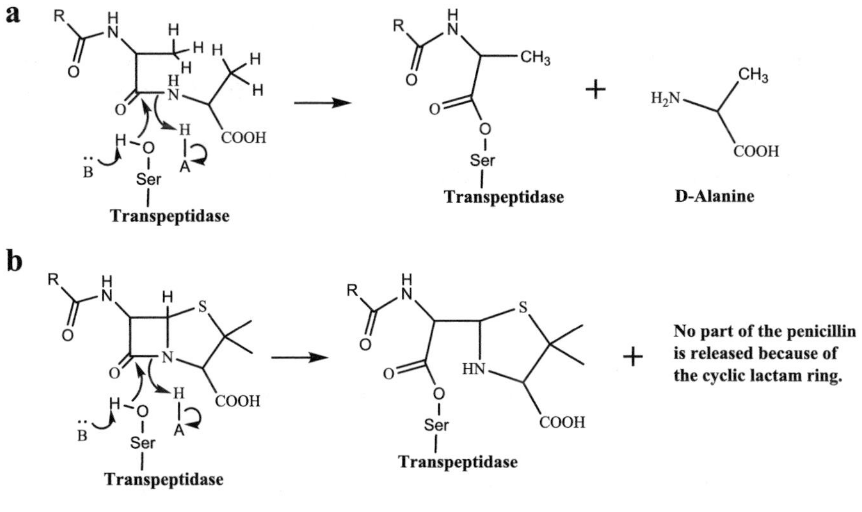

Fig. 3.16 Reaction of transpeptidase. (a) with the natural substrate, the D-Ala-D-Ala of the pentapeptide and (b) with penicillin

exist in multiple conformations formed by rotation around the C-C single bonds but a penicillin molecule has limited variation of conformation because of the rigidity of the four-membered lactam ring. Of the many conformations possible for the terminal dipeptide the one that binds to the enzyme resembles the structure of penicillin (Fig. 3.16), and thus, the two can compete for binding to the active site of the enzyme. The –C(O)-N bond of the β-lactam mimics the –C(O)-N of the peptide bond of the terminal dipeptide.

Penicillin is an effective antibiotic not just because it functions as a competitive inhibitor, but because it is a suicide (mechanism based) inhibitor (Sect. 1.7.3). Suicide inhibitors are more effective inhibitors because they stoichiometrically and irreversibly inactivate the enzyme and so are needed in very small amount. Since penicillin is a mechanism-based inhibitor, it is important to first understand the mechanism of the transpeptidation reaction. During normal transpeptidation reaction a peptide bond is formed between a glycine and the penultimate D-alanine of the pentapeptide while the terminal D-ala is released. Energy is needed to form the new peptide bond but no ATP is used in the reaction because energy is obtained from the terminal D-ala-D-ala peptide bond which is broken. Such reactions usually follow a general mechanism known as the ping pong mechanism as shown below.

$$\text{Enzyme} + \text{A-B} \rightarrow \text{Enzyme-B} + \text{A}$$
$$\text{Enzyme-B} + \text{C} \rightarrow \text{C-B} + \text{Enzyme} \tag{3.6}$$

The energy of the A-B bond is conserved in the intermediate species, Enzyme-B and then is used to form a bond between C and B. Such a mechanism is seen for numerous biochemical reactions. For transpeptidation reaction in bacterial cell wall, the OH group of a serine residue at the active site of the transpeptidase enzyme forms an ester linkage with the penultimate D-ala while displacing the terminal D-ala. Then the NH_2 group of the terminal glycine residue of the pentaglycine of the adjacent peptidoglycan strand displaces the serine residue of the enzyme while forming a peptide bond with the D-ala. Thus, the two adjacent peptidoglycan chains are cross-linked, making the cell wall strong and rigid (Eq. 3.7).

$$\text{Enz-Ser-O-H} + \text{Peptidoglycan-Ala-Ala}$$

$$\downarrow\!\!\!\!\searrow \text{Ala}$$

$$\text{Peptidoglycan-Ala-O-Ser-Enz} \tag{3.7}$$

$$\downarrow \text{Peptidoglycan-(Gly)}_4\text{-Gly}$$

$$\text{Peptidoglycan-Ala-(Gly)5-Peptidoglycan} + \text{Enz}$$

When penicillin competes with the natural substrate and binds to the active site, the serine–OH reacts at the β-lactam ring since its structure resembles that of the D-ala-D-ala of the peptidoglycan. However, unlike reaction with the normal substrate, in which case one alanine residue is released, no part of the penicillin molecule is released. Because of the ring structure of the β-lactam, the penicillin is still attached to the enzyme even after reaction. The normal transpeptidation reaction and the reaction with penicillin are shown in Fig. 3.16. Since the bond between the enzyme and the penicillin is a stable covalent bond, the enzyme becomes irreversibly inactivated. The acyl bond between the enzyme and the penicillin cannot be hydrolyzed to free the enzyme since the active site of the transpeptidase enzyme does not allow water molecules to enter and so the penicillin remains covalently bound to the transpeptidase enzyme.

Multiple Targets of Penicillin In order for any drug to inhibit the activity of a protein, the two must first bind to each other. Thus, the target of the drug can be identified by determining which protein binds to the drug. For this, scientists use radioactive drug as a marker. The protein bound to the drug can be purified by using various separation techniques while monitoring the presence of the drug by its radioactivity. One such technique is polyacrylamide gel electrophoresis, which can be used to separate proteins based their sizes. Exposing a photographic film on the gel determines the position and thus the size of the protein, which is radioactive due to the drug bound to it. Further characterization and identification of the protein can help to determine the mechanism of action of the drug. In a similar experiment, radioactive penicillin was added to bacterial cells and then all proteins were extracted from the cells and were analyzed by electrophoresis [12]. Surprisingly, the radioactivity was found bound to not one but six proteins indicating that they all bind to penicillin. Until all these proteins, which range in size from 40 to 90 kDa, were identified and characterized they were all named Penicillin Binding Proteins (PBP1, PBP2, PBP3, etc.). The names may be somewhat misleading. The function of these proteins is not to bind penicillin; they just happen to bind penicillin because it resembles the normal substrate that binds to these proteins. Thus, all these PBPs are expected to be involved in the formation of the cell wall.

Why do bacteria need more than one PBP? In fact, later more PBPs were discovered in *E. coli*, increasing the number to 12 (PBP1a, 1b, 1c, 2, 3, 4, 4b, 5, 6, 6b, 7, and AmpH). Similarly, multiple PBPs were discovered in many other bacteria. Although it was initially surprising to find so many proteins binding to penicillin, the reason was understood later. The purpose of cell wall synthesis is not just for growth of bacterial cell size, it is also needed for the formation of septum during cell division. Note that for septation the direction of cell wall synthesis is perpendicular to the existing cell wall and thus requires a different transpeptidase. Also, being a complex process, septation may require separate enzymes for initiation, elongation, and termination of the process. In *E. coli*, PBP1, 2, and 3 are all transpeptidases. PBP1 actually contains two similar size proteins 1A and 1B. However, PBP3, 4, and 5 are not transpeptidases but carboxypeptidases, whose function is to limit the extent of cross-linking in order to control rigidity/flexibility of the cell wall. The carboxypeptidases are not essential for survival of the cell. Mutations in these enzymes do not have any significant effect on cell growth or viability. The antibiotic activity of penicillin is dependent on their binding to the transpeptidase. Although it also binds to the carboxypeptidases, that binding is not related to its antibiotic property since cell growth or survival is not affected. Some of the PBPs have more than one activity in the same protein. One domain (part) of the protein has transpeptidation activity (for cross-linking) and another domain has glycosyltransferase activity that is needed for elongation of the glycan chain. Penicillin binds to only the domain with transpeptidase activity. Note that the numbers 1, 2, 3, etc., associated with the PBP is based on their relative molecular weights, PBP1 having the largest size. These numbers are for *E. coli*. The number of PBPs and their sizes in other bacteria may be different. Since many bacterial chromosomes have now been sequenced, their PBP's can now be compared and classified based on their sequences. Such studies have been extensively reviewed [13]. Such comparisons reveal that the PBP numberings (which are based on size) do not necessarily relate to function and thus create confusion. For example, PBP2 of *S. aureus* is similar to PBP1a of *E. coli* and PBP3 of *S. aureus* is similar to PBP2 of *E. coli* [13].

Why Is Penicillin Bactericidal? Attachment of penicillin to PBPs only results in the inhibition of growth since new cell wall is not synthesized. However, the question arises as to why this leads to cell death. Many reviews have been written on bactericidal effects of penicillin and other antibiotics [14–16]. Initially, it was thought that cell death was due to the buildup of internal pressure resulting from increase of cell mass by normal growth in the cytoplasm and lack of growth of the cell wall [17]. This was called the unbalanced growth hypothesis. However, according to a later model, cell death is due to hydrolysis of the cell wall by an enzyme called autolysin, also known as L-alanyl-*N*-acetylmuramic acid amidase, or murein hydrolase [18]. Mutant bacteria that were deficient in murein hydrolase activity were not killed by β-lactam antibiotics. So these bacteria are said to be penicillin tolerant. Such cells can be lyzed by the addition of exogenous purified autolysin enzyme, thus confirming the mechanism of cell death. Autolysins are enzymes that break bonds between and within peptidoglycan strands. Autolysins are present in all bacteria that have cell wall. The normal function of autolysins is to maintain a rate of cell wall turnover by hydrolyzing the β-(1 → 4) bond between *N*-acetylmuramic acid and N-acetyl glucosamine in the peptidoglycan chain (Sect. 3.1.2). The purpose of breaking the bonds in the cell wall is to make room for cell wall expansion. New peptidoglycan monomers are added to these points where the existing peptidoglycan strands have been broken.

Cell walls contain lipoteichoic acid (LTA), which is an inhibitor of autolysin. Inhibition of peptidoglycan synthesis by penicillin weakens the cell wall especially at the points where the cell wall is broken due to autolysin activity. This triggers the release of lipoteichoic acids present in the cell wall and this relieves the inhibition of autolysin and so more covalent bonds are broken in the peptidoglycan polymer. This results in loss of osmotic integrity and thus causes cell lysis.

Another model explaining the role of autolysins is the "inside to outside" model [19]. The cell wall consists of three zones: inner, middle, and outer [20]. New peptidoglycan is synthesized in the inner zone (closest to the cytoplasmic membrane) and is removed by autolysin action from the outer zone [21]. The inner zone contains new unstressed peptidoglycan. As the cell grows in size, the peptidoglycan of the inner zone passes outwards and stretches to become the stress-bearing middle zone. The process continues and with further growth, the peptidoglycan of the middle becomes the outer zone, where partial hydrolysis by autolysins allow the cell wall to become loose and is eventually removed, making room for new cell wall to be made. Inhibition of synthesis of new cell wall by penicillin causes the cell to lyze since the outer zone is removed due to the action of autolysin.

3.3.2.4 Resistance to β-Lactam Antibiotics

Some bacteria are naturally resistant to certain antibiotics (intrinsic resistance), while others develop resistance to the antibiotics (acquired resistance).

Intrinsic Resistance In order to inhibit cell wall biosynthesis, the penicillin has to first approach the cell wall. Cell wall of gram-positive bacteria is more accessible to the drug than is the cell wall of gram-negative bacteria for which the drug has to be first transported through the outer membrane. However, the outer membrane is not as selective a barrier as the inner membrane. Hydrophobic antibiotics such as macrolides can diffuse through the hydrophobic lipid bilayer of the outer membrane while small hydrophilic antibiotics such as β-lactams can easily pass through the porin channels present in the outer membrane [22]. However, although not highly selective, the outer membrane porins of some bacteria have a preference for positive charges (cations) while in some other bacteria they may have a preference for negative charges (anions). There is a carboxyl group in penicillin which at the neutral body pH will exist as carboxylate anion since its pKa is less than 7.0. The OmpC porin in *E. coli* has a preference for transporting cations and so are less accessible to the negatively charged β-lactams than are gram-positive bacteria. *Gonococci,* on the other hand, are more sensitive to penicillin because unlike other gram-negative bacteria, gonococcal porins have preference for anions. *Pseudomonas aeruginosa* are highly resistant to β-lactams because their outer membranes lack the classical porins found in enteric bacteria.

Acquired Resistance There are three ways by which bacteria can acquire resistance to β-lactams: (1) by decreasing permeability of the drug into the cell, (2) by developing mutations in the target protein (PBP) to decrease binding affinity for the drug, and (3) by acquiring gene for β-lactamase enzyme which can degrade the drug.

Reduced Permeability Penicillin does not need to cross the cytoplasmic membrane because the cell wall, which is the target of the drug is located outside the cytoplasmic membrane. The cell wall itself does not act as a permeability barrier. So the transpeptidase enzyme (penicillin binding protein), which is the target for the drug, is easily accessible in gram-positive bacteria. Therefore, for gram-positive bacteria, resistance to penicillin cannot occur by decreasing permeability or prevention of intracellular accumulation. In gram-negative bacteria the outer membrane acts as a barrier for the drug. The drug passes through porins present in the outer membrane. Since the channels of the porins are formed by protein(s), point mutations developed in the protein sequence may alter the size or charge of the channel and thus lower the permeability of the drug.

Mutations in PBPs, The Targets of Penicillins The target of penicillin is the transpeptidase enzyme, also known as penicillin binding protein (PBP) (Sect. 3.3.2.3). Antibiotic resistance may be due to an alteration in the target of the drug such that it binds poorly to the target. Since many of these alterations arise as a result of a point mutation in the genes encoding the target enzymes, resistance to these drugs may occur spontaneously as frequently as 1 in 10^6 to 10^7 cells (Sect. 2.4). Due to the high cell density, there can be a significant number of mutants in the bacterial population. With continued selection there can be more than one point mutation in the protein giving rise to greater degree of resistance.

3.3.2.5 β-Lactamase: An Enzyme That Inactivates β-Lactam Drugs

Since the discovery of penicillin, numerous improvements have been made in the field of β-lactams as antibiotics. These include the various modifications made in the structure of penicillin as well as discovery and creation of several generations of cephalosporins (Sect. 3.3.3.2). However, resistance developed rapidly to all of these drugs. The most effective mechanism of developing resistance to penicillin is by the production of an enzyme that inactivates the drug. Since it breaks down β-lactams the enzyme is called β-lactamase (aka penicillinase), which hydrolyzes the four-membered β-lactam ring thereby inactivating the antibiotic. Because of the low penetration of β-lactams into gram-negative bacteria it was found that only a small amount of β-lactamase enzyme could confer a high level of resistance. In contrast, gram-positive bacteria required a much higher amount of the enzyme [23]. The first report of penicillinase was made in 1940, a few years before the start of clinical use of penicillin in humans [24]. This suggests that existence of the gene was not a result of clinical use of the antibiotic. As explained before (Sect. 2.6), formation of a new gene cannot happen in a few years but probably takes thousands of years. Since penicillin is a natural antibiotic, it has been used by nature long before humans discovered penicillin and produced it for clinical use. During this time the enzyme had already evolved to its current form even before the first clinical use of the drug.

   There are more than a thousand unique protein sequences for β-lactamases discovered. Genes encoding these enzymes may be on the bacterial chromosome or in plasmids or in transposons. Based on conserved and distinguishing amino acid motifs β-lactamases can be classified into four different classes: A, B, C, and D. Enzymes of the A, C, and D classes contain an active serine residue, which

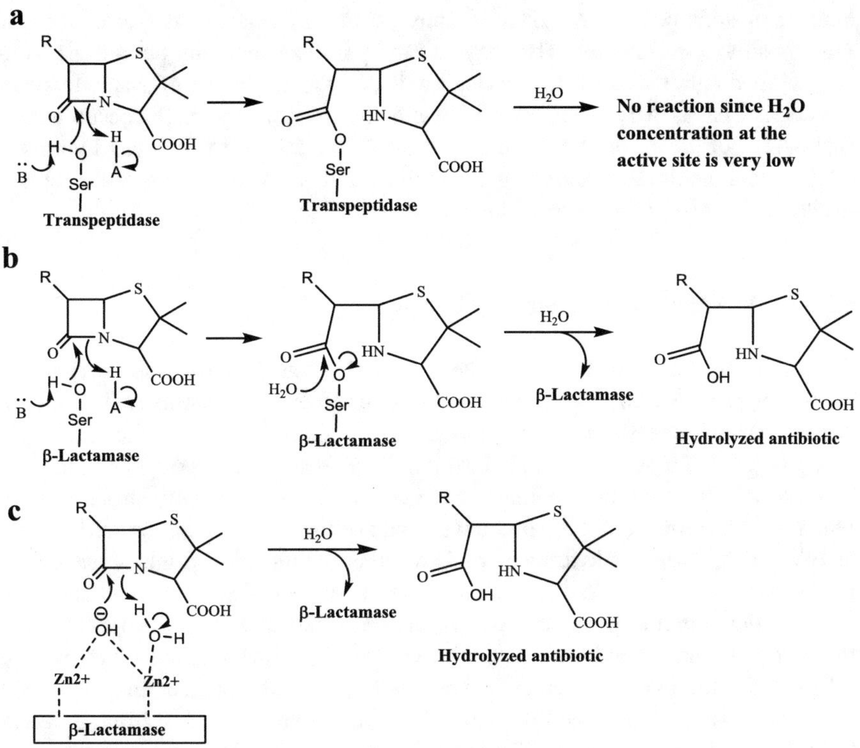

Fig. 3.17 Reactions of β-lactams. (**a**) Transpeptidase, (**b**) Serine β-lactamase, and (**c**) Metallo β-lactamase. All interactions between the Zn and the enzyme are not shown

means that the serine can act as a nucleophile. These enzymes, known as serine β-lactamases, hydrolyze the β-lactam antibiotic by first forming acyl bond between the active serine (Sect. 3.3.2.6) at its active site and the antibiotic and then hydrolyzing the bond (Fig. 3.17). β-lactamases of the B class are metalloenzymes that contain one or two zinc ions at the active site to facilitate β-lactam hydrolysis. Another way of classification of β-lactamases is based on their function. Group 1: cephalosporinases; Group 2: broad-spectrum, inhibitor-resistant, and extended-spectrum β-lactamases, and Group 3: metallo-β-lacatmases [25]. A β-lactamase is a penicillinase if it is more specific for penicillin, it is cephalosporinase if it is more specific for cephalosporins, it is a broad-spectrum β-lactamase if it hydrolyzes both penicillins and cephalosporins equally well. β-lactamase enzyme is usually found in gram-negative bacteria since the outer membrane is a permeability barrier for β-lactam antibiotics. So the concentration of the antibiotic in the periplasm, which is the site of cell wall synthesis, is very low. So gram-negative bacteria do not need to produce a large amount of β-lactamase and also do not need to secrete it outside the cell. Gram-positive bacteria, on the other hand, do not have an outer membrane and its cell wall is exposed to the outside environment. So a large amount of the β-lactamase has to be secreted to the outside environment to degrade the large amount of antibiotic present there. Staphylococcal β-lactamase is one of the few gram-positive β-lactamases. The enzyme can hydrolyze penicillin but not methicillin; so methicillin is an effective antibiotic for the treatment of staphylococcal infection. That is why emergence of methicillin-resistant Staphylococcus aureus (MRSA) is of great concern since they produce β-lactamase which can hydrolyze all β-lactam antibiotics including methicillin and so there is not much option available for treating infections with MRSA (Sect. 3.3.2.10).

3.3.2.6 Mechanism of Action of β-Lactamases

As mentioned above (Sect. 3.3.2.5), the A, C, and D classes of β-lactamases have an active Ser residue at the active site, similar to the PBPs (transpeptidases). The term "active serine" means the that there is enough negative charge on the serine OH group to make it a good nucleophile so that it can react at a carbonyl carbon. Note that alcohols are weak nucleophiles, whereas alkoxide ions are very strong nucleophiles. So all serines in a protein are weak nucleophiles but the active serine at the active site, by virtue of hydrogen bonds to other amino acids at the active site, is made a strong nucleophile that is somewhere in-between alcohols and alkoxides in strength. For both transpeptidases and β-lactamases, the Ser attacks the β-lactam ring to form a covalent bond with penicillin (Fig. 3.17). However, the outcomes are different for the two: the penicillin inactivates the transpeptidase enzyme (PBP) while β-lactamase destroys the penicillin. The difference arises from different stabilities of the enzyme-penicillin acyl−O−Ser complex. The acyl bond between the transpeptidase and penicillin is stable because water is excluded from the active site due to conformational change at the active site as explained for normal reaction of transpeptidase (Sect. 3.2.3). However, the structure of β-lactamase enzymes is such that water can easily enter the active site after acyl bond is formed between the β-lactamase and penicillin. This results in rapid hydrolysis of the acyl bond, thus releasing the penicillin with the hydrolyzed lactam ring. Stability of a complex is also expressed by its half-life ($t_{1/2}$) which is the time taken for half of the complex to be hydrolyzed. The $t_{1/2}$ of transpeptidase–penicillin complex is ~90 min, which is more than 10^6 times longer than the $t_{1/2}$ (~4 ms) for the β-lactamase–penicillin complex, which means that the latter is hydrolyzed more than a million times faster than the former. Note that half-life of 90 min is very high compared to usual doubling times of most bacteria and thus penicillin effectively stops the growth of cells. Cephalosporins and carbapenems are also hydrolyzed in the same way.

Metallo-β-lactamases (Class B) have one or two Zn^{2+} ions rather than a serine at the active site. These enzymes are known to inactivate all clinically important β-lactams except aztreonam [26]. Some β-lactamases (subclass B1 and B3) are most active as dizinc enzymes, while subclass B2 enzymes, such as *Aeromonas hydrophila* CphA, are inhibited by the binding of a second zinc ion [27]. Hydrolysis of the β-lactam ring requires reaction of a nucleophile at the carbonyl carbon of the β-lactam ring. Note that water is not a strong enough nucleophile as hydroxide (OH⁻) ion to carry out such a reaction. The Zn ions function by facilitating ionization of one water molecule to form hydroxide ion which functions as a nucleophile to react at the carbonyl carbon (Fig. 3.17). A similar involvement of zinc is seen in the enzyme carbonic anhydrase, which catalyzes the reaction of carbon dioxide and water to form bicarbonate. Another water molecule that is also bound to a zinc ion donates a proton to the nitrogen of the β-lactam ring and in the process becomes a hydroxide ion which carries out further catalytic cycles.

3.3.2.7 β-Lactamase Inhibitors

Since the discovery of penicillin scientists have made numerous other antibiotics. Some of these are natural ones, some are synthetic, but majority of these are minor modifications of existing antibiotics resulting in different variations of physical and chemical properties. Bacteria that are resistant to the existing antibiotics may not be resistant to these modified molecules because of the difference in structure. The most effective method that bacteria have for resistance to β-lactams is by secretion of the enzyme β-lactamase either to the outside by gram-positive bacteria or into the periplasmic space by gram-negative bacteria. In order to overcome the problem of resistance, scientists have developed new antibiotics that cannot be degraded by β-lactamases. One example of such antibiotic is methicillin, which is resistant to degradation by β-lactamase. This property makes the antibiotic methicillin a very successful antibiotic (Sect. 3.3.2.10).

Another strategy that scientists undertake for rational drug design is to make compounds that can inhibit or permanently inactivate enzymes that destroy antibiotics [28]. So the next big discovery in the field of antibiotics was that of small molecules that function as mechanism-based (suicide) inhibitors (Sect. 1.7.3) of β-lactamases. These compounds are used along with the antibiotics in order to prevent the destruction of the antibiotic by the β-lactamase enzyme. There are three β-lactamase inhibitors that have become clinically successful: Clavulanic acid, Sulbactam, and Tazobactam (Fig. 3.18). The first one is an enol ether lactam while the latter two are both sulfonyl derivatives of β-lactam. All these three compounds inhibit only serine β-lactamases and not the metallo-β-lactamases, which still continue to pose significant threat. Of special concern are the bacteria

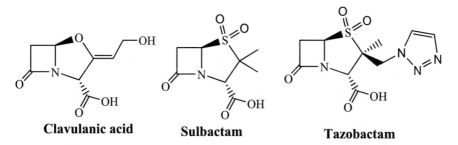

Clavulanic acid	**Sulbactam**	**Tazobactam**

Fig. 3.18 Mechanism-based (suicide) inhibitors of β-lactamases

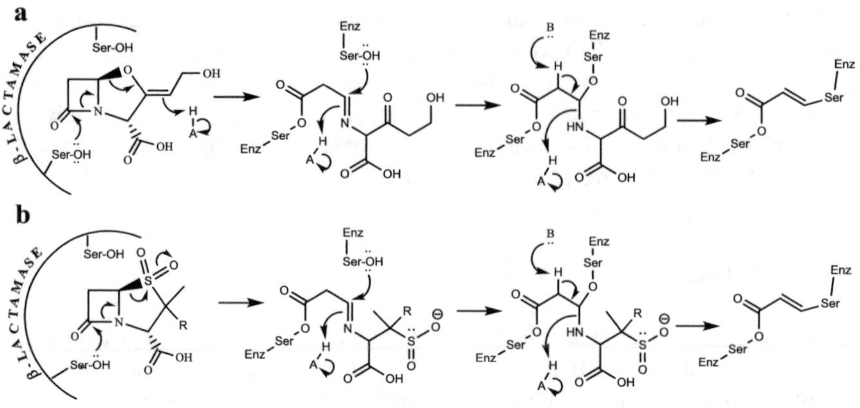

Fig. 3.19 Suicide inhibitors of β-lactamase. Reactions of (**a**) Clavulanic acid and (**b**) Sulbactam (R = methyl) and Tazobactam (R = triazole) with β-lactamase enzyme. The base, B, and the acid, A-H, represent various basic and acidic amino acids, respectively, at the active site of the enzyme

that are resistant to carbapenems by virtue of secretion of metallo β-lactamase. For further discussion of carbapenem-resistant bacteria, see Sect. 3.3.2.12. The good news is that recently several inhibitors for metallo-β-lactamase also been developed (Sect. 3.3.2.8) [29, 30].

Clavulanic Acid, Sulbactam, and Tazobactam Clavulanic acid is a β-lactamase inhibitor produced by *Streptomyces clavuligerus*. It is not an antibiotic (it has only weak antibacterial activity), its main function is to counteract the effect of β-lactamases. *S. clavuligerus* also produces and secretes the β-lactam antibiotic cephamycin C. The genes for biosynthesis of the antibiotic as well as the β-lactamase inhibitor are clustered together in the same region of the chromosome. Thus, this organism is unique in that the biosynthesis of an antibiotic and of a molecule to protect the antibiotic from enzymatic degradation is controlled by shared mechanisms [31]. This concept has been utilized clinically by prescribing β-lactam antibiotics in combination with clavulanic acid. The clavulanic acid inactivates the β-lactamase enzyme that is secreted by the infecting bacteria and thus allows penicillin to function. Some popular and successful combinations are Augmentin (contains penicillin and clavulanic acid), Timentin (contains ticarcillin and clavulanic acid), Zosyn (contains piperacillin and tazobactam), Unasyn (contains ampicillin and sulbactam).

The mechanism of action of clavulanic acid and the sulfones involves an "electron sink," which is a double bond at the right position for indirectly accepting the electrons of the nucleophile, serine that initiates the reaction. In case of transpeptidase reacting with peptidoglycan or with penicillin, and also in case of β-lactamases reacting with penicillin or with other β-lactams, the electrons of the serine go to the nitrogen of the amide bond that is broken, which then accepts a proton from an acidic amino acid at the active site of the enzyme (Figs. 3.16 and 3.19). In case of reaction of β-lactamase with clavulanic acid and the sulfones,

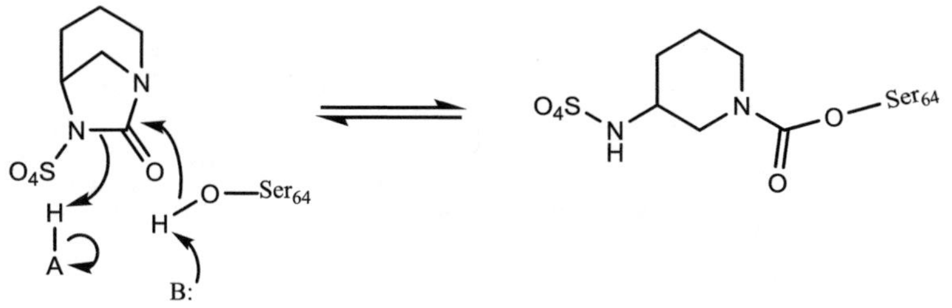

Fig. 3.20 Reaction of avibactam with β-lactamase enzyme. B and A-H are basic and acidic amino acids, respectively, at the active site

sulbactam and tazobactam, the electrons of the nucleophile go all the way to the alkenyl carbon or to the oxygen of the sulfone, which then accepts a proton from an acidic amino acid at the active site of the β-lactamase. The acyl enzyme bond formed between the β-lactamase and the inhibitor is different from the acyl bond formed with penicillin in that it is much slower to hydrolyze. What makes the release of free active enzyme even less probable is the formation of a second bond to another serine at the active site (Fig. 3.19). Because of the bonds of two serines of the enzyme to the fragmented inhibitor molecule, the β-lactamase becomes permanently inactivated.

Mechanism of action of sulbactam and tazobactam is similar to that of clavulanic acid. The main difference between them is that clavulanic acid has a carbon–carbon double bond while sulbactam and tazobactam have a sulfur–oxygen double bond to act as "electron sink" to accept the electrons of the nucleophilic serine. After the first step of the reaction the remaining steps are the same for the three inhibitors (Fig. 3.19).

Avibactam   The β-lactamase inhibitors clavulanic acid, sulbactam, and tazobactam are effective against class A but not class C β-lactamases, the genes for which are found in the chromosomes of many clinically important pathogens such as *Pseudomonas aeruginosa* and many *Enterobacteriaceae* spp. and also on plasmids that can be transferred to many other species of bacteria. One member of class C β-lactamase is AmpC, a cephalosporinase, which mediates resistance to cephalothin, cefazolin, cefoxitin, and most penicillins [32]. A newly discovered compound, Avibactam, is a β-lactamase inhibitor that inhibits both class A and class C and some class D enzymes [33]. Avibactam is different from the commonly used β-lactamase inhibitors, clavulanic acid, sulbactam, and tazobactam in that it is not a β-lactam. There is some difference in the mechanism of action also. The reaction starts with a ring opening as in case of the other three; however, the reaction is reversible, so the inhibitor is regenerated as opposed to being hydrolyzed (Fig. 3.20).

3.3.2.8   Inhibitors of Metallo-β-Lactamases

Clavulanic acid, sulbactam, tazobactam, and avibactam inhibit only serine β-lactamases and not the metallo-β-lactamases. There are no known clinical

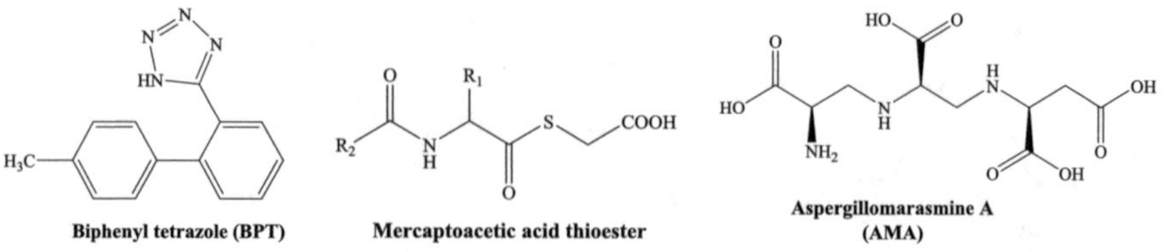

Fig. 3.21 Inhibitors of metallo-β-lactamase

inhibitors of the metallo-β-lactamases to date [28, 34]. However, several inhibitors of metallo-β-lactamases are being studied. Carbapenem resistance in *Bacteroides fragilis* is due to the secretion of a metallo-β-lactamase that hydrolyzes the antibiotic. Biphenyl tetrazoles (BPTs) (Fig. 3.21) are a structural class of potent competitive inhibitors of metallo-beta-lactamase identified through screening and using molecular modeling of the enzyme structure. The compound was shown to inhibit the metallo-β-lactamase in vitro and to convert imipenem-resistant *B. fragilis* from resistant to sensitive [29]. Based on crystallographic and kinetic studies, the authors proposed that the inhibitor functions by displacing a water molecule bound to the Zn^{2+} atom at the active site of the enzyme while the biphenyl moiety interacts with the hydrophobic amino acids of the flap extending above the active site.

Other metallo-β-lactamase inhibitors that are not related to biphenyl tetrazoles have also been reported. Mercaptoacetic acid thiol esters are irreversible inhibitors of metallo-β-lactamases and function via hydrolytic release of mercaptoacetic acid which subsequently covalently modifies a cysteine residue at the active site [26, 35]. Trifluoromethyl alcohol and ketone derivatives of L- or D-alanine have been reported to be competitive inhibitors of metallo-β-lactamase. These compounds are proposed to bind to one of the Zn atoms in the active site [28]. Recently, a fungal natural product, Aspergillomarasmine-A (AMA) has been identified as a potent inhibitor of metallo-β-lactamases such as New Delhi metallo-β-lactamase-1 (NDM-1) and Verona Integron encoded metallo-β-lactamase (VIM-2) [29]. The authors also demonstrated that a combination of AMA and the carbapenem antibiotic meropenem was able to significantly reduce the bacterial load in mice infected with meropenem-resistant bacteria. It has been previously shown that AMA can inhibit angiotensin converting enzyme (ACE), which is involved in increasing blood pressure by constricting blood vessels. Both ACE and metallo-β-lactamases are zinc containing enzymes and share some functional similarities, which suggests that the inhibitor AMA may have similar mechanism of action on these enzymes [36]. It has been shown that AMA is able to remove zinc ion from NDM-1 and thus inactivate the enzyme [30]. There has been growing alarm over recent reports of infection by carbapenem-resistant bacteria that contain the NDM-1 gene (see Sect. 3.3.2.12). The newly discovered inhibitors may be able to alleviate some of that concern.

3.3.2.9 Extended-Spectrum β-Lactamases (ESBLs)

Bacteria are constantly evolving in order to survive various environmental conditions. Since the introduction and subsequent frequent use of the β-lactamase inhibitors, bacteria have evolved to make their β-lactamases carry out their function even in the presence of the inhibitors. This is achieved by the bacteria in two ways: (1) Develop mutations in the β-lactamase promoter that increase the expression level of the enzyme and (2) develop mutations at the active site that allow the enzyme to bind to the antibiotic (β-lactam) with greater affinity than to the β-lactamase inhibitor. The new enzymes containing these mutations are called the Extended-Spectrum β-Lactamases (ESBLs). Thus, the ESBLs are derived from the usual broad-spectrum β-lactamases, TEM, which was named after the patient, Temoniera from whom it was first isolated and SHV, named so because sulfhydryl reagent had variable effect on activity [37]. The first report of ESBL was in 1979 [38]. The ESBL enzyme makes the bacteria resistant to most beta-lactam antibiotics, including penicillins, cephalosporins, and the monobactam aztreonam. There have been many reports of outbreaks of infections caused by bacteria-producing ESBL. These infections are very difficult to treat. However, the bacteria are usually sensitive to carbapenem antibiotics (Sect. 3.3.2.12).

3.3.2.10 Methicillin-Resistant *Staphylococcus aureus* (MRSA)

These gram-positive bacteria are part of the normal flora in 30% of the population but can also cause severe infectious diseases mainly of the skin. *S. aureus* should be sensitive to the β-lactam antibiotics such as penicillin since it has four penicillin binding proteins, PBP-1,2,3, and 4, which are needed for cell wall synthesis (Sect. 3.3.2.3). However, because of the uncontrolled use and misuse of penicillin the bacteria rapidly developed resistance to penicillin and other common β-lactams. The most common mechanism of resistance is the acquisition of the gene for the enzyme β-lactamase, which hydrolyzes the penicillin to penicilloic acid (Sect. 3.3.2.5, Fig. 3.17). In the 1950s, scientists at Beecham pharmaceutical company in the UK developed a derivative of penicillin that contained two bulky methoxy substituents in the side chain benzyl group. Because of these bulky groups, the antibiotic did not bind to β-lactamase and so were resistant to destruction by the enzyme [39]. This new antibiotic was initially introduced under the trade name Celbenin but was later changed to the current name methicillin (Fig. 3.14). This new antibiotic has since been highly effective especially for treating penicillin-resistant bacteria. However, unfortunately, within 1 year the first case of resistance to methicillin was reported [40]. This resistant strain, methicillin-resistant *Staphylococcus aureus* (MRSA), which has now become a household name is of immense concern and is widely referred to as a "superbug" because of the difficult-to-treat diseases that it causes. Initially it was responsible for hospital-acquired (nosocomial) infections but has now spread outside the hospitals causing numerous illnesses in the community [41].

Mechanism of Resistance to Methicillin Because of the binding and subsequent suicide inactivation of penicillin binding proteins (PBPs), methicillin can kill most bacteria even if they produce β-lactamase since methicillin does not bind to β-lactamase. However, some bacteria develop resistance to methicillin by expressing a different PBP protein named PBP2a (also called PBP2′) in addition to the usual PBP2. Note that this same protein also confers resistance to carbapenems which is discussed later (Sect. 3.3.2.12). This new protein does not bind to methicillin and thus is not inactivated by the antibiotic. The gene, *mecA* that codes for the protein PBP2a is present in the chromosome in a mobile genetic element which is a large 40–60 kb stretch of foreign DNA called the *mec* element. DNA sequence analysis suggests that the *mecA* gene has originated from *Staphylococcus sciuri*. *S. aureus* has four PBPs involved in cell wall synthesis (Sect. 3.3.2.3). In presence of methicillin, which binds to and inactivates the four usual PBPs, the PBP2a enzyme can take over the function of the four PBPs and continue with cell wall synthesis [42]. Since PBP2a is essential in conferring methicillin resistance, any factor that interferes with the expression of the *mecA* gene or with the activity of PBP2a will affect methicillin resistance. Since PBP2a has strict substrate requirements, factors that influence formation of the substrate have the potential to perturb or modulate methicillin resistance. Some of these factors have been discussed in a review article [43]. However, another research lab has reported that the enzyme PBP2A is very adaptable to the conditions of the host cell. The *mecA* gene (which codes for PBP2A) of methicillin-resistant *S. aureus* is originally from *S. sciuri,* however, the cell wall structures of the two wild-type strains, *S. sciuri* and *S. aureus* are quite different. Transpeptidase enzymes in the two strains use different substrates for the cross-linking reaction and thus synthesize cell walls with different structures. However, when the *mecA* gene from *S. sciuri* was cloned into *S. aureus* it was found that the cell wall structure of *S. aureus* did not change which suggests that the protein product of the *S. sciuri mecA* can efficiently participate in cell wall biosynthesis and build a cell wall using the cell wall precursors that are characteristic of the *S. aureus* host [44].

3.3.2.11 Unusual β-Lactams: Monobactams

Search for more β-lactams have resulted in discovery of many other antibiotics with similar mechanisms of action. A range of diverse β-lactams, that are neither penicillins nor cephalosporins, have been discovered. Two such antibiotics will be discussed: aztreonam and carbapenem. Aztreonam (Fig. 3.22) is an unusual β-lactam because it is a monobactam (a monocyclic β-lactam), in which the β-lactam ring is alone and not fused to another ring. Aztreonam is the only commercially available monobactam and is active against aerobic gram-negative bacteria such as Neisseria [45] and Pseudomonas and was approved by the FDA in 1986. Commercial aztreonam is made synthetically but was originally isolated from the bacterium *Chromobacterium violaceum*. It is a useful antibiotic since it is resistant to some β-lactamases including metallo-β-lactamases (Sect. 3.3.2.6) but are sensitive to

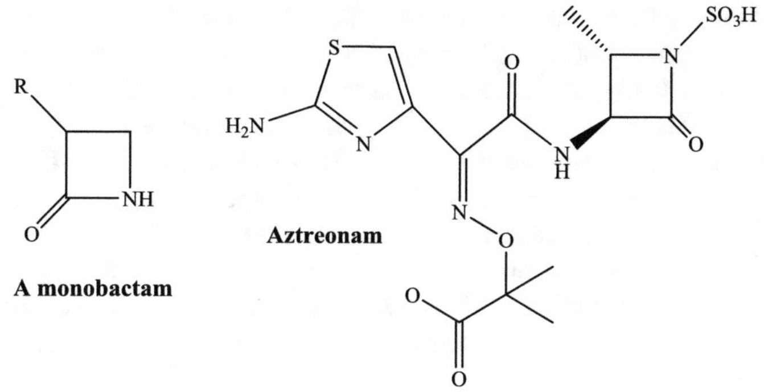

Fig. 3.22 Structure of aztreonam, a monobactam

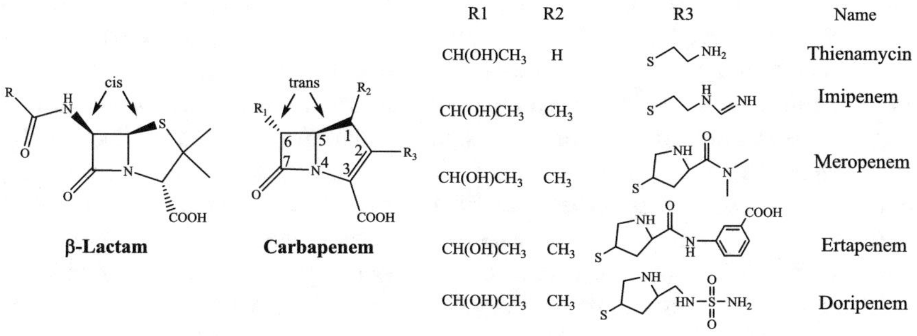

Fig. 3.23 Structures of β-lactam and some carbapenems

extended-spectrum β-lactamases (Sect. 3.3.2.9). The drug is especially useful for treatment of patients who are allergic to penicillin, cephalosporin, or carbapenem. This is because monobactams, in spite of being lactams, are structurally different enough from the usual β-lactam antibiotics, so as to not induce allergic reactions.

3.3.2.12 Unusual β-Lactams: Carbapenems

Another antibiotic, thienamycin has a carbon in place of the sulfur in the five-membered ring, but does have a sulfur containing side chain (Fig. 3.23). It is not inactivated by most β-lactamases including the extended-spectrum β-lactamases (ESBL). Thienamycin was the first naturally occurring member of the class of antibiotic called carbapenems. It was isolated from the soil bacteria, *Streptomyces cattleya* and its structure was determined [46]. Other carbapenems were derived based on the structure of thienamycin. As the name suggests, carbapenems differ from conventional penicillins (penams) in two ways: they have a carbon instead of a sulfur atom and have a double bond (an "e" instead of "a" in the name) in the molecule (Fig. 3.23). Carbapenems have a very broad spectrum of activity. They are active against a broader range of both gram-negative and gram-positive bacteria than are penicillins and cephalosporins including bacteria that have acquired the extended-spectrum β-lactamases (ESBL). The spectrum of activity of carbapenems encompasses virtually all bacterial pathogens except mycobacteria, cell wall-deficient organisms, and a few infrequent non-fermenters and aeromonads [47]. The structures of some carbapenems are shown in Fig. 3.23. The important structural features of carbapenems that are responsible for their broad spectrum of activity as well as their resistance to β-lactamases are the *trans* configuration of the substituents of the β-lactam ring, the hydroxyethyl substituent (R1) at C6, and the methyl substituent (R2) at C1 [48]. The hydroxyethyl side chain at C6 has a chiral center and the stereochemistry of the carbon (C8) should be R for the carbapenem to be active.

Older carbapenems, such as imipenem, were susceptible to degradation by the enzyme dehydropeptidase-1 (DHP-1) located in renal tubules but newer ones such as meropenem (approved by FDA in 1996), ertapenem (approved in 2001), and doripenem (approved in 2007) are more stable [49]. One great advantage of carbapenems over β-lactams such as penicillins is that they are not degraded by most beta-lactamases except metallo-beta-lactamases. *Pseudomonas aeruginosa* is frequently the cause of nosocomial (hospital acquired) pneumonia. Infections by *P. aeruginosa* have very high mortality rates due to widespread antibiotic resistance, both intrinsic and acquired. One antibiotic that is frequently used effectively against these bacteria is carbapenem. Because of its broad spectrum of activity and its stability against β-lactamases, carbapenems are one of the most valuable antibiotics available today. In order to prevent development of resistance to the antibiotic, its use has been limited and is used as an antibiotic of last resort. Still, frequent use of the antibiotic has resulted in the development of resistance to carbapenem.

Resistance to Carbapenems   Since the antibiotic has not yet been used excessively, there is less resistance development. However, many cases of resistant bacteria have been reported. There can be several methods of resistance development.

By Point Mutations   One method of developing resistance to carbapenems is by acquiring point mutations in the target proteins, which in this case are the PBPs such that they bind less tightly to the drug. Such examples are seen in *Enterococcus faecium* and methicillin-resistant Staphylococci [50]. However, the most common cause of resistance to carbapenem is mutation in the outer membrane protein OprD which is known to form outer membrane porins that are needed for transport of basic amino acids or peptides containing basic amino acids. Unlike the porin protein OmpF of *E. coli*, the OprD protein of *P. aeruginosa* forms a narrower porin. Carbapenems enter the cell through these porins. Since basic amino acids compete with carbapenem antibiotics for passage through the porins, MIC of carbapenems depends on the presence of basic amino acids in the growth medium. Mutations in the *oprD* gene cause conformational changes in the porin and thus give rise to resistance to carbapenems [51].

By Insertion of DNA   Besides acquiring point mutations in a gene, another common method of mutation is by insertion of some DNA sequence in a gene. The inserted

DNAs are called a transposons or insertion sequence (IS) elements, which are commonly known as "jumping genes" (Sect. 2.6). Insertion of DNA into a gene can give rise to antibiotic resistance in two ways. The insertion sequence itself may contain the sequence of an antibiotic resistance gene thus conferring resistance to that antibiotic. Another possibility is that the insertion into a gene inactivates a protein that is essential for functioning of an antibiotic. One such example is seen in *P. aeruginosa* which can become resistant to carbapenem by insertion of IS elements in the *oprD* gene. In *P. aeruginosa* the OprD protein is used to form the outer membrane porins whose normal function is to transport basic amino acids. Since carbapenems use these porins to enter the cell, insertions in the *oprD* gene make the bacteria resistant to antibiotics. Many such instances of carbapenem-resistant *P. aeruginosa* have been reported from various countries [52].

By Acquiring Resistance Gene   As mentioned above, carbapenems are resistant to most β-lactamases, and this makes them highly reliable antibiotics. However, carbapenems are not resistant to the metallo-β-lactamases (MBL) which are capable of degrading the drugs and are also known as carbapenemases. There have been several reports of carbapenem-resistant bacteria that contain a metallo-β-lactamase gene. However, it was a matter of great concern when a new carbapenem-resistant *Klebsiella pneumoniae* was detected in a patient. These bacteria contained the New Delhi metallo-β-lactamase 1 (NDM-1) gene [53]. A high incidence of NDM-1 producing carbapenem-resistant *Enterobacter cloacae* has been reported from China [34]. Such spread of the resistance gene is becoming a major public health threat. As discussed before (Sect. 3.3.2.7), several inhibitors of serine β-lactamases are known and are in clinical use in combination with β-lactam antibiotics. No such inhibitors of metallo β-lactamases are in clinical use as of today although, there are several promising candidates that have been discovered for inhibiting the metallo-β-lactamases (Sect. 3.3.2.8).

A New Multistep Method of Acquiring Resistance   Another very unique mechanism of resistance to carbapenem that is independent of carbapenemase (metallo-β-lactamase) enzyme has been identified recently by Levy and his colleagues at Tufts University [54]. They showed that the infecting bacteria, *E. coli* had mutated four separate times in order to develop resistance to carbapenems. The results support the concept that continued exposure to sublethal concentrations of an antibiotic will select for multiple mutations that can have a cumulative effect to eventually provide resistance to the antibiotic (Sect. 2.5). Mutations in the bacteria in a regulatory protein MarR decreased the expression of outer membrane porin proteins OmpF and OmpC, thus decreasing the permeability of carbapenem antibiotics, and increased the expression of a multidrug efflux pump resulting in pumping out any antibiotic that enters the cell. The mutations also resulted in the expression of a new protein YedS, which is usually a non-translatable protein due to a large gap in the open reading frame (ORF). Expression of the outer membrane protein YedS also increases resistance to carbapenems.

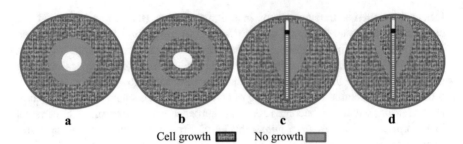

a   b   c   d

Cell growth [image]   No growth [image]

Fig. 3.24 Diagrammatical representation of zone of inhibition. Kirby–Bauer test (**a, b**) and E-test (**c, d**). (**a**) and (**c**) represent the use of most antibiotics. (**b**) and (**d**) show carbapenem resistance and demonstrate the target bypass mechanism. Note that the cells and background can be stained differentially with methylene blue to make the zone of inhibition appear more prominent [55]

Target Bypass Mechanism In some bacteria resistance takes place by a target bypass mechanism. These strains produce a second PBP called PBP2′ (also known as PBP2a), which is also a peptidoglycan transpeptidase as are other PBPs. The same mechanism applies to methicillin resistance also and has been discussed in more detail before (Sect. 3.3.2.10). In the presence of carbapenem the usual transpeptidases (PBP2 and 3) are inactivated. However, the presence of the carbapenem induces the expression of PBP2′ which now continues the cross-linking reaction. PBP-2′ has decreased affinity for all β-lactams, so the resistance is developed against a broad spectrum of antibiotics including carbapenems.

As discussed before (Sect. 2.2), antibiotic resistance can be determined by the Kirby–Bauer disk diffusion test. Antibiotic sensitive strains will give a zone of inhibition while resistant strains will grow uniformly throughout the plate. Disk diffusion test done with the antibiotic imipenem (a carbapenem) on a lawn of *Streptococcus haemolyticus* gives an interesting result: a double zone of growth around the disk is observed. There is growth around the disk as well as at the edge of the plate with a zone of no growth in between (Fig. 3.24, compare A and B). A similar pattern can be seen in the E-test (Sect. 2.2), which also gives a double zone of growth (Fig. 3.24, compare C and D). Such a double zone of growth is a characteristic of *Streptococcus haemolyticus* in the presence of carbapenem [56]. A likely explanation for the double zone of growth is that induction of the PBP2′ gene requires a certain minimum concentration of the carbapenem antibiotic. The concentration is high enough around the antibiotic disk and so PBP2′ is synthesized and the bacteria are resistant immediately around the disk. Beyond a certain distance from the disk, the concentration of the antibiotic is not high enough to induce the expression of the gene for PBP2′ and so the bacteria are killed by the antibiotic. At even farther distance the concentration of antibiotic is below the MIC and so does not kill the bacteria and results in the second zone of growth.

An example of multiple zones of inhibition has also been observed with an unidentified antibiotic present in the aqueous extract from rhubarb stalk [57]. The appearance of multiple zones of inhibition depends on the number of cells spread on the plate for demonstrating zone of inhibition.

3.3.3 Antibiotics Targeting the Membrane Phase of Synthesis

As explained before (Sect. 3.3.2), although the membrane phase comes before the cell wall phase in the sequence of reactions for the biosynthesis of the cell wall, the antibiotics acting in the membrane phase of cell wall biosynthesis are being discussed here after the β-lactams which affect the cell wall phase of biosynthesis of the cell wall.

3.3.3.1 Bacitracin

Bacitracin, named after a patient, Treacy, is a peptide antibiotic produced by some strains of *Bacillus licheniformis* and *Bacillus subtilis*, and inhibits the cell wall formation in gram-positive as well as some gram-negative bacteria [58].

Structure and Use   Bacitracin is a mixture of several cyclic polypeptides containing 12 amino acids (Fig. 3.25) many of which are unusual amino acids. Four of them are D-amino acids instead of the usual L-ones, one is ornithine and two are modified amino acids. In fact, there is something unusual about every peptide bond of bacitracin. This is a common biochemical strategy that organisms often take to protect against degradation by proteases which can cleave only normal peptide bonds in proteins. Remember, bacterial cell wall also has several unusual peptide bonds (Fig. 3.4). Another unusual feature of bacitracin is that an extra peptide bond is formed between the γ-carboxyl group of aspartate and the ε-amino group of lysine,

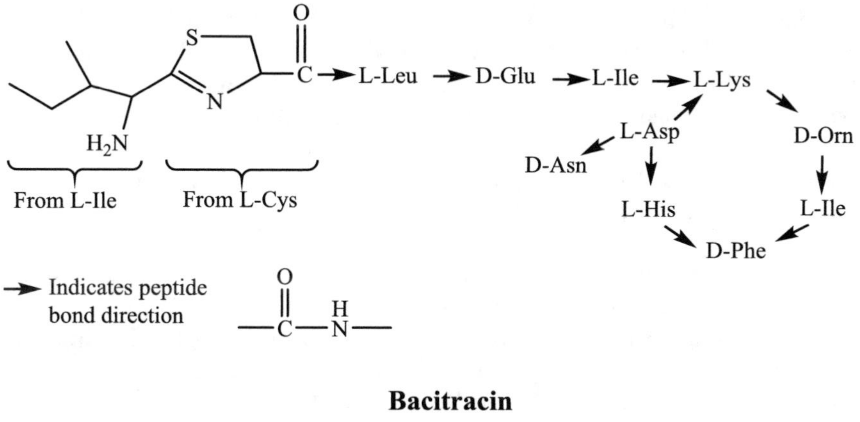

Bacitracin

Fig. 3.25 Structure of bacitracin

thus forming a ring. The Lys and Asp, each participates in more than two peptide bonds.

Bacitracin is too toxic for systemic use and so is used topically for skin infections. At high concentration it is nephrotoxic if ingested. However, it is an ingredient in many over-the-counter first aid ointments. Topical application on pets may result in toxic effects including death since pets have a tendency to lick the topically applied medication. At low concentrations bacitracin is widely used as a growth-promoting food additive for animals (see subtherapeutic use, Sect. 2.10.2). Bacitracin is not absorbed through the intestinal walls and so is one of the most commonly used growth-promoting antibiotic in animals. Of all the bacitracin manufactured, 90% is used for growth promotion in farm animals.

Mechanism of Action Although the mechanism of subtherapeutic effect is not clearly understood, the mechanism of therapeutic effect of bacitracin has been well studied. Bacitracin inhibits the hydrolysis of C55-P-P by pyrophosphatase (Sect. 3.2.2). Since the product C55-P is not formed, the next cycle of reaction cannot take place, thus stopping cell wall synthesis. Mechanism of action of bacitracin is somewhat unusual. Most antibiotics function by binding to an enzyme and inhibiting its activity. Bacitracin functions by binding not to the enzyme but to the substrate, C55-P-P in the presence of Zn^{2+} or other divalent cations [59]. Two other antibiotics that function by binding to their respective substrates are mersacidin (Sect. 3.3.3.4) and vancomycin (Sect. 3.3.3.6).

Resistance Development One common mechanism of resistance development to any antibiotic is by point mutations in the enzyme catalyzing an essential reaction. However, in this case that is not possible since bacitracin does not bind to any protein. So resistance development to bacitracin is rare in spite of the widespread use since its discovery in the early 1940s. Nevertheless, several methods of resistance to bacitracin have been reported [60].

Some bacteria are resistant to the antibiotic because of the presence and expression of the BacA enzyme. It was initially proposed that BacA is a kinase catalyzing the phosphorylation of the lipid C55 (undecaprenol) to produce C55-P, thus compensating for the deficiency of the undecaprenyl (C55) pyrophosphatase, which also produces C55-P but is inhibited by bacitracin [61]. However, when BacA was purified it was found to be an undecaprenyl pyrophosphatase enzyme [62]. In fact there are multiple genes encoding membrane proteins with undecaprenyl pyrophosphate phosphatase (UppP) activity. All these proteins outcompete bacitracin for binding to C55-P-P to produce C55-P and provide low-level resistance to bacitracin. This is an example of a unique phenomenon. Usual enzyme inhibitors compete with the substrate for binding to the enzyme. Here, enzyme molecules compete with the inhibitor (bacitracin) for binding to the substrate.

Another mechanism of resistance to bacitracin is by the BcrABC transporter system which pumps out the bacitracin molecules from the interior of the membrane, where the C55-P-P resides. The hydrophobic proteins BcrB and BcrC form a transmembrane channel and two identical subunits of BcrA protein function as

ATPases that hydrolyze ATP to provide energy for the transport [63]. The presence of bacitracin is sensed by the bacitracin sensor BcrR, which activates the expression of the BcrABC transport system. The original source of the BcrABC transport system is *Bacillus licheniformis*, which produces the antibiotic bacitracin and protects itself from the antibiotic by pumping it out using the transport system. The system can be acquired by a wide range of bacterial species to gain resistance.

3.3.3.2 Antibiotics That Inhibit Transglycosylation Reaction

As discussed before (Sect. 3.2.3), transglycosylation reaction involves both the plasma membrane and the cell wall as its location. Three antibiotics that inhibit the transglycosylation reaction will be discussed: Moenomycin, Mersacidin (a lantibiotic), and Vancomycin.

3.3.3.3 Moenomycin

Moenomycin complex of antibiotics (other names flavomycin, bambermycin) is a mixture of five components A, A12, C1, C3, and C4 isolated in the 1960s from several strains of *Streptomyces*: *Streptomyces bambergiensis, ghanaensis, ederensis,* and *geysiriensis*. Unlike the many antibiotics that inhibit the transpeptidase enzyme, moenomycin is the only antibiotic known that inhibits the transglycosylase enzyme (also known as PBP1b or glycosyltransferase). The structure of moenomycin A is shown in Fig. 3.26. The various components of moenomycin complex differ from each other in the substituents of the sugar rings marked in blue. The molecules

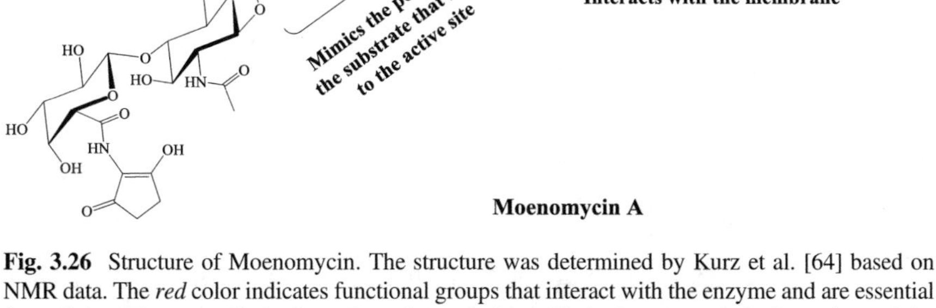

Moenomycin A

Fig. 3.26 Structure of Moenomycin. The structure was determined by Kurz et al. [64] based on NMR data. The *red* color indicates functional groups that interact with the enzyme and are essential for activity. The two carbohydrate rings shown in *blue* mimic the two carbohydrate rings of the substrate

contain a pentasaccharide chain attached to a lipid (menocinol) via a phosphodiester linkage. The structure was determined by Kurz et al. [64]. The red color indicates functional groups that interact with the enzyme and are essential for activity. The two carbohydrate rings shown in blue mimic the two carbohydrate rings of the substrate. Moenomycin has a broad spectrum of activity against gram-positive bacteria. It is not as potent against gram-negative bacteria because of the outer membrane barrier. However, it can lyze growing *E. coli* cell but not stationary phase cells [65]. Some gram-negative bacteria including *Neisseria, Helicobacter, Brucella, Pasteurella,* and *Pseudomonas* are more sensitive to moenomycins than others suggesting differences in permeability of the outer membrane [66].

Some transglycosylases can be bifunctional enzymes which also have transpeptidase activity. The enzyme has a transmembrane domain and thus, is anchored to the membrane. Binding of moenomycin was studied by Cheng et al. [67] who made deletions in the gene for the enzyme to obtain truncated proteins. They observed that the transmembrane domain was important for binding to moenomycin.

In spite of its unique mechanism of action, moenomycin is not used in humans because of toxicity and poor bioavailability. Due to the hydrophobic lipid tail and hydrophilic sugars and phosphate, moenomycin behaves as a detergent and so is very stable but forms aggregates in aqueous solution. This prevents absorption through the intestinal walls and if injected into the bloodstream, can cause hemolysis, which explains its toxicity [66]. So its main use is as a growth promoter in animal feed (Subtherapeutic use, Sect. 2.10.2). However, appropriate moenomycin derivatives should be further explored as promising antibiotics. There are several advantages of such antibiotics [68]. This is the only antibiotic known that functions by inhibiting glycosyltransferase enzyme. Function of glycosyltransferase is conserved among all bacteria and does not have any eukaryotic counterpart [69]. Similar to penicillin, it inhibits cell wall synthesis and so it has bactericidal activity. Since the antibiotic has never been used in humans and when added to animal feed, it is not absorbed but excreted, there is no known case yet of significant resistance to the drug. In an effort to obtain derivatives of moenomycin that can be of human use, Yuan et al. [70] have first identified six amino acid residues at the active site of the glycosyltransferase enzyme that make contact with moenomycin. Mutational analysis showed that all six residues are important for enzymatic activity. The authors also designed the biosynthesis of a smaller version of moenomycin which can still maintain these six contacts with the enzyme and the new compound had comparable activity as the natural drug. This can facilitate the design of new moenomycin analogs that can be used in humans.

3.3.3.4 Lantibiotics: Mersacidin

One type of antibiotics that functions by inhibiting cell wall synthesis is called a lantibiotic (an abbreviation for lanthionine containing antibiotic). Lantibiotics are antimicrobial peptides (AMPs) but unlike other peptide antibiotics such as bacitracin and gramicidin, lantibiotics are ribosomally synthesized and then post-translationally modified. Lantibiotics can be of two types: Type A lantibiotics are long and flexible molecules (e.g., nisin) while Type B lantibiotics are globular (e.g., mersacidin). Both types of lantibiotics function by binding to the C55-P (Lipid II), which plays an important role in cell wall synthesis (Sect. 3.2.2). However, since Type A lantibiotics also function by disrupting the cell membrane, these are discussed further in more detail in the chapter on membrane-acting antibiotics (Sect. 7.2.2.8). The focus here is on the Type B lantibiotic, mersacidin.

Mersacidin was first discovered and isolated from a Bacillus species [71]. It is the smallest lantibiotic known (molecular mass 1825). Mersacidin has antibiotic activity comparable to that of vancomycin against gram-positive bacteria such as strepto-cocci, bacilli, and staphylococci including methicillin-resistant *Staphylococcus aureus* (MRSA, Sect. 3.3.2.10) strains [72]. Mechanism of action of mersacidin involves binding to membrane-bound Lipid II (C55-P-P-NAM-NAG-pentapeptide, Sect. 3.2.2). Structure of mersacidin and its interaction with Lipid II has been studied using NMR [73]. Mersacidin is a 20-residue peptide with nine post-translationally modified amino acids and a single negatively charged residue, Glu-17. Unlike type A lantibiotics, which are positively charged and are mostly extended and flexible in structure, mersacidin is globular and compact and have a negative charge or no net charge.

Mersacidin inhibits peptidoglycan biosynthesis, at the level of transglycosylation [74]. Of the three types of antibiotics that inhibit transglycosylation, moenomycin functions as a competitive inhibitor of the transglycosylase enzyme, vancomycin binds to the terminal D-ala-D-ala portion of the substrate and thus inhibits the reaction. Mersacidin also functions by binding to the substrate but not to the terminal D-ala-D-ala part. Binding of mersacidin to the substrate has been studied and shown to an ionic interaction. Mutagenesis or modification of Glu-17 of mersacidin was shown to inactivate the antibiotic activity. However, since both mersacidin and Lipid II are negatively charged, interaction between them is proposed to be between Glu-17 of mersacidin and the sugar phosphate head group of Lipid II through positively charged calcium ions [73].

3.3.3.5 Vancomycin

The glycopeptide antibiotic vancomycin (from the word "vanquish") was first isolated from *Streptomyces orientalis* in the 1950s at Eli Lilly and Company [75]. It was found to be active against most penicillin-resistant staphylococci [76] and some anaerobic organisms [77] as well as *Neisseria gonorrhoeae* [78]. Although vancomycin was quickly approved for clinical use in 1958 soon after its discovery, it was not a favored antibiotic because of its toxicity. Vancomycin has several disad-vantages such as poor tissue penetration, difficulty of administration as well as other side effects. People became interested in it again in the early 1980s for the treatment of pseudomembranous enterocolitis. Another reason for the increased interest in vancomycin was the spread of methicillin-resistant *Staphylococcus aureus* (MRSA).

Infections caused by MRSA are of major concern since the bacteria are resistant to all β-lactam antibiotics. Since vancomycin is effective against MRSA, it is considered to be a very valuable antibiotic and so is usually used as an antibiotic of last resort when all β-lactams fail or when the patient is allergic to β-lactams. Because of the limited use of vancomycin and because of its unique mechanism of action (Sect. 3.3.3.5), resistance development to vancomycin has been rare. Vancomycin is one of the very few antibiotics that can be used against infection by multidrug-resistant enterococci. Enterococci are the major cause of hospital-acquired infections. Of these, *E faecalis* accounts for 80–90% of all clinical isolates of *Enterococci* while the remaining 10–15% is due to infection by *E. faecium* [79].

3.3.3.6 Mechanism of Action of Vancomycin

The structure of vancomycin consists of a heptapeptide ring attached to a disaccharide via a glycosidic bond (hence the name glycopeptide). The site of action is the cell wall and so the vancomycin does not enter the cytoplasm of the bacteria. Because of its large structure vancomycin does not pass through the outer membrane and so is not effective against gram-negative bacteria except *Neisseria gonorrhoeae,* which has a *m*ultiple *t*ransferable *r*esistance (MTR) pump that is used to translocate drugs across the membrane [80]. Another glycopeptide antibiotic with very similar mechanism of action as vancomycin is teicoplanin. The structures of the two antibiotics are shown in Fig. 3.27 in which the differences between the two are shown in red. Because of the hydrophobic tail, teicoplanin can be described as a lipoglycopeptide. Teicoplanin has lower toxicity than vancomycin [81]. Teicoplanin and several derivatives of it were approved by the FDA in 2014.

Vancomycin blocks the transglycosylase (aka glycosyltransferase) reaction in which the monomer unit is transferred to the broken points of the cell wall. Vancomycin is an unusual type of antibiotic that does not bind to the

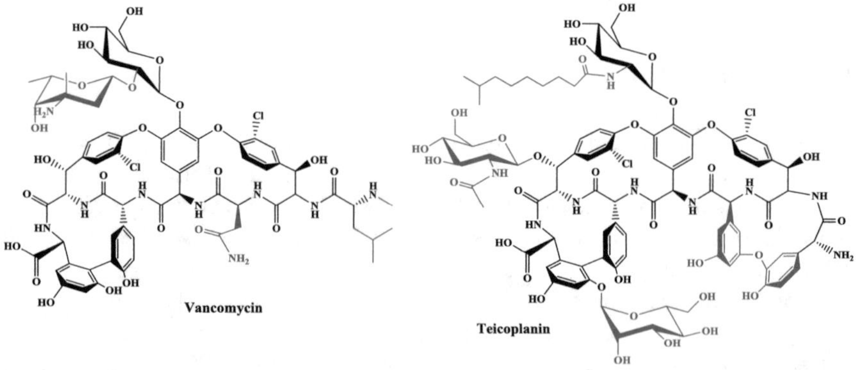

Fig. 3.27 Structures of vancomycin and teicoplanin. *Red* color indicates differences between the two

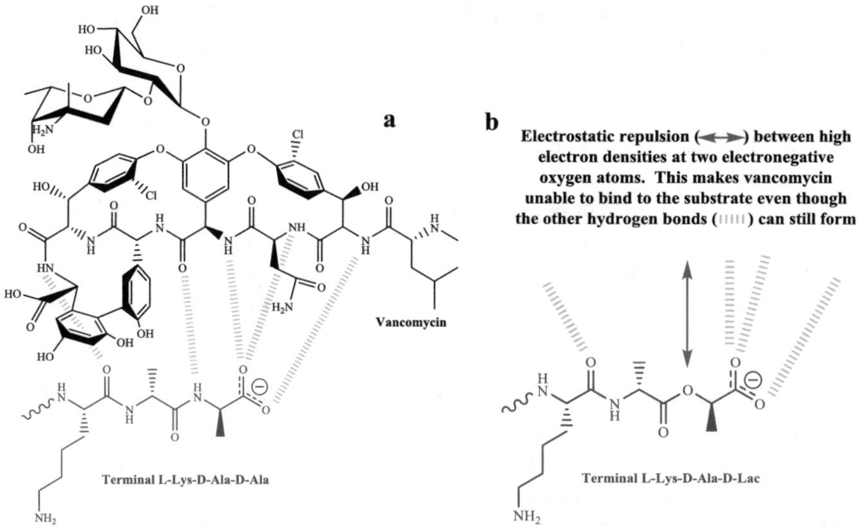

Fig. 3.28 Mechanism of vancomycin resistance. (**a**) Structure of Vancomycin (*black*) bound to the terminal L-Lys-D-Ala-D-Ala (*blue*) of the peptidoglycan strand by hydrogen bonds (*green*). The H-bond lengths are not drawn to scale. (**b**) Interaction with modified substrate in vancomycin-resistant mutants

transglycosylase enzyme or to any other protein. Instead, it binds to the terminal D-ala-D-ala dipeptide residue of pentapeptide part of the monomer unit, C55-P-P-NAM-(pentapeptide-pentaglycine)-NAG. By blocking the substrate, it inhibits the transglycosylation reaction. Since the terminal D-ala-D-ala part of the peptidoglycan strand also serves as a substrate for the transpeptidase enzymes (Sect. 3.2.3), vancomycin can inhibit both the transglycosylation and the transpeptidation reaction and have a bactericidal effect. Unlike the β-lactam antibiotics and all other suicide inhibitors (Sect. 1.7.3), which become covalently bonded to the enzyme, vancomycin functions in a unique way. Instead of binding to an enzyme, it binds non-covalently to the substrate and so is also described a "supramolecular antibiotic." The binding between the antibiotic and the substrate takes place through five hydrogen bonds between the carbonyl and secondary amino groups present in both vancomycin and the terminal L-Lys-D-ala-D-ala of the peptidoglycan monomer (Fig. 3.28). Another contributing interaction that is not shown in the figure is the hydrophobic interaction between the alanine methyl group of the monomer and the aromatic groups of the vancomycin [82]. The binding of vancomycin to its target was determined by NMR and X-ray diffraction studies [83, 84]. Since it specifically binds to the D-ala-D-ala moiety of the glycopeptide molecule (actually lysine, the third amino acid from the end also fits into the binding pocket of vancomycin), other intermediates that contain D-ala-D-ala can also bind to the vancomycin. So the D-ala-D-ala dipeptide formed by D-ala-D-ala synthetase (Sect. 3.2.1) can also bind to vancomycin in vitro. However, no binding is observed in vivo since the antibiotic is unable to cross the membrane and enter the cytoplasm where the D-ala-D-ala intermediate is present.

3.3.3.7 Resistance Development to Vancomycin

As discussed above, vancomycin was first approved for clinical use in 1958 but was not widely used because of its toxicity. Then in the 1980s, when methicillin-resistant *Staphylococcus aureus* (MRSA, Sect. 3.3.2.10) strains began to emerge, vancomycin was brought back into use. By the late 1980s, it was the only reliable antibiotic against MRSA infections. Today the use of vancomycin has been made limited so that too much use does not result in selection for bacteria resistant to the drug. Vancomycin is used as an antibiotic of last resort only when all other antibiotics have failed. Although vancomycin has been in use since the 1960s, significant resistance has not developed to it. One reason for this is the relatively infrequent use of the drug for reasons explained above. However, the major reason for the lack of resistance development is the unusual mechanism of action of vancomycin in that it binds to the substrate and not to an enzyme. As discussed before (Sect. 2.4), if an antibiotic binds to a protein, resistance can develop by acquiring point mutations in the gene for the protein. But this is not possible if the antibiotic binds to a substrate. In order to prevent binding of the antibiotic to the substrate, the structure of the substrate will have to be changed, which cannot be done by acquiring a simple point mutation in any protein. Synthesis of a substrate requires several steps and several enzymes. In order for any one of these enzymes to catalyze a different reaction, there will have to be a drastic change in the enzyme. Also, in order to synthesize a new molecule a whole set of new enzymes will have to be made. Moreover, even if a different molecule is synthesized, there is no guarantee that it can function as a substrate for the synthesis of the cell wall. However, in spite of all these barriers to resistance development, there are many reports of vancomycin-resistant *enterococci* (VRE) in hospital-acquired infections; the first reports coming in 1988 in Europe [85, 86] and similar strains were detected in the USA [87]. Between 1989 and 1993, the percentage of enterococcal tests that were positive for VRE in the USA rose from 0.3 percent to 7.9 percent according to the Center for Disease Control and Prevention (CDC). Enterococcal infections can be fatal, particularly those caused by strains of vancomycin-resistant *enterococci* (VRE). During 2004, VRE caused about one of every three infections in hospital intensive-care units, according to the CDC.

Resistance development to vancomycin has been possible not because the resistance mechanism has evolved due to extensive use of the antibiotic. In fact, the antibiotic has hardly been used in the last 50 years. Instead, resistance to vancomycin has developed because the resistance genes already existed in nature. In an evolutionary time scale, it can take millions of years for bacteria to develop a whole set of genes for the synthesis of an alternative substrate that will not bind to the drug and so 50 years of use is not sufficient to develop these genes. Resistance is possible because such resistant bacteria already existed in nature and contain the cassette of genes for the synthesis of the alternative substrate. Widespread use of the antibiotic only facilitates the horizontal transfer of the genes to other bacteria by conjugation, transformation, or transduction (Sect. 2.6) and further selection of resistant bacteria. *S. aureus* strains that are resistant to vancomycin can be classified into three categories: vancomycin-resistant strains (VRSA; MIC, ≥ 16 μg/ml); vancomycin-intermediate strains (VISA; MIC, ≥ 4 μg/ml); and heterogeneous vancomycin-intermediate strains (hVISA; MIC <4 μg/ml with subpopulations with higher MIC) [88]. VRSA strains are extremely rare, whereas the others have been reported numerous times.

As mentioned above, genes for resistance to vancomycin already existed in some bacteria long before humans started using the antibiotic [89]. Examples include Enterococcal species such as *Enterococcus gallinarum* and *E. casseliflavus* and gram-positive bacteria such as *Leuconostoc*, *Lactobacillus*, and *Pediococcus* which have intrinsic resistance to vancomycin because the monomer units that they synthesize for making their cell wall have a terminal D-ala-D-lac or D-ala-D-ser instead of the usual D-ala-D-ala (lac = lactate, ser = serine). This makes only a minor change in the structure of the substrate, which is the change of an amide bond to an ester bond. This does not change the structure of the cell wall since the terminal D-lac or D-ala is removed during transpeptidation reaction (Sect. 3.2.3). However, this small change in the monomer has a strong effect on binding to vancomycin. As discussed above and shown in Fig. 3.28, vancomycin binds to the L-lys-D-ala-D-ala terminus of the monomer. The binding is through five hydrogen bonds of the carbonyl and N-H groups of the peptide linkages in vancomycin to the N-H and carbonyl groups of the D-lys and D-ala as well as the free carboxyl group of the terminal D-ala of the monomer substrate (Fig. 3.28a). One such hydrogen bond is lost if one D-ala is replaced by D-lac in the monomer. The situation is made even worse because the hydrogen bonding is replaced by a repulsive force between two electronegative oxygen atoms (Fig. 3.28b), which further decreases the binding of vancomycin to the cell wall monomer and thus confers resistance to the antibiotic. This explains the intrinsic resistance of these bacteria to the drug. Since these bacteria do not cause any disease in humans, their intrinsic resistance may not have any direct effect on the effectiveness of vancomycin in human diseases. However, they indirectly facilitates the development of resistance in pathogens affecting humans through transfer of the resistance genes by conjugation, transformation, or transduction (Sect. 2.6).

Synthesis of alternate substrate for resistance to vancomycin can be catalyzed by any of the seven known clusters of genes (also known as operons): VanA, VanB, VanC, VanD, VanE, VanG, and VanL [89]. Of these the VanA type resistance gene cluster is the most common and the most studied and is the one that is described here. The VanA type resistance gene cluster is originally found on the transposon (Sect. 2.6) Tn1546, which has a size of 10,581 bp and contains nine genes and is present within a plasmid. Two of these genes are for transposition function and the other seven genes, *vanR*, *vanS*, *vanH*, *vanA*, *vanX*, *vanY*, and *vanZ* provide resistance to vancomycin and similar antibiotics (Fig. 3.29). The most important of these is the protein vanA which codes for a D-ala-D-lac ligase, which joins D-ala to D-lac to form a depsipeptide (peptides in which one or more amide bonds are replaced by

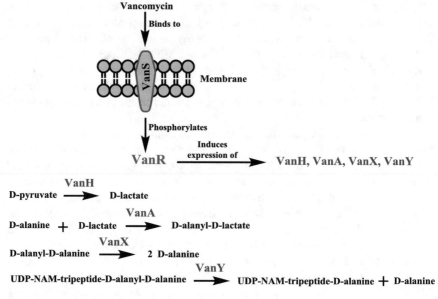

Fig. 3.29 Expression of VanA type resistance

ester bonds). It has sequence similarity to the usual D-ala-D-ala ligase enzyme but it specifically ligates D-ala to D-lac. Binding of the alternate substrate to VanA ligase has been studied in detail [90]. The specific functions of the other genes are as follows. VanR and VanS proteins constitute a two-component regulatory system that regulates the transcription of the *vanHAXYZ*. A typical two-component regulatory system contains a transmembrane receptor protein that senses an environmental signal such as the presence of vancomycin. It then sends the signal to a response regulator protein in the cytoplasmic side of the membrane. This protein then facilitates expressions of some specific genes that respond to the signal present on the other side of the membrane. In this case, VanS is the receptor protein for vancomycin. It is a histidine kinase that autophosphorylates a histidine residue in its sequence when it senses the presence of vancomycin in the environment. The phosphoryl group is then transferred to the response regulator protein, VanR which then binds to DNA to activate the synthesis of the proteins encoded by *vanHAXYZ*. Although VanA catalyzes the ligation of D-ala to D-lac which is the most important reaction for the formation of the depsipeptide monomer, VanA alone cannot confer resistance to vancomycin. This is because D-lac is not normally produced by enterococci. VanH, which is a lactate dehydrogenase enzyme, performs the function of producing D-lac by reducing pyruvate. Even in the presence of these resistance genes, the normal cell wall monomer containing the terminal D-ala-D-ala linkage, as well as the precursor of the monomer, the D-ala-D-ala dipeptide (Sect. 3.2.1 and Fig. 3.5) will continue to be produced. In order to prevent these from interfering with resistance, the VanA gene cluster contains two more genes, *vanX* and *vanY*. VanX is a D-ala-D-ala dipeptidase, which hydrolyzes any D-ala-D-ala normal dipeptide that may be present. It has no activity against D-ala-D-lac, the altered substrate precursor. VanY is a D-D-carboxypeptidase which hydrolyzes the D-ala-D-ala bond in any monomer molecule that may already be formed before Van X can degrade all of the D-ala-D-ala dipeptide. This way the four enzymes VanH, VanA, VanX, and VanY together synthesize alternate depsipeptide substrate, which cannot bind to vancomycin (Fig. 3.29).

The function of VanZ is not yet known. It is not necessary for resistance to vancomycin but may play a role in resistance to teicoplanin since it moderately increases the MIC of teicoplanin [79].

Other Types of Vancomycin Resistance   Besides the VanA type of resistance, there are reports of other types such as VanB, VanC, VanD, VanE, VanG, and VanL types. They all have significant sequence similarities to VanA type cluster. However, there are also differences between the various types. Some are present on plasmid and so can be easily transferred to other bacteria by conjugation while others are in the chromosome and cannot be transferred so easily. They can also differ in the gene expression levels. Some clusters are inducible by vancomycin (genes expressed only when vancomycin is present in the environment) while others are constitutive (genes always expressed). The alternate monomer substrate for some resistance type may have a D-ala-D-serine linkage (Van C, E, G, and L) instead of D-ala-D-lac (Van A, B, and D) [91].

Vancomycin-Dependent Bacteria   One interesting observation related to vancomycin is that some bacteria are not just vancomycin resistant, they can actually become vancomycin-dependent, which means that they cannot grow in the absence of vancomycin. Such bacteria are a major concern since they require vancomycin for growth and so cannot be detected or cultured under usual growth conditions. So the disease will not even be diagnosed as a bacterial infection until it becomes too late. The first example of vancomycin-dependent enterococcus was reported in 1994 [92]. This was followed by many other reports of vancomycin-dependent bacteria [93–95]. Other antibiotic-dependent bacteria have been observed long before vancomycin dependence was discovered. For example, streptomycin-dependent bacteria were known since the 1940s [96].

Mechanism of Vancomycin Dependence   Vancomycin-dependent bacteria are actually vancomycin-resistant bacteria that have developed mutations that inactivate the wild-type D-ala-D-ala ligase gene. So they cannot make cell wall using the wild-type genes and so cannot grow under normal growth conditions. In the presence of vancomycin the vancomycin resistance genes are expressed which allow it to make cell wall using D-ala-D-lactate ligase as discussed above (Sect. 3.3.3.6). Note that there is no change in the final cell wall structure since the terminal D-ala or D-lactate is removed in the cross-linking step of cell wall synthesis. One evidence that supports this mechanism is that the bacteria are able to grow in the absence of vancomycin if they are supplied with premade D-ala-D-ala dipeptide which bypasses the need for D-ala-D-ala ligase.

Reversion to vancomycin independent phenotype is also possible [97]. This can happen in two ways: (1) The bacteria can develop mutation(s) that negates the effect of the earlier mutation and now make active D-ala-D-ala ligase enzyme again. (2) The bacteria can develop mutations in the promoter of the VanA or VanB resistance operon such that the expression of the resistance genes becomes constitutive and does not require the presence of vancomycin as inducer.

3.3.4   Teixobactin, A Newly Discovered Antibiotic

As mentioned before (Sect. 1.4), 99% of bacterial species on this planet have not yet been cultured. Kim Lewis and coworkers have developed a novel method to culture many of these organisms in soil and in the process, have discovered several new potential antibiotics [98]. One such antibiotic is teixobactin which was obtained from a screen of uncultured bacteria grown in diffusion chambers in situ. Teixobactin is effective against gram-positive pathogens, including drug-resistant strains. In fact, it showed far superior activity against *S. aureus* than did vancomycin. Teixobactin is ineffective against gram-negative bacteria, because it cannot cross the outer membrane barrier. When tested against an *E. coli* strain with a defective outer membrane permeability barrier, it did show strong activity. Besides being a highly effective antibiotic teixobactin has another advantage. No mutant of *S. aureus* or *M. tuberculosis* resistant to teixobactin could be obtained when they were plated in the presence of low dose of the antibiotic. Long-term growth at sub-MIC levels of the antibiotic also failed to produce any resistant mutants. Such an observation usually would suggest a nonspecific mode of action with accompanying toxicity against the host. However, the antibiotic was found to have no toxicity against mammalian cells.

The authors also studied the mechanism of action of the antibiotic. Teixobactin inhibited the synthesis of peptidoglycan, but had no effect on DNA, RNA, or protein synthesis, suggesting that teixobactin is a new peptidoglycan synthesis inhibitor. The fact that no resistant mutant could be obtained against the antibiotic suggests that the target of the antibiotic is not a protein. In this respect it is similar to vancomycin which also does not have a protein target but binds to lipid II, the precursor of the cell wall peptidoglycan. The authors proposed that teixobactin also has the same target. They demonstrated that teixobactin did interact with the peptidoglycan precursor. Teixobactin is superior to vancomycin in another way. Bacteria can develop resistance to vancomycin by synthesizing a modified lipid II in which the terminal D-ala is replaced by a D-lac or D-Ser (Sect. 3.3.3.6). It was shown that teixobactin was active against vancomycin-resistant enterococci that make these modified targets suggesting that teixobactin is able to bind to these modified forms of lipid II while vancomycin cannot (Fig. 3.28). So it can be expected that resistance to teixobactin will not develop easily. After its introduction into the clinic, it took 30 years for vancomycin resistance to appear. The authors predict that it will probably take even longer for resistance to teixobactin to emerge.

References

1. Scheffers D-J, Pinho MG (2005) Bacterial cell wall synthesis: new insights from localization studies. Microbiol Mol Biol Rev 69:585–607
2. van Heijenoort J (2001) Formation of the glycan chains in the synthesis of bacterial peptido-glycan. Glycobiology 11:25R–35R
3. White D (2007) The physiology and biochemistry of prokaryotes, 3rd edn. Oxford University Press, New York, NY
4. Mengin-Lecrelux D, van Heijenoort J (1993) Identification of the *glmU* gene encoding N-acetylglucosamine-1-phosphate uridyltransferase in *Escherichia coli*. J Bacteriol 175: 6150–6157
5. Du W, Brown JR, Sylvester DR, Huang J, Chalker AF, So CY, Holmes DJ, Payne DJ, Wallis NG (2000) Two active forms of UDP-*N*-acetylglucosamine enolpyruvyl transferase in grampositive bacteria. J Bacteriol 182:4146–4152
6. Brown ED, Vivas EI, Walsh CT, Kolter R (1995) MurA (MurZ), the enzyme that catalyzes the first committed step in peptidoglycan biosynthesis, is essential in *Escherichia coli*. J Bacteriol 177:4194–4197
7. Eschenburg S, Kabsch W, Healy ML, Schonbrunn E (2003) A new view of the mechanisms of UDP-*N*-acetylglucosamine enolpyruvyl transferase (MurA) and 5-enolpyruvylshikimate3-PHOSPHATE synthase (AroA) derived from X-ray structures of their tetrahedral reaction intermediate states. J Biol Chem 278:49215–49222
8. Kim DH, Lees WJ, Kempsell KE, Lane WS, Duncan K, Walsh CT (1996) Characterization of a Cys115 to Asp substitution in the *Escherichia coli* cell wall biosynthetic enzyme UDPGlcNAc enolpyruvyl transferase (MurA) that confers resistance to inactivation by the antibiotic Fosfomycin. Biochemistry 35:4923–4928
9. Fillgrove KL, Pakhomova S, Newcomer ME, Armstrong RN (2003) Mechanistic diversity of fosfomycin resistance in pathogenic microorganisms. J Am Chem Soc 125:15730–15731
10. Reitz RH, Slade HD, Neuhaus FC (1967) The biochemical mechanisms of resistance by streptococci to the antibiotics D-cycloserine and O-carbamyl-D-serine. Biochemistry 6:2561–2570
11. Wargel RJ, Shadur CA, Neuhaus FC (1971) Mechanism of D-cycloserine action: transport mutants for D-alanine, D-cycloserine, and glycine. J Bacteriol 105:1028–1035
12. Spratt BG (1977) Properties of the penicillin-binding proteins of Escherichia coli K12. Eur J Biochem 72:341–352
13. Sauvage E, Kerff F, Terrak M, Ayala JA, Charlier P (2008) The penicillin-binding proteins: structure and role in peptidoglycan biosynthesis. FEMS Microbiol Rev 32:234–258
14. Kohanski MA, DePristo MA, Collins JJ (2010) Sublethal antibiotic treatment leads to multi-drug resistance via radical-induced mutagenesis. Mol Cell 37:311–320
15. Bayles KW (2000) The bactericidal action of penicillin: new clues to an unsolved mystery. Trends Microbiol 8:274–278
16. Tomasz A (1979) The mechanism of the irreversible antimicrobial effects of penicillins: how the beta-lactam antibiotics kill and lyse bacteria. Annu Rev Microbiol 33:113–137
17. Tipper DJ, Strominger JL (1965) Mechanism of action of penicillins: a proposal based on their structural similarity to acyl-D-alanyl-D-alanine. Proc Natl Acad Sci U S A 54:1133–1141
18. Tomasz A, Albino A, Zanati E (1970) Multiple antibiotic resistance in a bacterium with suppressed autolytic system. Nature 227:138–140
19. Smith TJ, Blackman SA, Foster SJ (2000) Autolysins of *Bacillus subtilis*: multiple enzymes with multiple functions. Microbiology 146:249–262
20. Graham LL, Beveridge TJ (1994) Structural differentiation of the *Bacillus subtilis* 168 cell wall. J Bacteriol 176:1413–1423
21. Merad T, Archibald AR, Hancock IC, Harwood CR, Hobot JA (1989) Cell wall assembly in *Bacillus subtilis*: visualisation of old and new wall material by electron microscopic

examination of samples stained selectively for teichoic acid and teichuronic acid. J Gen Microbiol 135:645–655

22. Delcour AH (2009) Outer membrane permeability and antibiotic resistance. Biochim Biophys Acta 1794:808–816

23. Percival A, Brumfitt W, De Louvois J (1963) The role of penicillinase in determining natural and acquired resistance of Gram-negative bacteria to penicillins. J Gen Microbiol 32:77–89

24. Abraham EP, Chain E (1940) An enzyme from bacteria able to destroy penicillin. Nature 46: 837–837

25. Bush K, Jacoby GA (2010) Updated functional classification of β-lactamases. Antimicrob Agents Chemother 54:969–976

26. Liu X-L, Shi Y, Kang JS, Oelschlaeger P, Yang K-W (2015) Amino acid thioester derivatives: a highly promising scaffold for the development of metallo-β-lactamase 11 inhibitors. ACS Med Chem Lett 6:660–664

27. Bebrone C, Delbruck H, Kupper MB, Schlomer P, Willmann C, Frere J-M, Fischer R, Galleni M, Hoffmann KMV (2009) The structure of the dizinc subclass B2 metallo–βlactamase CphA reveals that the second inhibitory zinc ion binds in the histidine site. Antimicrob Agents Chemother 53:4464–4471

28. Drawz SM, Bonomo RA (2010) Three decades of β-lactamase inhibitors. Clin Microbiol Rev 23:160–201

29. Toney JH, Fitzgerald PM, Grover-Sharma N, Olson SH, May WJ, Sundelof JG, Vanderwall DE, Cleary KA, Grant SK, Wu JK, Kozarich JW, Pompliano DL, Hammond GG (1998) Antibiotic sensitization using biphenyl tetrazoles as potent inhibitors of Bacteroides fragilis metallo-beta-lactamase. Chem Biol 5:185–196

30. King AM, Reid-Yu SA, Wang W, King DT, De Pascale G, Strynadka NC, Wals TR, Coombes BK, Wright GD (2014) Aspergillomarasmine A overcomes metallo-β-lactamase antibiotic resistance. Nature 510:503–506

31. Paradkar A (2013) Clavulanic acid production by Streptomyces clavuligerus: biogenesis, regulation and strain improvement. J Antibiot (Tokyo) 66:411–420

32. Jacoby GA (2009) AmpC β-Lactamases. Clin Microbiol Rev 22:161–182

33. Lahiri SD, Johnstone MR, Ross PL, McLaughlin RE, Olivier NB, Alma RA (2014) Avibactam and class c β-lactamases: mechanism of inhibition, conservation of the binding pocket, and implications for resistance. Antimicrob Agents Chemother 58:5704–5713

34. Liu C, Qin S, Xu L, Zhao D, Liu X, Lang S, Feng X, Liu HM (2015) New Delhi metallo-β lactamase 1(NDM-1), the dominant carbapenemase detected in carbapenem-resistant enterobacter cloacae from Henan province, China. PLoS One. https://doi.org/10.1371/journal.pone.0135044

35. Payne DJ, Bateson JH, Gasson BC, Proctor D, Khushi T, Farmer TH, Tolson DA, Bell D, Skett PW, Marshall AC, Reid R, Ghosez L, Combret Y, Marchand-Brynaert J (1997) Inhibition of metallo-β-lactamases by a series of mercaptoacetic acid thiol ester derivatives. Antimicrob Agents Chemother 41:135–140

36. Meziane-Cherif D, Courvalin P (2014) Antibiotic resistance: to the rescue of old drugs. Nature 510:477–478

37. Bradford PA (2001) Extended-spectrum β-lactamases in the 21st century: characterization, epidemiology, and detection of this important resistance threat. Clin Microbiol Rev 14:933–951

38. Sanders CC, Sanders WE (1979) Emergence of resistance to cefamandole: possible role of cefoxitin-inducible beta-lactamases. Antimicrob Agents Chemother 15:792–797

39. Rolinson GN, Stevens S, Batchelor FR, Wood JC, Chain EB (1960) Bacteriological studies on a new penicillin—BRL.1241. Lancet 276:564–567

40. Barber M (1961) Methicillin-resistant staphylococci. J Clin Pathol 14:385–393

41. Morell EA, Balkin DM (2010) Methicillin-resistant Staphylococcus aureus: a pervasive pathogen highlights the need for new antimicrobial development. Yale J Biol Med 83:223–233

42. Fuda C, Suvorov M, Vakulenko SB, Mobashery S (2004) The Basis for resistance to β-lactam antibiotics by Penicillin-binding Protein 2a of methicillin-resistant *Staphylococcus aureus*. J Biol Chem 279:40802–40806

43. Stapleton PD, Taylor PW (2002) Methicillin resistance in *Staphylococcus aureus*: mechanisms and modulation. Sci Prog 85:57–72

44. Severin A, Wu SW, Tabei K, Tomasz A (2005) High-level β-lactam resistance and cell wall synthesis catalyzed by the *mecA* homologue of *Staphylococcus sciuri* introduced into *Staphylococcus aureus*. J Bacteriol 187:6651–6658

45. Miller LK, Sanchez PL, Berg SW, Kerbs SB, Harrison WO (1983) Effectiveness of aztreonam, a new monobactam antibiotic, against penicillin-resistant gonococci. J Infect Dis 148:612

46. Kahan JS, Kahan FM, Goegelman R, Currie SA, Jackson M, Stapley EO, Miller TW, Miller AK, Hendlin D, Mochales S, Hernandez S, Woodruff HB, Birnbaum J (1979) Thienamycin, a new β-lactam antibiotic discovery, taxonomy, isolation and physical properties. J Antibiot 32: 1–12

47. Livermore DM, Woodford (2000) Carbapenemases: a problem in waiting? Curr Opin Microbiol 3:489–495

48. Papp-Wallace KM, Endimiani A, Taracila MA, Bonomo RA (2011) Carbapenems: past, present, and future. Antimicrob Agents Chemother 55:4943–4960

49. Zhanel GG, Wiebe R, Dilay L, Thomson K, Rubenstein E, Hoban DJ, Noreddin AM, Karlowsky JA (2007) Comparative review of the carbapenems. Drugs 67:1027–1052

50. Spratt BG (1994) Resistance to antibiotics mediated by target alterations. Science 264:388–393

51. Li H, Luo YF, Williams BJ, Blackwell TS, Xie CM (2012) Structure and function of OprD protein in *Pseudomonas aeruginosa*: from antibiotic resistance to novel therapies. Int J Med Microbiol 302:63–68

52. Al-Bayssari C, Valentini C, Gomez C, Reynaud-Gaubert M, Rolain J-M (2015) First detection of insertion sequence element *ISPa1328* in the *oprD* porin gene of an imipenem-resistant *Pseudomonas aeruginosa* isolate from an idiopathic pulmonary fibrosis patient in Marseille, France. New Microbes New Infect 7:26–27

53. Yong D, Toleman MA, Giske CG, Cho HS, Sundman K, Lee K, Walsh TR (2009) Characterization of a new metallo-β-lactamase gene, *bla*$_{NDM-1}$, and a novel erythromycin esterase gene carried on a unique genetic structure in *Klebsiella pneumoniae* sequence type 14 from India. Antimicrob Agents Chemother 53:5046–5054

54. Warner DM, Yang Q, Duval V, Chen M, Xu Y, Levy SB (2013) Involvement of MarR and YedS in carbapenem resistance in a clinical isolate of *Escherichia coli* from China. Antimicrob Agents Chemother 57:1935–1937

55. Bhattacharjee MK (2015) Better visualization and photodocumentation of zone of inhibition by staining cells and background agar differently. J Antibiot 68:657–659

56. Schwalbe RS, Ritz WJ, Verma PR, Barranco EA, Gilligan PH (1990) Selection for vancomycin resistance in clinical isolates of *Staphylococcus haemolyticus*. J Infect Dis 161:45–51

57. Bhattacharjee MK, Bommareddy PK, DePass AL (2021) A water-soluble antibiotic in rhubarb stalk shows an unusual pattern of multiple zones of inhibition and preferentially kills slow-growing bacteria. Antibiotics 10:951. https://doi.org/10.3390/antibiotics10080951

58. Johnson BA, Anker H, Meleney FL (1945) Bacitracin: a new antibiotic produced by a member of the *B. subtilis* group. Science 102:376–377

59. Stone KJ, Strominger JL (1971) Mechanism of action of bacitracin: complexation with metal ion and C55-isoprenyl pyrophosphate. Proc Natl Acad Sci U S A 68:3223–3227

60. Cao M, Helmann JD (2002) Regulation of the *Bacillus subtilis* bcrC bacitracin resistance gene by two extracytoplasmic function σ factors. J Bacteriol 184:6123–6129

61. Cain BD, Norton PJ, Eubanks W, Nick HS, Allen CM (1993) Amplification of the *bacA* gene confers bacitracin resistance to *Escherichia coli*. J Bacteriol 175:3784–3789

62. El Ghachi M, Bouhss A, Blanot D, Mengin-Lecreulx D (2004) The bacA gene of *Escherichia coli* encodes an undecaprenyl pyrophosphate phosphatase activity. J Biol Chem 279:30106–30113

63. Podlesek Z, Comino A, Herzog-Velikonja B, Grabnar M (2000) The role of the bacitracin ABC transporter in bacitracin resistance and collateral detergent sensitivity. FEMS Microbiol Lett 188:103–106

64. Kurz M, Guba W, Vertesy L (1998) Three-dimensional structure of moenomycin A—a potent inhibitor of penicillin-binding protein 1b. Eur J Biochem 252:500–507

65. Baizman ER, Branstrom AA, Longley CB, Allanson N, Sofia MJ, Gange D, Goldman RC (2000) Antibacterial activity of synthetic analogues based on the disaccharide structure of moenomycin, an inhibitor of bacterial transglycosylase. Microbiology 146:3129–3140

66. Ostash B, Walker S (2010) Moenomycin family antibiotics: chemical synthesis, biosynthesis, biological activity. Nat Prod Rep 27:1594–1617

67. Cheng TJR, Sung MT, Liao HY, Chang YF, Chen CW, Huang CY, Chou LY, Wu YD, Chen YH, Cheng YSE, Wong CH, Ma C, Cheng WC (2008) Domain requirement of moenomycin binding to bifunctional transglycosylases and development of high-throughput discovery of antibiotics. Proc Natl Acad Sci U S A 105(2):431–436

68. Shah NJ (2015) Reversing resistance: the next generation antibacterials. Indian J Pharmacol 47: 248–255

69. Derouaux A, Sauvage E, Terrak M (2013) Peptidoglycan glycosyltransferase substrate mimics as templates for the design of new antibacterial drugs. Front Immunol 4:78–83

70. Yuan Y, Fuse S, Ostash B, Sliz P, Kahne D, Walker S (2008) Structural analysis of the contacts anchoring moenomycin to peptidoglycan glycosyltransferases and implications for antibiotic design. ACS Chem Biol 3:429–436

71. Chatterjee S, Chatterjee S, Lad SJ, Phansalkar MS, Rupp RH, Ganguli BN, Fehlhaber HW, Kogler H (1992) Mersacidin, a new antibiotic from Bacillus. Fermentation, isolation, purification and chemical characterization. J Antibiot 45:832–838

72. Chatterjee S, Chatterjee DK, Jani RH, Blumbach J, Ganguli BN, Klesel N, Limbert M, Seibert G (1992) Mersacidin, a new antibiotic from *Bacillus, in vitro* and *in vivo* antibacterial activity. J Antibiot 45:839–845

73. Hsu S-TD, Breukink E, Bierbaum G, Sahl H-G, de Kruijff B, Kaptein R, van Nuland NAJ, Bonvin AMJJ (2003) NMR study of Mersacidin and Lipid II interaction in Dodecylphosphocholine Micelles: conformational changes are a key to antimicrobial activity. J Biol Chem 278:13110–13117

74. Brotz H, Bierbaum G, Markus A, Molitor E, Sahl H-G (1995) Mode of action of the lantibiotic mersacidin: inhibition of peptidoglycan biosynthesis via a novel mechanism? Antimicrob Agents Chemother 39:714–719

75. Levine DP (2006) Vancomycin: a history. Clin Infect Dis 42:S5–S12

76. Anderson RCGR, Higgins HM Jr, Pettinga CD (1961) Symposium: how a drug is born. Cinci J Med 42:49–60

77. Geraci JE, Heilman FR, Nichols DR, Ross GT, Wellman WE (1956) Some laboratory and clinical experiences with a new antibiotic, vancomycin. Mayo Clin Proc 31:564–582

78. Geraci JE, Wilson WR (1981) Vancomycin therapy for infective endocarditis. Rev Infect Dis 3 (Suppl):S250–S258

79. Cetinkaya Y, Falk P, Mayhall CG (2000) Vancomycin-resistant enterococci. Clin Microbiol Rev 13:686–707

80. Janganan TK, Zhang L, Bavro VN, Matak-Vinkovic D, Barrera NP, Burton MF, Steel PG, Robinson CV, Borges-Walmsley MI, Walmsley AR (2011) Opening of the outer membrane protein channel in tripartite efflux pumps is induced by interaction with the membrane fusion partner. J Biol Chem 286:5484–5493

81. Svetitsky S, Leibovici L, Paul M (2009) Comparative efficacy and safety of vancomycin versus teicoplanin: systematic review and meta-analysis. Antimicrob Agents Chemother 53:4069–4079

82. Smith DK (2005) A supramolecular approach to medicinal chemistry: medicine beyond the molecule. J Chem Educ 82:393–400

83. Rao J, Whitesides GM (1997) Tight binding of a dimeric derivative of Vancomycin with Dimeric L-Lys-D-Ala-D-Ala. J Am Chem Soc 119:10286–10290

84. Schafer M, Schneider TR, Sheldrick GM (1996) Crystal structure of vancomycin. Structure 4: 1509–1515

85. Uttley AHC, Collins CH, Naidoo J, George RC (1988) Vancomycin-resistant enterococci. Lancet 1:57–58

86. Leclercq R, Derlot E, Duval J, Courvalin P (1988) Plasmid-mediated resistance to vancomycin and teicoplanin in *Enterococcus faecium*. N Engl J Med 319:157–160

87. Friden TR, Munsiff SS, Low DE, Willey BM, William G, Faur Y, Eisner W, Warren S, Kreiswirth B (1993) Emergence of vancomycin resistant enterococci in New York City. Lancet 342:76–79

88. Rasmussen RV, Fowler VG Jr, Skov R, Bruun NE (2011) Future challenges and treatment of *Staphylococcus aureus* bacteremia with emphasis on MRSA. Future Microbiol 6:43–56

89. Nelson RRS (1999) Intrinsically vancomycin-resistant gram-positive organisms: clinical relevance and implications for infection control. J Hosp Infect 42:275–282

90. Lessard IAD, Healy VL, Park I-S, Walsh CT (1999) Determinants for differential effects on D-Ala-D-Lactate vs D-Ala-D-Ala formation by the VanA ligase from vancomycin-resistant enterococci. Biochemistry 38:14006–14022

91. Perichon B, Courvalin P (2009) VanA-type vancomycin-resistant *Staphylococcus aureus*. Antimicrob Agents Chemother 53:4580–4587

92. Fraimow HS, Jungkind DL, Lander DW, Delso DR, Dean JL (1994) Urinary tract infection with an *Enterococcus faecalis* isolate that requires vancomycin for growth. Ann Intern Med 121:22–26

93. Kirkpatrick BD, Harrington SM, Smith D, Marcellus D, Miller C, Dick J, Karanfil L, Perl TM (1999) An outbreak of vancomycin-dependent *Enterococcus faecium* in a bone marrow transplant unit. Clin Infect Dis 29:1268–1273

94. Majumdar A, Lipkin GW, Eliott TS, Wheeler DC (1999) Vancomycin-dependent enterococci in a uraemic patient with sclerosing peritonitis. Nephrol Dial Transplant 14:765–767

95. Tambyah PA, Marx JA, Maki DG (2004) Nosocomial infection with Vancomycin-dependent enterococci. Emerg Infect Dis 10:1277–1281

96. Vanderlinde RJ, Yegian D (1948) Streptomycin-dependent bacteria in the identification of streptomycin producing microorganisms. J Bacteriol 56:357–361

97. van Bambeke F, Chauvel M, Reynolds PE, Fraimow HS, Courvalin P (1999) Vancomycin dependent *Enterococcus faecalis* clinical isolates and revertant mutants. Antimicrob Agents Chemother 43:41–47

98. Ling LL, Schneider T, Peoples AJ, Spoering AL, Engels I, Conlon BP, Mueller A, Till F, Schäberle TF, Hughes DE, Epstein S, Jones M, Lazarides L, Steadman VA, Cohen DR, Felix CR, Fetterman KA, Millett WP, Nitti AG, Zullo AM, Chen C, Lewis K (2015) A new antibiotic kills pathogens without detectable resistance. Nature. https://doi.org/10.1038/nature14098

Chapter 4
Antimetabolites: Antibiotics That Inhibit Nucleotide Synthesis

Abstract Antimetabolites as antibiotics that inhibit the synthesis of nucleotides needed for nucleic acid synthesis. Background biochemistry on folic acid metabolism is included. Antibiotics presented are sulfa drugs, p-aminosalicylic acid, dapsone, trimethoprim, and fluorouracil. Methotrexate is discussed as an anticancer drug. Mechanisms of action of all these antibiotics are also discussed.

4.1 Antimetabolites

Unlike the peptidoglycan cell wall, which is present in the infecting bacteria but not in the host, nucleic acids, which include both DNA and RNA, are present in all living cells of all species and synthesis of nucleic acids is essential for growth of all living cells. Enzymes that are essential for synthesis of nucleic acids can be effective targets for designing antibiotics. However, the ideal antibiotics that target nucleic acid synthesis should be selective, which means that it should specifically inhibit the enzymes in the bacteria but not in the host even though both follow very similar mechanisms for the synthesis of their nucleic acids. Similar selectivity is also needed for antibiotics that inhibit protein synthesis. For a discussion on selectivity, see Sect. 1.3. Lack of selectivity of a drug results in its side effects.

One potential target of antibiotics can be reactions for the synthesis of metabolites (Sect. 1.7.4) such as nucleotides (both purines and pyrimidines) that are needed for the synthesis of DNA or RNA (Chap. 5). Of all the nucleotides, the synthesis of deoxythymidine is a preferred target for the development of antibiotics because deoxythymidine is present only in DNA and not in RNA (Sect. 4.3.7). Such antibiotics that inhibit synthesis of metabolites are also known as antimetabolites. So, an antimetabolite is a compound that resembles a certain metabolite and thus interferes with the normal metabolism involving that metabolite. Since the antimetabolite resembles the normal substrate, it functions by binding to the active site of an enzyme that catalyzes a reaction with the substrate. This can interfere with the normal metabolic process in two different ways: (a) It can function as a competitive inhibitor of the enzyme thereby slowing the process. (b) If the resemblance with the metabolite is significant, the antimetabolite can function as a substrate and form a product; however, the product will be unable to function as a substrate for the next step of the metabolic pathway. Although the term antimetabolite can apply for any metabolic step of any biochemical pathway, it usually refers to metabolites of the nucleotide synthesis pathways. Depending on the specificity of the antimetabolite, it can function as antibacterial, antiparasitic, or anticancer drug.

4.2 Background Biochemistry Information: Folic Acid

Folate, or folic acid (Fig. 4.1) is a coenzyme that is essential for cell growth of all organisms including the pathogen (bacteria) and the host (human). For a discussion on coenzymes, see Sect. 1.7.1. Humans are unable to make some coenzymes, known as vitamin-derived coenzymes, and so must obtain them in their diet in the form of vitamins. Vitamins are usually further modified in humans to convert them to coenzymes. The vitamin, folic acid is further modified in two ways: (1) the pterin ring is reduced to 7,8-dihydropterin to form dihydrofolate (DHF) or to 5,6,7,8-tetrahydopterin to form tetrahydrofolate (THF) (Fig. 4.2) and (2) several (five to six) glutamate residues are added to the glutamate of folic acid. Note that the glutamyl residues are linked to each other by peptide bonds to the γ-carboxyl group instead of to the α-carboxyl group that is usually seen in proteins. The anionic polyglutamate chain is important for binding of the coenzyme to the enzyme. Another function of the charged polyglutamate chain is to keep the coenzyme inside the cell since charged molecules are unable to cross the hydrophobic cell membrane. Folate, present in food can enter the human cells through a folate transport machinery present in the cell membrane. Bacteria can synthesize the folic acid that they need (Fig. 4.2). They do not have a transporter to take in folate even if it is provided in their food (growth medium).

Tetrahydrofolate (THF) is an important coenzyme, required by many enzymes that catalyze the transfer of one carbon unit (usually as methyl group). First the THF is converted to various one-carbon derivatives of folate that are then used to donate the one carbon in various reactions (Fig. 4.3). For example, 5-methyl THF is used for the synthesis of the amino acid methionine, 5,10-methylene THF for the synthesis of

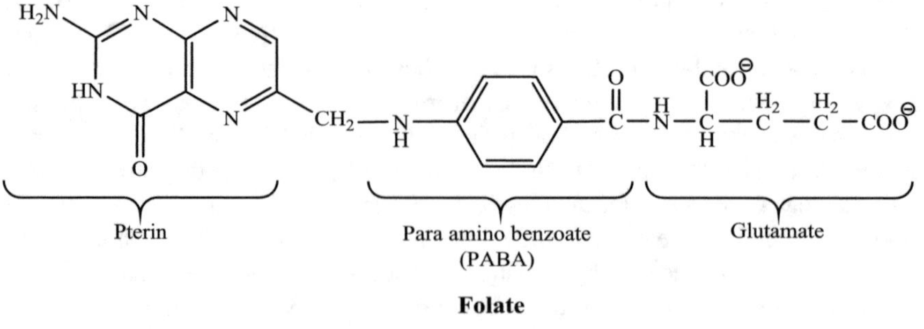

Fig. 4.1 Structure of folic acid

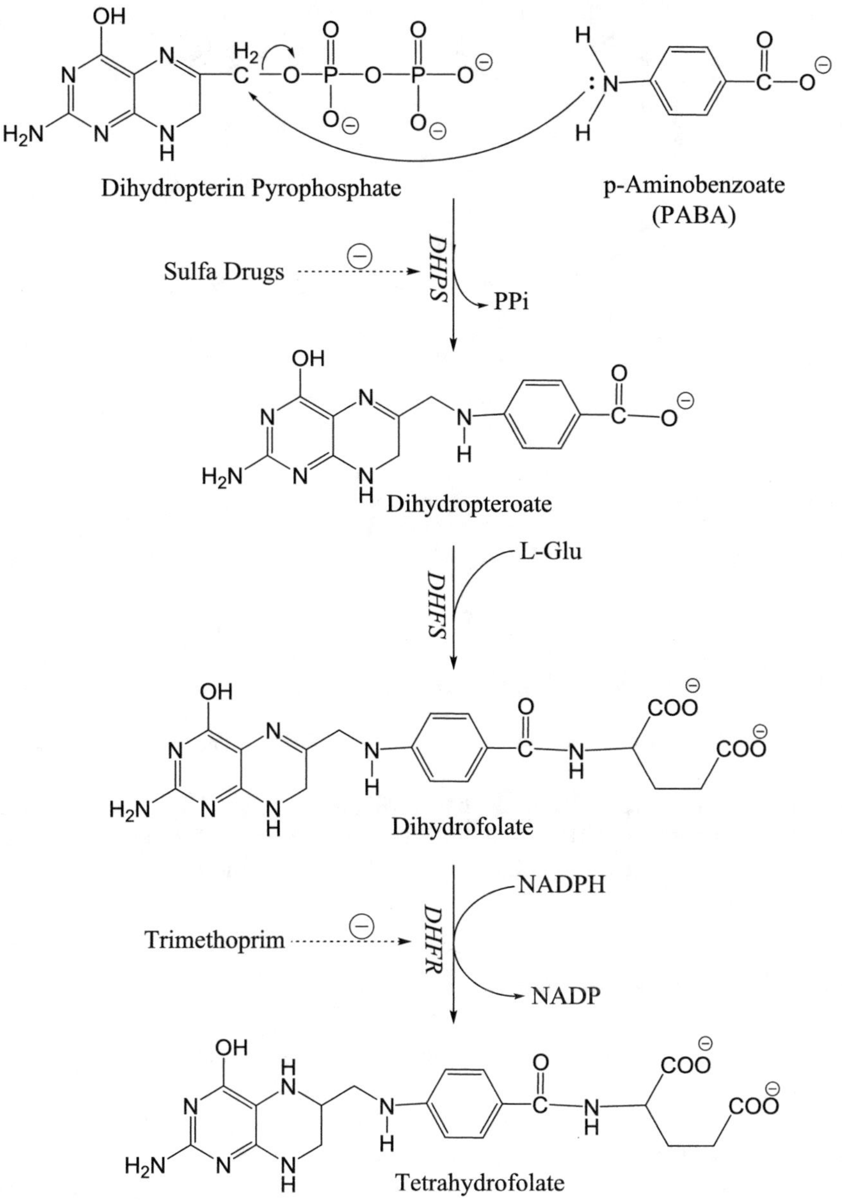

Enzymes: DHPS (Dihydropteroate Synthase)
DHFS (Dihydrofolate Synthase)
DHFR (Dihydrofolate Reductase)

Fig. 4.2 Folate synthesis pathway in bacteria. *Broken arrows* with (−) sign indicate inhibition of enzyme activity by the antibiotic

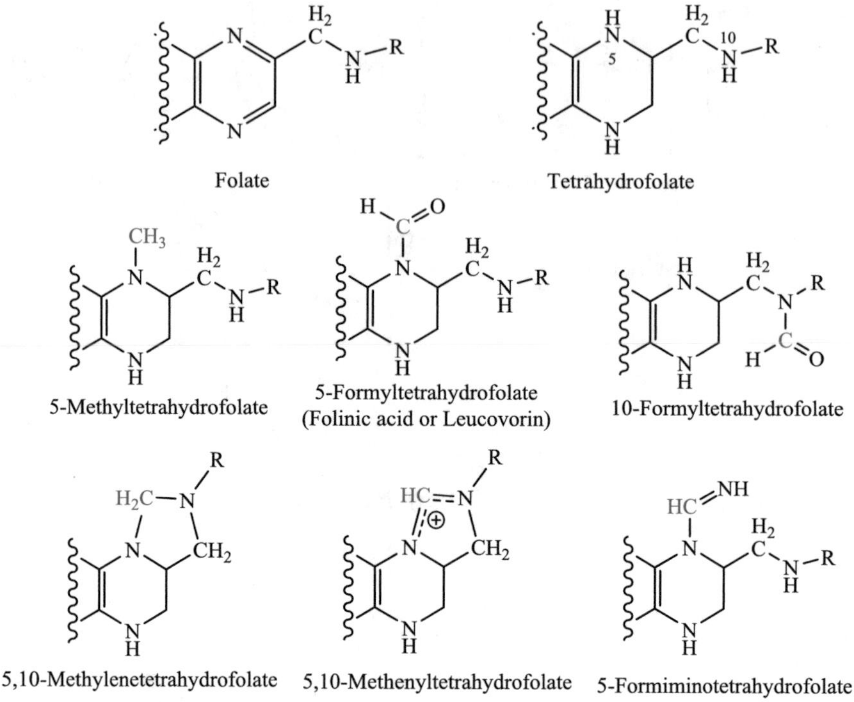

Fig. 4.3 Folates and their one-carbon derivatives. For complete structure of folates, see Fig. 4.1

the amino acid serine and the nucleotide dTMP, 5,10-methenyl THF for the synthesis of purine bases, and 10-formyl THF for the synthesis of *N*-formylmethionine-tRNA which is required for initiating bacterial protein synthesis (Sect. 6.1).

4.3   Antibiotics That Inhibit Folate Metabolism

As discussed before (Sect. 1.3), one requirement of an antibiotic is selectivity, which means that at the concentrations used, it should inhibit growth of the infecting species but not of the host. This is possible if the metabolic reaction that is inhibited takes place only in the infecting species (bacteria) but not in the host (human). One such example is the group of antibiotics called the *sulfa drugs* which inhibit folic acid synthesis which takes place in bacteria but not in humans (Sect. 4.2, Fig. 4.2). Even if a reaction that is inhibited does take place in both the bacteria and the human, the inhibitor can still act as an antibiotic if it specifically inhibits the bacterial enzyme but not the human enzyme due to differences in protein sequence. Examples of such antibiotics include those that inhibit further modifications of folic acid to form THF.

4.3.1 Sulfa Drugs

One group of antibiotics called the sulfa drugs, which includes the sulfonamides (Sect. 2.8), function by inhibiting folic acid biosynthesis in bacteria. Sulfa drugs are broad-spectrum bacteriostatic antibiotics because folic acid synthesis is essential for all species of bacteria. Human cells are not affected since they do not synthesize the folic acid. As discussed before (Sect. 2.8), Gerhard Domagk had demonstrated that the dye Prontosil functioned as antibiotic when injected into infected mice but not when tested against bacteria growing in a test tube. The reason for this unusual behavior was understood later. Prontosil was reductively cleaved in the liver of the mice to sulfanilamide (Fig. 2.2), which had antibiotic activity. The dye property of prontosil is unimportant for its antibacterial activity. Later scientists made several other derivatives of sulfanilamide to obtain the sulfonamide series of antibiotics (Fig. 4.4) all of which have been used to cure a wide spectrum of bacterial infections.

There are many other sulfonamide-containing drugs that have no antibiotic activity but have other medical uses. There are two main structural features that differentiate sulfonamide antibiotics from sulfonamide non-antibiotics: (1) All sulfonamide antibiotics have a free amino group at the para position from the sulfonyl group on the benzene ring (N4). In non-antibiotics the primary amino group may be missing or be replaced by a secondary or tertiary amino group. (2) The sulfonamide antibiotics usually contain a 5- or 6-membered nitrogen containing ring attached to the sulfonylamino group (N1) while the sulfonamide non-antibiotics do not contain such rings [1].

4.3.2 Mechanism of Action of Sulfonamides

As shown in Fig. 4.5, the structure of sulfonamide resembles that of para aminobenzoic acid (PABA), which is a precursor and integral part in the structure

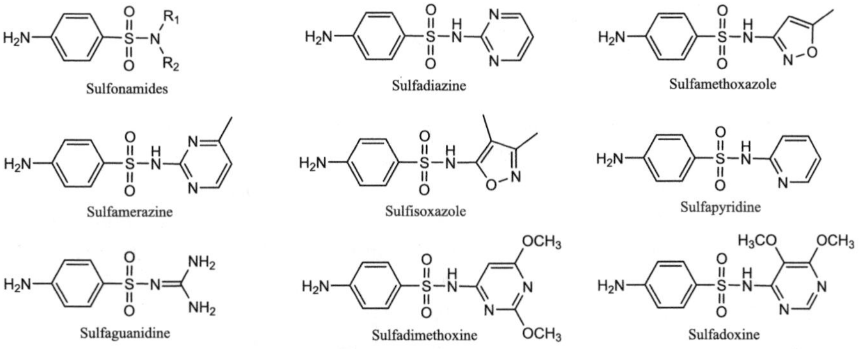

Fig. 4.4 Structures of some sulfonamide antibiotics

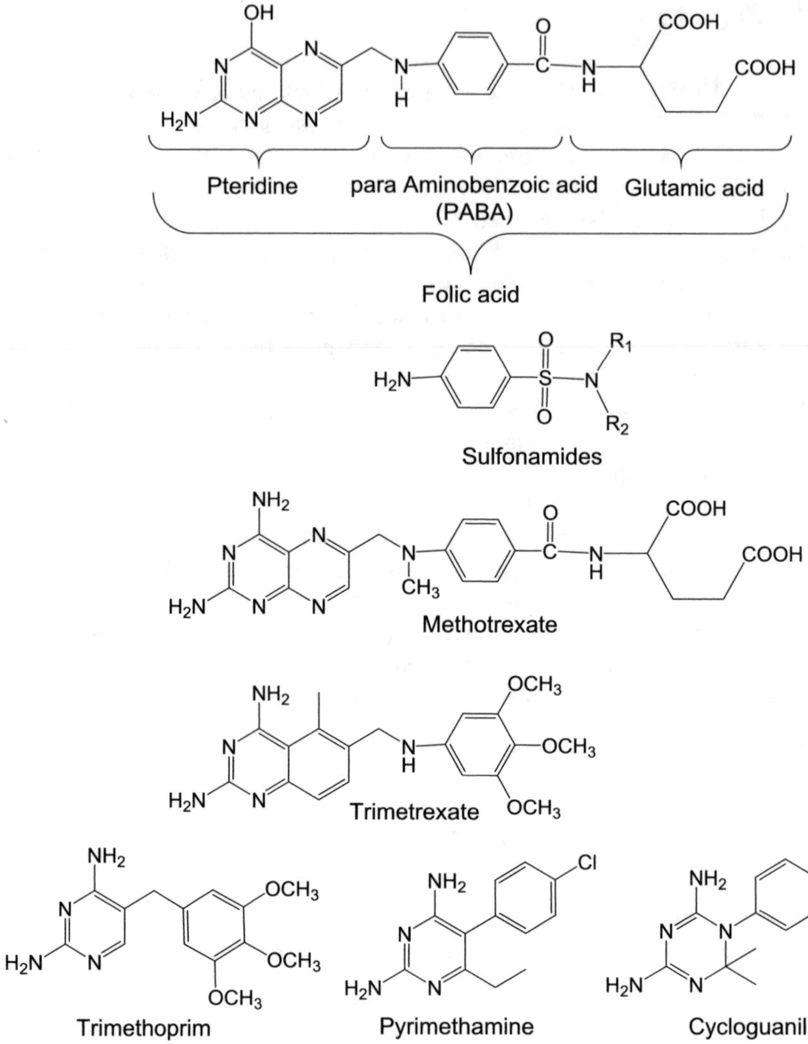

Fig. 4.5 Folic acid and its analogs

of folic acid. Folic acid (DHF) is synthesized in two stages. First, PABA reacts with a pteridine derivative to form dihydropteroic acid. This is followed by reaction with glutamic acid. Sulfonamide antibiotics function by inhibiting the enzyme dihydropteroate synthase (DHPS) which catalyzes the reaction of p-aminobenzoic acid (PABA) with dihydropterin pyrophosphate to form dihydropteroate (Fig. 4.2). Because of the similarity in structure, PABA and sulfonamides compete for binding to the active site of the enzyme. So the antibiotic effect of sulfonamide can be reversed by adding high concentration of exogenous PABA. However, this fact is only of academic interest; clinically, sulfonamides are highly effective antibiotics.

As will be discussed later (Sect. 4.3.5), one function of folic acid is to participate in the synthesis of the nucleotide dTMP. Thus, sulfonamides can be said to eventually inhibit synthesis of dTMP which is an essential component of DNA. Thus, sulfonamides prevent replication of DNA and that is mainly responsible for their bactericidal activity. So the bactericidal effect is also known as "thymine-less death."

Bacterial Resistance to Sulfa Drugs Sulfa drugs have been highly effective till today even after more than eight decades of continuous use. One reason for this is that resistance development to sulfa drugs is rare because they are synthetic antibiotics (Sect. 2.8). However, some examples of resistance to sulfonamides have been reported. One mechanism of resistance is by development of mutations that affect the binding of the drug to the active site of the enzyme dihydropteroic acid synthetase. Another mechanism seen in some bacteria is by increasing the expression of enzymes that are used to synthesize PABA. So the high concentration of PABA is made by the cells which can effectively compete with the sulfa drug for binding to the active site of the enzyme.

4.3.3   Negative Aspects of Sulfonamides

There is usually a time lag before the effect of the sulfa drug can be seen. This delay is because the bacterial cell will already have a certain concentration of folic acid made before the administration of the drug. Also, other metabolites that require folic acid for their synthesis, such as purines, pyrimidines, and amino acids, will also be already present in sufficient quantity in the cell when the drug is administered. So the time delay for antimetabolite antibiotics to become effective corresponds to the time taken for the stock of the metabolites to be depleted, which is approximately equal to about five cell divisions [2]. Another drawback of sulfa drugs is that about 3% (which is a high percentage) of the general population is allergic to sulfonamide-containing drugs and that number can be as high as 60% among AIDS patients [1]. Some patients also experience nonallergic response to the drugs such as nausea, diarrhea, and headaches.

4.3.4   Non-sulfonamide Antimetabolites of Folic Acid

Besides sulfonamides, other analogs of PABA have also been used as antibiotics. Two such antibiotics are diaminodiphenylsulfone (*Dapsone*) for leprosy and para-amino salicylic acid (*PAS*) for curing tuberculosis.

Dapsone (diaminodiphenylsulfone) (not to be confused with daptomycin) is a sulfone with a structure similar to that of sulfa drugs (Fig. 4.6). It is used as an antibacterial agent and its mechanism of action is similar to that of sulfa drugs, which inhibit the synthesis of folic acid (Sect. 4.3.2). It is commonly used in combination

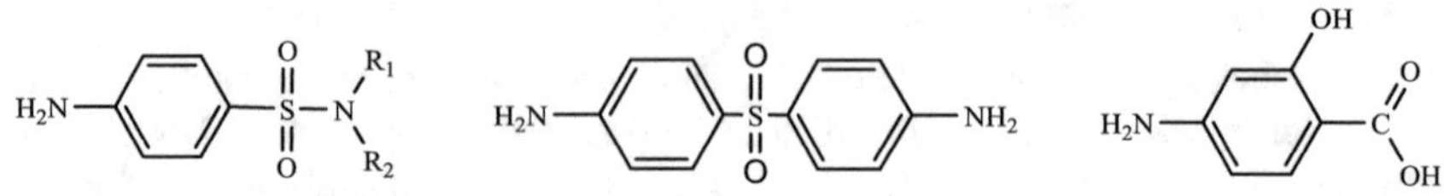

Fig. 4.6 Folic acid antimetabolites

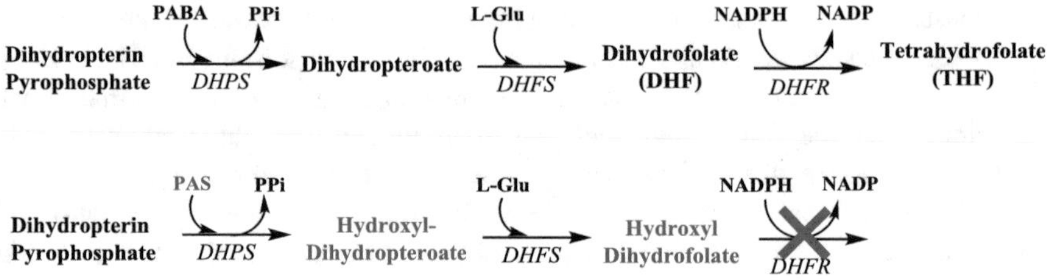

Fig. 4.7 Mechanism of action of para-aminosalicylic acid

with other antibiotics such as rifampicin (Sect. 5.6) for the treatment of leprosy. Dapsone was first used as an antibiotic in 1937 and was used for the treatment of leprosy in 1945 [3]. Besides acting as an antibacterial agent, dapsone also works as an anti-inflammatory agent and is also used to treat several dermatological disorders.

The application of PAS (p-amino salicylic acid) for the treatment of tuberculosis was discovered by the Swedish chemist Jorgen Lehmann based on the reasoning that *Mycobacterium tuberculosis*, the bacteria that cause the disease, could metabolize salicylic acid [4]. Within 2 years p-aminosalicylic acid came into clinical use and was the second antibiotic that was effective against tuberculosis, the first one being streptomycin. PAS is always used in combination with other anti-TB drugs such as isoniazid, rifampicin, ethambutol, pyrazinamide, and streptomycin. Although PAS has been used clinically for more than six decades, its mechanism of action has been confusing. Initially it was believed to have the same mechanism of action as sulfonamides because both sulfonamides and PAS have structures that resemble that of PABA. Since sulfonamides function by inhibiting the enzyme dihydropteroate synthetase (DHPS), which uses PABA as a substrate (Fig. 4.2), it was assumed that PAS also followed the same mechanism. However, it was then found that PAS does not actually inhibit DHPS [5]. Recently the mechanism of action of PAS has been determined [6]. It was shown that PAS does actually bind to the active site of DHPS but instead of inhibiting the enzyme, it acts as a substrate to form the product hydroxyl dihydrofolate, which can also be recognized and used as a substrate for the next enzyme of the pathway, Dihydrofolate synthetase (DHFS). The product, hydroxyl dihydrofolate that is formed however, cannot be used as a substrate for the next enzyme of the pathway, Dihydrofolate reductase (DHFR) (Fig. 4.7), thereby inhibiting the pathway for the synthesis of THF. Since it is hydroxyl dihydrofolate and not PAS that is the actual antimetabolite that inhibits the enzyme DHFR, PAS can be described as a prodrug (Sect. 2.8).

4.3.5   Antimetabolites as Dihydrofolate Reductase (DHFR) Inhibitors

Sulfonamides inhibit DNA synthesis by inhibiting the formation of DHF. Another target of antimetabolites is the enzyme DHFR, which catalyzes a later step in the pathway in which DHF is reduced to THF as shown in Figs. 4.2, 4.7, and 4.8. The two hydrogens that are added to DHF come from NADPH, which is the usual reducing agent for most biosynthetic reactions. The enzyme DHFR is required in all species irrespective of whether they synthesize their folic acid (e.g., bacteria, or protozoa) or obtain it as a vitamin (e.g., humans). The function of THF is to combine with a methylene group from serine to form methylene-THF (Fig. 4.9), which is one of the two main methylating agents commonly used in biochemical reactions (the

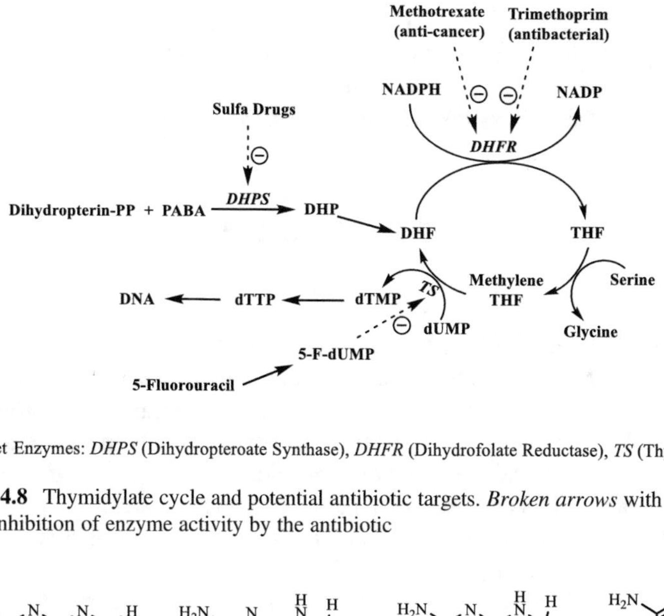

Target Enzymes: *DHPS* (Dihydropteroate Synthase), *DHFR* (Dihydrofolate Reductase), *TS* (Thymidylate Synthase)

Fig. 4.8 Thymidylate cycle and potential antibiotic targets. *Broken arrows* with (−) sign indicate the inhibition of enzyme activity by the antibiotic

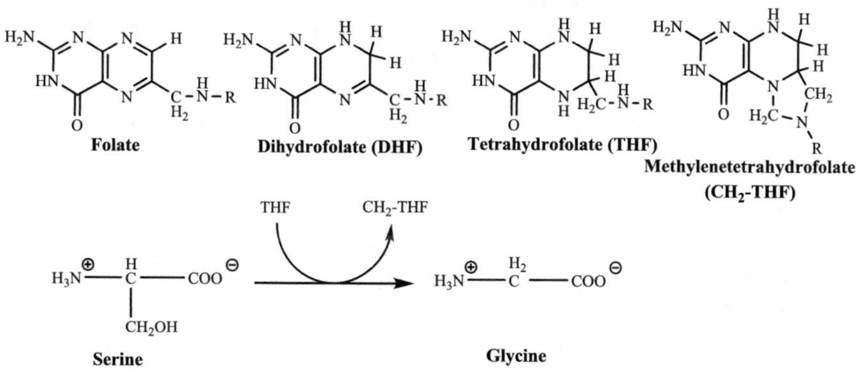

Fig. 4.9 Formation of methylene tetrahydrofolate

other one being S-adenosylmethionine or, SAM). For structure of methylene-THF and other one-carbon derivatives of folic acid, see Fig. 4.3. The methyl group is then transferred to deoxy uridine monophosphate (dUMP) to form deoxy thymidine mono phosphate (dTMP) and in the process the THF is oxidized to DHF. Formation of dTMP is an essential reaction for all species for the synthesis of DNA and so a constant concentration of THF should be maintained. This is done by reducing the DHF back to THF by the enzyme DHFR. So an active DHFR is essential for multiplication of cells and this makes DHFR an obvious target for antibiotics (Fig. 4.8).

Since the reduced form of folic acid (THF) is an essential coenzyme for the synthesis of nucleotides, analogs of folic acid may be expected to inhibit nucleotide synthesis and thus function as antibiotics. Various analogs of folic acid were tested for antibacterial properties. However, they did not function as antibiotics because the bacterial cell membranes are not permeable to folic acid or its derivatives. On the contrary, these folic acid analogs are highly toxic for human cells since their cell membranes are permeable to these molecules. So, to be an antibiotic it should be an analog of intermediates in the synthesis of folic acid and not analogs of the large molecule, folic acid which is not taken in by the bacteria.

4.3.6   Antimetabolites as Antibacterial, Antimalarial, and Anticancer Agents

Nucleotides are needed by both the bacteria and the human. As explained above, many antimetabolites that inhibit synthesis of nucleotides are not effective as antibiotics because they are too toxic for human cells. However, this same property makes them useful as anticancer agents. Since cancer cells multiply more frequently than normal cells in adults, the inhibition of nucleotide synthesis affects the cancer cells more than normal cells and that is the basis of anticancer chemotherapy. One target of antimetabolites is the enzyme dihydrofolate reductase (DHFR), which catalyzes the formation of tetrahydrofolate (THF). The enzyme is essential in both bacteria and humans. However, separate inhibitors have been designed to specifically inhibit the bacterial enzyme and thus act as an antibacterial agent or to inhibit protozoal enzyme and thus act as antimalarial agent or to inhibit the human enzyme and thus act as anticancer agent. One example of such anticancer drug is methotrexate (Fig. 4.5), which resembles folic acid and inhibits the enzyme DHFR and thus results in decreased synthesis of thymine (Fig. 4.8). Although DHFR is an essential enzyme in most species, its structure is different in different species. After testing many synthetic analogs of a portion of folic acid as inhibitors, scientists have identified 2,4 diaminopyrimidines as a series of drugs that selectively inhibit bacterial DHFR. This includes the compound trimethoprim as an inhibitor of bacterial DHFR and pyrimethamine as inhibitor of plasmodia, which cause malaria (Fig. 4.5) and are used for the respective infections.

Table 4.1 Inhibition by antimetabolites of DHFR from various sources

| | Molar concentration needed for 50% inhibition | | |
| | Source of DHFR | | |
	Rat liver	*Escherichia coli*	*Plasmodium berghei*
Trimethoprim	2.6×10^{-4}	5×10^{-9}	7×10^{-8}
Pyrimethamine	7×10^{-7}	2.5×10^{-6}	5×10^{-10}
Methotrexate	2.1×10^{-9}	1×10^{-9}	7×10^{-10}

Adapted from Ferone et al. [7] with permission

Although the host (human) also has the same enzyme, the structure of DHFR enzyme of humans is sufficiently different from that in bacteria or plasmodia such that both trimethoprim and pyrimethamine bind to mammalian DHFR 10,000- and 1000-fold less tightly, respectively, (Table 4.1) resulting in high specificity as antibiotics (antibacterial and antimalarial) [7]. Similarly, methotrexate, another analog of folic acid is transported into human (but not bacterial) cells and inhibits the human DHFR and so is used as an anticancer drug. Since both normal and cancer cells contain the same DHFR enzyme, methotrexate inhibits both cancerous and normal cells. However, the cancer cells multiply at a much faster rate than normal cells and so are affected more. THF is also needed for other pathways besides DNA synthesis (Sect. 4.2). Thus, there is a significant toxicity of methotrexate even for normal cells which are not proliferating rapidly. As an antidote for methotrexate therapy folinic acid is often administered [8]. Folinic acid, which is also known as Leucovorin or 5-formyltetrahydrofolate, bypasses the DHFR enzyme to make tetrahydrofolate and thus decreases the toxicity of methotrexate.

As shown in Table 4.1, methotrexate binds equally well to bacterial, plasmodial, and mammalian DHFR. So one may expect it to be also useful as antibacterial and antimalarial antibiotics. However, methotrexate is not transported into bacterial or plasmodial cells and so cannot be used as antibacterial or antimalarial drug. Note that the structure of methotrexate is very similar to that of folate and enters mammalian cells using the same transport system as that for folate. Since bacteria and plasmodia synthesize their folic acid, they do not have a transporter for folate and thus methotrexate cannot enter those cells. On the other hand, trimethoprim and pyrimethamine are lipid-soluble and enter the bacterial or plasmodial cells rapidly without requiring specific transport mechanisms. These antibiotics may also enter mammalian cells but will not have any effect due to their poor binding to mammalian DHFR as explained above. Even if there is some slight effect due to some binding to the mammalian enzyme, that effect can be reversed by administering folinic acid, which bypasses the DHFR enzyme to make tetrahydrofolate as discussed above. So folinic acid is prescribed along with high doses of DHFR inhibitors such as trimethoprim and pyrimethamine. Folinic acid is not transported into bacterial or plasmodial cells due to lack of transporter and so they are not rescued by folinic acid administration.

Whereas sulfonamide is only bacteriostatic, trimethoprim is bactericidal above a certain concentration. In addition to its antibacterial activity, trimethoprim also has activity against *Pneumocystis carinii*, a fungus that causes opportunistic lung infection in people with weak immune system such as AIDS patients and cancer patients undergoing chemotherapy [9]. Although the trimethoprim and methotrexate separately have useful applications as antibacterial and anticancer drugs, the two should not be taken in combination as that can lead to severe toxicity [10, 11].

4.3.6.1   Combination Antibiotics

Sulfa drugs and trimethoprim function as antibacterial antibiotics by inhibiting the synthesis of folic acid but they inhibit different steps in the synthetic pathway. Thus, the use of both drugs in combination is expected to have a synergistic effect, which means that the combination of the two drugs has a greater effect than either drug alone. So the two antibiotics have been marketed as a combination. One such combination of trimethoprim with sulfamethoxazole is known as *cotrimoxazole* (not to be confused with clotrimazole) and is sold under the brand name Septra. The combination drug also has a broader spectrum of activity than either drug alone. The combination drug has been prescribed for many years. However, the greater effectiveness of the combination drug has been questioned since the synergism is only observed in vitro and may actually be toxic in vivo [12].

Another combination antimetabolite is Fansidar which is a combination of pyrimethamine and sulfadoxine and is used for prophylaxis and treatment of malaria (See also Sect. 8.2.3).

4.3.7   Thymidylate Synthase Inhibitor: 5-Fluorouracil

As shown in Fig. 4.8, the third potential target in the thymidylate cycle is the thymidylate synthase enzyme. Since the enzyme carries out the essential function of DNA synthesis in all species, inhibition of this step will result in not just inhibition of the infecting bacteria but also will cause toxicity for the host. Because of this property thymidylate synthase is an important target for anticancer drugs. Fluorinated pyrimidines were first reported in 1957 to be effective antitumor agents [13]. The substrate for this reaction is dUMP and the substrate analog 5-fluoro-dUMP functions as a substrate analog that binds to the active site. It is considered to be a very effective anticancer drug because it functions as a suicide inhibitor and thus is required in a very small, stoichiometric amount. For a general discussion on how suicide inhibitors (mechanism-based inhibitors) function, see Sect. 1.7.3. As shown in Fig. 4.8, instead of 5-fluoro-dUMP, one can use 5-fluorouracil as the anticancer agent since the cell can convert it to a variety of metabolites, one of which is 5-fluoro-dUMP, which then inhibits the synthesis of thymidine (Note that technically, the correct name should be deoxythymidine but it is commonly referred to as thymidine because the base is present only in DNA and not in RNA. The prefix deoxy- should be used for the other three deoxynucleotides to differentiate between

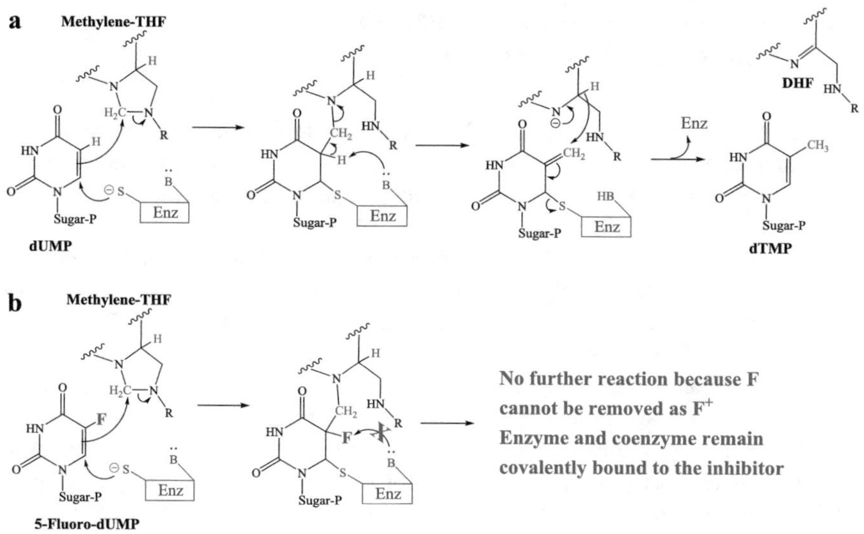

Fig. 4.10 Reaction mechanism of methyl transfer. (**a**) with normal substrate and (**b**) with suicide inhibitor. Only the relevant portion of the coenzyme structure is shown

DNA and RNA). 5-Fluorouracil is frequently used for cancer chemotherapy since the inhibition of thymidylate synthase affects DNA synthesis which is the most crucial for rapidly proliferating cells such as cancer cells. Cell death resulting from lack of thymidine is commonly called "thymineless death." 5-Fluorouracil is often used in combination with methotrexate since the two drugs inhibit two different steps of the same pathway.

The mechanism of action of fluoropyrimidines is shown in Fig. 4.10. In one of the steps of the reaction with the normal substrate, a basic amino acid at the active site removes a H^+ from position C5 of thymidine. Since that H is replaced by an F in 5-fluoropyrimidine, the F, which is highly electronegative, cannot be removed as an F^+ by the base. So the reaction stops at this stage with the inhibitor remaining covalently bound at C6 to a cysteine at the active site of the enzyme. Thus, the enzyme is inactivated in a stoichiometric ratio (1:1 ratio of inhibitor to enzyme). Since enzymes are catalysts and are present in small amounts, only a small amount of suicide inhibitor is needed to inactivate all the enzyme molecules.

The anticancer drug 5-fluorouracil is frequently used along with folinic acid which is also known by other names such as leucovorin or citrovorum factor. The folinic acid increases the effectiveness of 5-fluorouracil since it stabilizes the complex formed between 5-fluoro-dUMP and the enzyme thymidylate synthase. Clinically, response rate to 5-fluorouracil/leucovorin combination is significantly higher than that of fluorouridine alone [14, 15]. As discussed above (Sect. 4.3.6), folinic acid is also administered along with methotrexate.

4.3.8 Other Antimetabolites: Azaserine and Diazo-Oxo-Norleucine

The most commonly used antimetabolites are methotrexate, trimethoprim, and sulfonamides which inhibit the synthesis of dTMP, which is a pyrimidine. Each monomer unit of DNA and RNA contains either a purine or a pyrimidine as the base (Sect. 5.1.1). Both purines and pyrimidines contain nitrogen in their structures. During synthesis of these nucleotides, the source of the nitrogen is the amino acid glutamine. Thus, analogs of glutamine can potentially inhibit the synthesis of nucleotides. Two such analogs of glutamine are Azaserine and Diazo-oxo-norleucine (DON) both of which inhibit the reaction step in which glutamine acts as a nitrogen donor (Fig. 4.11). Besides the α-amino group which all amino acids have, glutamine also has another nitrogen as an amide. This nitrogen is usually transferred in the synthesis of nucleotides. Besides the synthesis of nucleotides, glutamine is used as a source of nitrogen for the biosynthesis of many biological compounds such as some amino acids. The enzymes that catalyze these N transfer reactions can be inhibited by analogs of glutamine. However, since amino group transfer reactions are universal in all species, these drugs will also act as competitive inhibitors for similar reactions in the host and thus, will be toxic. So although these drugs can be used as antibiotics against bacteria in vitro in the laboratory, they cannot be used as antibiotics to treat infections. Another important use of these drugs is for cancer chemotherapy and as antiviral agents.

Fig. 4.11 Structures of glutamine and its analogs

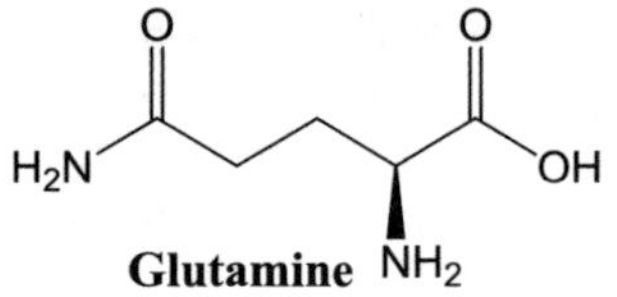

Glutamine

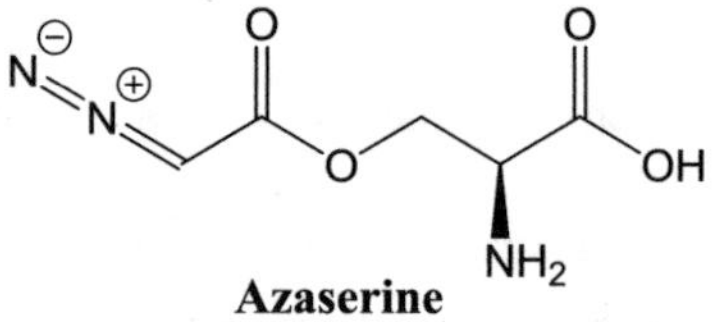

Azaserine
[O-(2-Diazoacetyl)-L-serine]

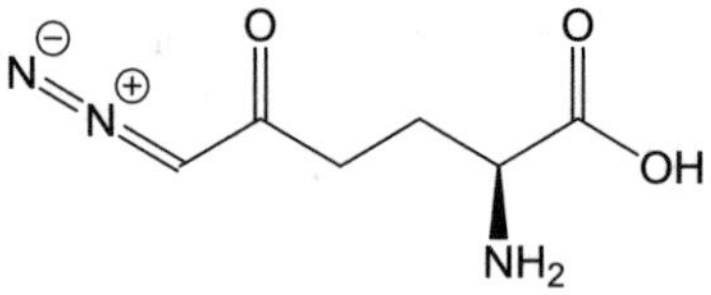

DON (6-diazo-5-oxo-L-norleucine)

References

1. Brackett CC, Singh H, Block JH (2004) Likelihood and mechanisms of cross-allergenicity between sulfonamide antibiotics and other drugs containing a sulfonamide functional group. Pharmacotherapy 24:856–870
2. Seydel JK, Wempe E, Miller GH, Miller L (1973) Quantification of the antibacterial action of trimethoprim alone and in combination with sulfonamides by bacterial growth kinetics. J Infect Dis 128:S463–S469
3. Zhu YI, Stiller MJ (2001) Dapsone and sulfones in dermatology: overview and update. J Am Acad Dermatol 45:420–434
4. Lehmann J (1946) Para-aminosalicylic acid in the treatment of tuberculosis. Lancet 1:15
5. Nopponpunth V, Sirawaraporn W, Greene PJ, Santi DV (1999) Cloning and expression of *Mycobacterium tuberculosis* and *Mycobacterium leprae* dihydropteroate synthase in *Escherichia coli*. J Bacteriol 181:6814–6821
6. Zheng J, Rubin EJ, Bifani P, Mathys V, Lim V, Au M, Jang J, Nam J, Dick T, Walker JR, Pethe K, Camacho LR (2013) *para*-Aminosalicylic acid is a prodrug targeting dihydrofolate reductase in *Mycobacterium tuberculosis*. J Biol Chem 288:23447–23456
7. Ferone R, Burchall JJ, Hitchings GH (1969) *Plasmodium berghei* dihydrofolate reductase isolation, properties, and inhibition by antifolates. Mol Pharmacol 5:49–59
8. Flombaum CD, Meyers PA (1999) High-dose leucovorin as sole therapy for methotrexate toxicity. J Clin Oncol 17:1589–1594
9. Allegra CJ, Kovacs JA, Drake JC, Swan JC, Chabner BA, Masur H (1987) Activity of antifolates against Pneumocystis carnii dihydrofolate reductase and identification of a potent new agent. J Exp Med 165:926–931
10. Al-Quteimat OM, Al-Badaineh MA (2013) Methotrexate and trimethoprim-sulphamethoxazole: extremely serious and life-threatening combination. J Clin Pharm Ther 38:203–205
11. Cudmore J, Seftel M, Sisler J, Zarychanski R (2014) Methotrexate and trimethoprimsulfamethoxazole toxicity from this combination continues to occur. Can Fam Physician 60:53–56
12. Ball P (1986) Toxicity of sulphonamide-diaminopyrimidine combinations: implications for future use. J Antimicrob Chemother 17:694–696
13. Heidelgberger C, Chaudhuri NK, Danneberg P, Mooren D, Griesbach L, Duschinsky R, Schnitzer RJ, Scheiner PE (1957) Fluorinated pyrimidines, a new class of tumour-inhibitory compounds. Nature 179:663–666
14. Rustum YM (1990) Biochemical rationale for the 5-fluorouracil leucovorin combination and update of clinical experience. J Chemother Suppl 1:5–11
15. Moran RG, Keyomarski K (1987) Biochemical rationale for the synergism of 5-fluorouracil and folinic acid. NCI Monogr 5:159–163

Chapter 5
Antibiotics That Inhibit Nucleic Acid Synthesis

Abstract Antibiotics that inhibit the synthesis of nucleic acids including DNA and RNA are presented. Background biochemistry information on the structure of DNA, replication, and transcription is provided. Antibiotics discussed include DNA intercalators, topoisomerase inhibitors such as nalidixic acid and fluoroquinolones, and nitroheterocycles such as nitroimidazoles and nitrofurans, RNA synthesis inhibitors such as actinomycin D and rifamycins are also included. Mechanisms of action of these drugs and resistance development against them are discussed.

5.1   Background Biochemistry Information

5.1.1   Structure of Nucleotides

Nucleic acids include DNA and RNA, which are polymers of nucleotides. Nucleotides have three components: ribose, one or more phosphate groups, and a base, which can be a purine or a pyrimidine (Fig. 5.1). The bases A, G, and C are found in both ribonucleotides and deoxyribonucleotides. U is found only in ribonucleotides, while deoxyribonucleotides have a T instead of U. Deoxyribose lacks the oxygen (O) at C2'.

All of the five bases can participate in hydrogen bonding either as hydrogen donor or as acceptor. Figure 5.2 shows the possible H-bonds at three corresponding positions of the four bases of DNA. The arrows point from H-bond donors (always a H that is covalently bonded to N or O or F) and to H-bond acceptor (any of the three electronegative atoms N, O, or F). The G, C, and T can form 3 H-bonds, while A can form only two. Note that more H-bonds (not shown) can be formed with water when the free nucleotides are dissolved in water. DNA is double-stranded. The two strands interact by hydrogen bonds. In double-stranded DNA, G always base pairs with C and A with T. In order to form base pairs, the donor arrow must be next to an acceptor arrow. Although T is capable of forming three H-bonds, it cannot pair with G or C since the acceptor and donor positions do not match. Similarly, T can form only two H-bond with A by matching the donor and acceptor positions. Since G

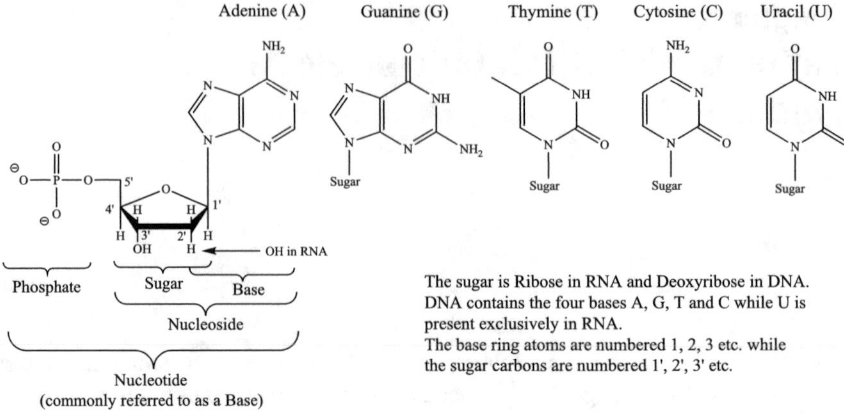

Fig. 5.1 Structures of nucleotides (nucleoside monophosphates)

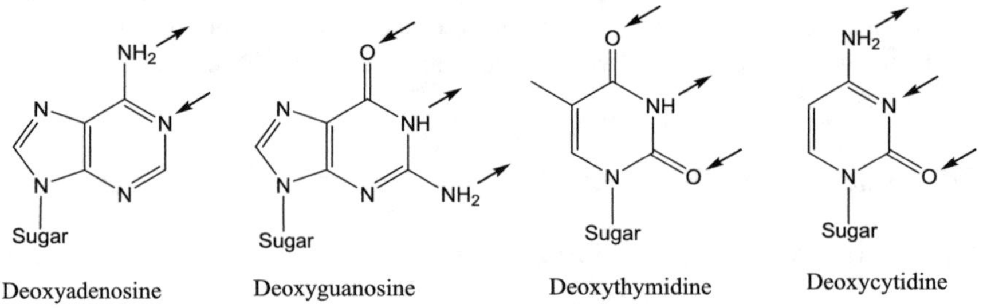

Fig. 5.2 Possible H-bonds at the three corresponding positions of the bases of DNA

pairs with C and A with T, the percentage of G is equal to that of C, and of A is equal to that of T.

5.1.2   Watson–Crick Model of DNA

The 3′-5′ Phosphodiester Linkage In a nucleoside monophosphate, the 5′-CH$_2$-P is perpendicular to the ribose ring, and the base is also perpendicular to the ribose ring. In a single strand of DNA the nucleotides are linked to each other by phosphoester bonds between 3′-OH of one and the 5′-CH$_2$-P of the other (Fig. 5.3). The backbone of a DNA chain consists of the phosphoryl groups and the 5′, 4′, 3′ carbon atoms and the 3′ oxygen atoms. All the nucleotides within a chain have the same orientation. Thus DNA chain has a directionality. The end with the free P on C5′ is called the 5′-end, and the end with a free-OH at C3′ is called the 3′-end. By convention, the sequence of one strand of DNA is always read in the 5′ to 3′ direction. When shown as double stranded, the sequences of the two strands are shown together with one in the 5′ to 3′ direction and the other in the 3′ to 5′ direction.

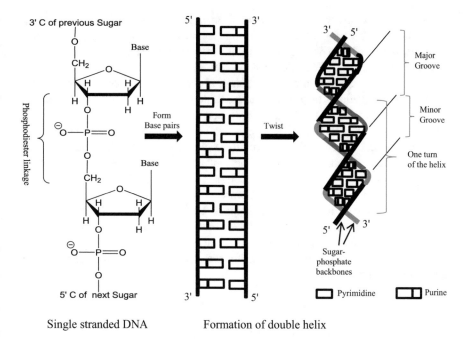

Single stranded DNA   Formation of double helix

Fig. 5.3 Formation of DNA double helix

Double-Helical Structure of DNA   In order to form H-bonds between two strands, they have to be in antiparallel orientations. One strand will be 5′-3′ and the other 3′-5′. The two strands are said to be complementary to each other. The two strands are held together by H-bonding between complementary bases (A = T or G ≡ C). In the double-stranded DNA the distance between the two sugar phosphate backbones is the same for each base pair (1 purine +1 pyrimidine). So all DNA molecules have the same regular structure in spite of their different sequences (Fig. 5.3).

Due to the H-bonds and the equal distance between the two backbones through-out, the structure should look like a ladder. Note that the ladder is not in one plane, but the bases are perpendicular to the length of the ladder. However, we know that the actual structure of ds DNA as described by Watson and Crick is not like a ladder, but the two strands wrap around each other to form a double-stranded helical structure (Fig. 5.3). The double-helical structure is formed because the ladder twists at an angle such that one helical turn contains about 10 (10.4 to be exact) nucleotides (bases) in each strand. Why does it form a helical structure? Base pairing alone does not explain the helical structure. In fact, H-bonds are not the stabilizing factor for the ds DNA structure. The reason it forms a helical structure and not a linear ladder is that in the helical form, the adjacent bases come close to each other and thus exclude all water from the interior of the DNA double helix. This hydrophobic interaction that stacks one base pair over another is called "stacking interaction." The stability obtained because of this hydrophobic interaction causes the DNA ladder to twist. By forming a helical structure, less vertical distance is achieved for the same length of covalent bonds between nucleotides, so the bases can stack against each other.

Thus, the interior of the double helix is very hydrophobic. As explained above, much of the stability of double-stranded DNA is due to the stacking interactions between base pairs. The hydrophobic interior also makes the H-bonds stable because there is no competition from water. The double helix has two grooves of unequal width because of the way the base pairs stack and the backbones twist (Fig. 5.3). These are called the major groove (bigger one) and the minor groove (smaller one). Since the helical structure gives thermodynamic stability, any DNA automatically forms a helical structure without the help of any enzyme or requiring any extra energy. Unwinding and separation of complementary strands are called "denaturation" and will require added energy. Complete denaturation can occur only in vitro. This can be done by heating, which helps to break the H-bonds between the two strands. In vivo, there can be localized denaturation in a short stretch of the DNA, a process that uses energy from ATP and is catalyzed by the enzyme, DNA helicase. Such localized denaturation is essential for the process of replication, transcription, and conjugative transfer.

5.1.3   Superhelical Structure of DNA

In order to understand the function of antibiotics which inhibit DNA synthesis, it is necessary to understand the structure of DNA. In normal unstrained double-helical DNA, which is known as relaxed DNA, there are about 10 bases per turn of the helix. The total number of turns in a certain piece of DNA is called its linking number. For example, E. coli chromosome has 4×10^6 bp in a closed circle and thus has a linking number of $4 \times 10^6 \div 10 = 4 \times 10^5$. If a linear helical DNA is again twisted in the same direction or against the direction of the helix, and then the two ends are joined to form a closed circle, the DNA will now have an increased or decreased linkage number, respectively. Both situations will increase the strain in the DNA molecule, and the strain can be relieved in the cell in three different ways: (1) The DNA can wrap around proteins known as histones in eukaryotes and histone-like proteins in prokaryotic cells. (2) The DNA double helix can coil around itself to form a supercoiled conformation. Supercoils can be of two types: (a) positive supercoil, which means coiled in the same direction as helical turn, which means there are less than 10 bases per turn, leading to a linking number that is greater than that of relaxed DNA and (b) negative supercoil, which means coiled in the opposite direction as the helical turn and thus has more than 10 bases per turn and a linking number lower than that of relaxed DNA. Supercoiling makes the DNA more compact. If E. coli chromosomal DNA existed as relaxed DNA, its total length would be a thousand times longer than the length of the cell. However, because of supercoiling, the DNA occupies only a small volume inside the cell. The concept of positive and negative supercoiling can be demonstrated using a telephone cord that has been twisted in the same or opposite direction as its normal helical structure (Fig. 5.4). Most of the DNA

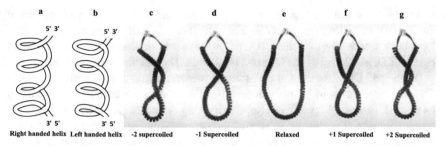

Fig. 5.4 Demonstration of double-helical and superhelical structure of DNA. (**a**) Right-handed and (**b**) left-handed helices are shown. A coiled telephone cord has been used to demonstrate supercoiling. Although normal DNA forms right-handed helix, a left-handed helical telephone cord has been used here due to availability. Two ends of a telephone cord are joined to get the relaxed form (**e**). One end of the cord was twisted once or twice by 360° each in the direction of the coil to obtain +1 (**f**) and + 2 (**g**) supercoiled forms and in the direction opposite to the coil to obtain −1 (**d**) and − 2 (**c**) supercoiled forms. Note that the directions of the positive and negative supercoils are opposite

in the cell is negatively supercoiled. In order to denature a DNA (completely or locally), the linking number of the DNA has to be first made normal (corresponding to 10.4 bases per turn), and then the two strands can be separated by breaking the hydrogen bonds between them. (3) The third method of relieving the strain of DNA applies to only negative supercoil (decreased linking number) and takes place by separation of the base pairs in one small localized region of the DNA (a process known as "localized melting") in order to decrease the number of bases per turn (same as increasing the linking number) in the rest of the DNA to a value closer to the normal 10 bp per turn. Thus, a large stretch of the DNA will be in relaxed form, while a small region will have the two single strands separated from each other. Such localized melting usually takes place in A/T rich regions, which can be melted more easily since A and T form two hydrogen bonds with each other while G and C form three. These A/T rich regions are usually present at the sites where replication, transcription, or conjugative transfer is initiated. Localized melting at these sites is essential for binding of the proteins that are involved in these processes that use single-stranded (not double-stranded) DNA to make a copy of DNA or RNA.

5.1.4   DNA Replication

As discussed above, DNA replication starts when two copies of the replication machinery (known as replisome) bind to an A/T rich site called the origin of replication (oriC for chromosome replication and oriV for plasmid replication). Localized melting at the origin of replication is possible because of the overall negative supercoiling of the DNA (Sect. 5.1.3). After replication is initiated at the origin, the two replisomes move in opposite directions while making copies of both strands simultaneously. One strand is replicated continuously in the $5' \to 3'$ direction and is called the leading strand. Since DNA polymerase can only synthesize DNA in the $5' \rightarrow 3'$ direction, the second strand, which is called the lagging strand, has to loop back frequently into the DNA polymerase and thus is synthesized in small pieces of DNA known as the Okazaki fragments. The junction of replicated and not yet replicated DNA, where the replisome is present, is known as a replication fork. Thus two replication forks move in opposite directions while each replisome copies its share of half of the total DNA. Replication stops when a replication fork reaches a termination site. The process of replication involves binding to the right nucleotide that is complementary to the template strand followed by joining of the nucleotide to the previous one, a reaction that is catalyzed by the enzyme DNA polymerase. The *E. coli* chromosome replicates in 40 minutes. Other bacteria also replicate at a similarly fast rate. Since only single-stranded DNA can be replicated, the two strands must be first separated as the replication fork moves as the two strands are copied. In order to replicate the whole chromosome in 40 minutes, the DNA must be unwound at the rate of $4 \times 10^5 \div 40 = 10,000$ turns per minute. As the DNA is unwound at the replication fork, a positive supercoiled region forms ahead of the fork to compensate for the unwinding. The supercoiling makes the DNA more compact and difficult to separate into two strands for replication. In spite of this, a fast rate of unwinding is achieved because cells have an enzyme called topoisomerase, whose function is to decrease the degree of supercoiling in the DNA.

Topoisomerases are essential for the survival of all cell types. These enzymes function by cutting the DNA, then twisting the cut end to lower the degree of supercoiling and then resealing the two ends. Doing this decreases the linking number. Topoisomerases can be of two types. Type I enzymes cut one strand, while type II enzymes cut both strands. After cutting either one or both strands, the DNA is twisted (strand passage) to change the linking number and then the cut ends are religated. All topoisomerases can relax supercoiled DNA, but DNA gyrase, which is a Type II enzyme, can also introduce negative supercoils using energy from ATP. Its most important function is in removing positive supercoils formed ahead of the replication fork as explained above. DNA gyrase is essential in all bacteria for replication and transcription but is not present in higher eukaryotes, including humans. This makes the enzyme an ideal target for the development of antibiotics.

Eukaryotic type II topoisomerase is a homodimer of a large subunit (~170 kDa), while prokaryotic DNA gyrase contains two subunits A and B, which form an A_2B_2 complex. *E. coli* gyrase is the most studied topoisomerase and contains the two subunits GyrA (97 kDa) and GyrB (90 kDa). The A subunit is responsible for binding to DNA and for cutting and religating DNA while the B subunit has ATPase activity. Since DNA gyrase is unique for bacteria, antibiotics can be developed that interfere with any of these steps, including DNA binding, DNA cleavage, strand passage, ATP hydrolysis, and DNA religation. Structures of some antibiotics that inhibit DNA gyrase are shown later in Fig. 5.6.

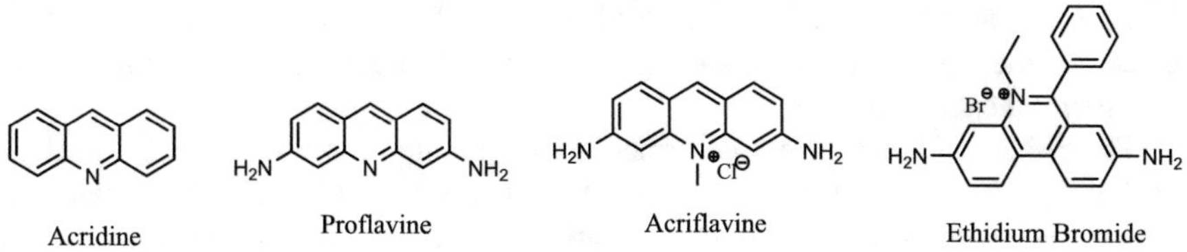

Fig. 5.5 Structures of some intercalators

5.2 Intercalators as Antibiotics

The double-stranded helical structure of DNA is a result of stacking interaction that places the planar aromatic rings of the base pairs stacked one on top of another. Aromatic molecules, which can insert themselves between the two layers of base pairs in DNA are known as "intercalators." These intercalators are planar aromatic molecules containing three rings. Some examples of intercalators are acridine and its derivatives Proflavine and Acriflavine (Fig. 5.5). The structure and function of these and several other acridine derivatives have been reviewed in detail [1]. The tricyclic aromatic structure of the intercalators is ideally suited to intercalate between two base pairs since each base pair also contains three aromatic rings, two for a purine and one for a pyrimidine.

Intercalators function as antibiotics for two main reasons. (1) Since the tricyclic intercalating agent binds to the three aromatic rings of the bases of both strands, it holds the two strands tightly, making it difficult for them to separate. This has a negative effect on replication as well as transcription since both these processes are initiated by first separating the two strands locally at the respective sites of initiation. Such separation of two strands in a short stretch of DNA is known as localized melting. Even if replication is initiated, the process cannot take place efficiently because of the second effect of the intercalator. (2) Since the intercalator is present between two layers of base pairs, the distance between the layers increases, and so the helix becomes partially unwound, thus disrupting the normal double-helical structure of the DNA. (Note that the helical structure is a result of stacking interaction in which the base pairs are brought closer to each other as discussed above in Sect. 5.1.2). The distortion of the helical structure can result in either deletion or addition of one or more bases during the copying of the DNA. This will lead to frameshift mutations resulting in either wrong or truncated protein sequence (Sect. 6.1). Since this will have a detrimental effect on the bacteria, these intercalators can function as antibiotics. A mechanism of frameshift mutation caused by intercalators has been suggested by Streisinger et al. [2]. According to the authors, frameshift mutations occur because of localized pairing out of register, more commonly known as slippage, during replication. Such mutations usually occur in regions containing repetitive base sequences such as CCCCCC. Because of the repetitive sequence and distortion of the helix by the intercalator, slippage occurs during replication of the region, thus creating bulges in either the template strand or the primer strand and leading to deletion or insertion, respectively. The intercalating agent further enhances the possibility of strand slippage by binding to the DNA bulges and stabilizing the structures [3].

The effects of intercalators described above are applicable for both the host as well as the infecting bacteria. In fact mutations happen more in the mammalian host than in the infecting bacteria because mutations take place when chromosomes cross over, an event that rarely happens in bacteria except during conjugation. So intercalators are too toxic for the host if used systemically and thus their use is limited to topical application only. Acridines are the active ingredients in most yellow ointments used for burns and small cuts. It is to be noted that all intercalators do not cause mutations and that DNA intercalation is necessary but not sufficient for genotoxicity [4].

Before the discovery of penicillin, acridines were widely used as antibiotics for the treatment of protozoal diseases such as malaria and trypanosomiasis. For systemic use, the acridine must be made soluble in water. This is made possible by alkylating the ring nitrogen to form a quaternary ammonium salt, with a positive charge on the nitrogen in Acriflavin (Fig. 5.5). This concept of adding quaternary ammonium group to make molecules water soluble is very common in many modern-day drugs.

Another intercalator is Ethidium Bromide, which is not normally used as an antibiotic since it is highly toxic for the host, although it has been used to treat trypanosomiasis in cattle [5]. Intercalating agents more selectively inhibit replication of small closed circular DNA such as plasmids in bacteria. For the same reason, they also inhibit replication of mitochondrial DNA in eukaryotes, which explains their toxicity. This is the basis of the activity of ethidium bromide against trypanosomes as well as its use as an anticancer chemotherapeutic agent. In the research laboratory, ethidium bromide is used to stain DNA for detection after electrophoresis. This is because in UV light, the dye fluoresces with an orange color, the intensity of which is about 20-fold higher if it is bound to DNA.

Another DNA binding antibiotic, Actinomycin D, can inhibit DNA replication, but its main use is as an antibiotic inhibiting transcription and so is discussed later (Sect. 5.6).

5.3   Inhibitors of DNA Gyrase: Quinolones

Quinolones are the highly effective broad-spectrum antibiotics that target DNA gyrase as their site of action. Representative members of the quinolone family of antibiotics include Nalidixic acid and Ciprofloxacin (Fig. 5.6). Nalidixic acid, the first quinolone antibiotic, was discovered by accident as a byproduct during the synthesis of the antimalarial drug chloroquine [6]. Later other quinolones with much better antibacterial activity were developed. The most popular and highly effective of these is ciprofloxacin (see below). Since there are many quinolones available, a new four-generation classification system has been described for quinolones

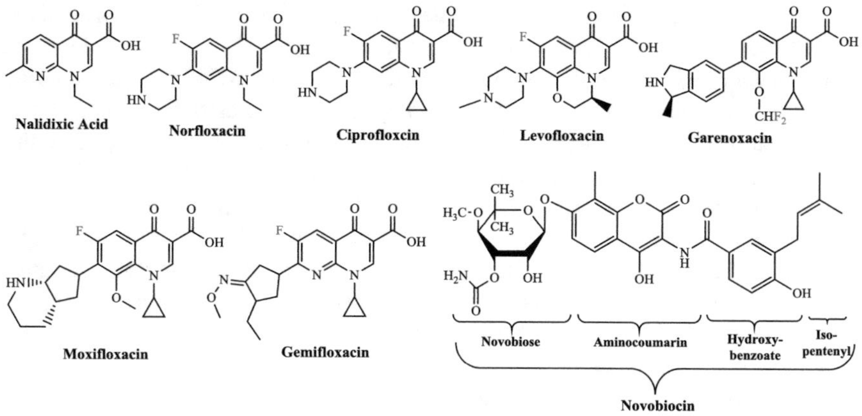

Fig. 5.6 Inhibitors of DNA gyrase

[7, 8]. The first-generation drugs, which includes the first quinolone antibiotic nalidixic acid, achieve only minimal serum concentration and so is not of much use. Later generation quinolones can reach high serum levels, have good tissue penetration and have broader spectrum of activity. Second-generation quinolone antibiotics, including norfloxacin and ciprofloxacin, are effective against gram-negative bacteria. Third-generation quinolones include levofloxacin and are effective against both gram-negative and gram-positive bacteria, while fourth-generation quinolones include trovafloxacin, moxifloxacin, and gemifloxacin and have the broadest spectrum of activity. In addition to all the activities of the third-generation quinolones, they also have activity against anaerobes.

Nalidixic acid is easily absorbed through the intestinal walls, but its tissue penetration is poor because much of it remains bound to plasma protein. Also, the drug is rapidly cleared from the body. Thus more frequent doses of nalidixic acid are needed to maintain the required concentration for inhibiting bacteria. Also, because of the rapid clearance from the body, the serum concentration of nalidixic acid is not high enough in any tissue to kill the infecting bacteria. However, since nalidixic acid is cleared rapidly from the body, its concentration in the urine reaches high levels. So nalidixic acid is mainly used to treat urinary tract infections. However, such use has also decreased due to the availability of less toxic quinolone derivatives called fluoroquinolones.

Fluoroquinolones Quinolones containing a fluorine substituent were developed as better gyrase-targeting antibiotics than the non-fluorinated quinolone, nalidixic acid. Norfloxacin, the first fluoroquinolone that was approved by the FDA in 1986, is a widely used antibiotic. Similarly, ciprofloxacin, another fluoroquinolone, was introduced in 1987 and within a few years became the most frequently used antibiotic in the world. Later other fluoroquinolones were introduced in the USA: levofloxacin and sparfloxacin in 1996, trovafloxacin in 1997, gatifloxacin and moxifloxacin in 2000.

There are several reasons why fluoroquinolones, especially ciprofloxacin, are considered to be better antibiotics than nalidixic acid: (1) Ciprofloxacin has a much lower MIC (minimum inhibitory concentration) and MBC (minimum bactericidal concentration) than nalidixic acid. So ciprofloxacin is effective at much lower concentration than nalidixic acid. An analysis of 375 gram-negative bacterial strains causing urinary tract infection for susceptibility to six quinolone derivatives showed that ciprofloxacin had a much higher antibacterial activity against all bacterial strains tested than the other quinolones, except for norfloxacin, which also showed comparable effectiveness. The MIC90 and MBC90 (the number 90 refers to 90% inhibition or bactericidal activity) of ciprofloxacin were less than 1 μg/ml for all bacterial species tested, and MBC/MIC ratios were very low, making it a very effective antibiotic [9]. Ciprofloxacin was also reported to have a very low ($\leq$ 1 μg/ml) MIC and MBC for methicillin-resistant *Staphylococcus aureus* (MRSA, Sect. 3.3.2.10) [10]. It was also found to be the most active quinolone against genital isolates of *Chlamydia trachomatis* (MIC and MBC 1 μg/ml) [11]. (2) Ciprofloxacin has a much higher selectivity ratio than nalidixic acid and so has a lower toxicity for the host. Selectivity ratio is the ratio of IC_{50} for mammalian topoisomerase II to the IC_{50} for bacterial DNA gyrase. (3) Fluoroquinolones are cleared from the body at a much slower rate than nalidixic acid. So less frequent doses of fluoroquinolones are needed to maintain the required serum concentration. (4) Resistance development to ciprofloxacin is much less than to nalidixic acid. Because of the low serum concentration attainable for nalidixic acid, bacteria develop resistance to the drug very easily (Sect. 2.4). These factors, along with their low cost, have made fluoroquinolones a widely used drug.

Other Fluoroquinolone Derivatives Ofloxacin is a synthetic fluoroquinolone antibiotic that was developed as a broader spectrum analog of norfloxacin and was approved by the FDA in 1990. Ofloxacin has one chiral center and is a racemic mixture of the two enantiomers, levofloxacin, the active component (Fig. 5.6), and dextrofloxacin. Ofloxacin and levofloxacin are broad-spectrum antibiotics that are active against both gram-negative and gram-positive bacteria. They function by inhibiting DNA gyrase, an enzyme essential for replication and Type IV topoisomerase, an enzyme needed for separating replicated DNA into the daughter cells. Garenoxacin (previously known as BMS284756, a Bristol Meyers Squibb drug) is a Des-F(6)-quinolone (Fig. 5.6) and is more effective as an antibiotic in vitro than ciprofloxacin [12]. The compound lacks the classical C-6 fluorine of fluoroquinolones but has fluorine as a difluoromethyl ether linkage at C8.

Mechanism of Action of Quinolones Antibacterial activity of drugs targeting DNA gyrase enzyme follows one of two mechanisms. They either inhibit the catalytic activity of the enzyme by binding to the active site, or they stabilize the covalent enzyme–DNA complex that is formed during the reaction. The latter mechanism makes a more effective antibiotic. The antibiotic Novobiocin is of the former type and inhibits the ATPase activity of the gyrase enzyme. Fluoroquinolones such as ciprofloxacin work by stabilizing the enzyme–DNA complex and thus interrupting the religation step [13].

The targets of the fluoroquinolones are the two enzymes, DNA gyrase and DNA topoisomerase IV, both belonging to the type II topoisomerases [14]. The target of the quinolones is DNA gyrase in gram-negative bacteria, while in gram-positive bacteria, the target is topoisomerase IV [15, 16]. Function of DNA gyrase is to introduce negative supercoils ahead of the replication fork (Sect. 5.1.5). It accomplishes this by wrapping the DNA into a positive supercoil followed by cutting the DNA, then passing one region of the duplex through another and then religating the cut ends [17]. Topoisomerase IV, a tetrameric protein consists of two subunits of ParC and two subunits of ParE proteins (referred to as GrlA and GrlB in *Staphylococcus aureus*). The names come from "partition" since these proteins help to separate the replicated chromosome into the two daughter cells during cell division. After replication, the two new copies of the DNA are interlinked. The two copies of DNA should be delinked so that one copy goes to each daughter cell after cell division, which is the primary function of topoisomerase IV. Another function is to introduce negative supercoils similar to the function of DNA gyrase enzyme.

As discussed before (Sect. 5.1.4), prokaryotic DNA gyrase contains two subunits GyrA and GyrB. The GyrA subunit is responsible for binding to DNA and for cutting and religating DNA while the B subunit has ATPase activity. In the absence of quinolone antibiotic, usually after cutting the two strands, the $5'$ terminus is covalently linked to Tyr 122 of the GyrA subunit. After twisting the two strands by one turn, the cut ends are again religated. When quinolone is present, it binds to the cut ends of the DNA in the enzyme–DNA complex and thus prevents rotation of the DNA ends and religation of the cut ends. Quinolones cannot bind to DNA alone or to the DNA gyrase enzyme alone; they only bind to the complex of the enzyme and the cut DNA.

In vivo, quinolones can have both bacteriostatic and bactericidal actions. At low concentrations, they are bacteriostatic, while at high concentrations, they are bactericidal. At low concentration of the antibiotic, the cut ends of the DNA remain bound to the antibiotic, and so the DNA remains supercoiled, and this inhibits replication as well as transcription since the replication fork, or the transcription complex is unable to proceed due to the supercoiling of DNA and the presence of the antibiotic-DNA-enzyme complex. The binding of the quinolone to the complex of gyrase enzyme and the cut ends of the DNA is reversible and thus, explains the bacteriostatic activity of the quinolones. If the cells are treated with high concentration of quinolone, the DNA ends are released from the complex, which results in fragmentation of the chromosomal DNA and eventual cell death [12, 18]. However, it has been argued that cell death cannot be due to DNA fragmentation because it is possible to repair fragmented chromosome. It has been proposed that the fragmentation is followed by the formation of reactive oxygen species such as hydroxyl radicals which are responsible for cell death [19–21]. It is possible for the cell to repair the chromosome fragmentation but not the effect of ROS, which explains the ability of inhibitors of ROS to almost completely block cell death.

Antibiotics That Target GyrB Subunit of DNA Gyrase GyrB subunit of prokaryotic DNA gyrase is an ATPase and can also be target for antibiotic development.

Aminocoumarins, which include novobiocin, clorobiocin, and coumermycin A1, are natural products isolated from Streptomyces species and can inhibit the supercoiling activity of DNA gyrase. It was shown that these antibiotics compete with ATP binding to gyraseB subunit and thus inhibit its ATPase activity. Novobiocin, also known as albamycin or cathomycin, was first reported in the mid-1950s when it was called streptonivicin. Novobiocin can be effectively used to treat MRSA infection [22].

5.3.1 Mechanism of Resistance to Quinolones

Intrinsic Resistance One explanation for the intrinsic resistance of pseudomonas to many antibiotics such as fluoroquinolones is that the antibiotics are unable to cross the outer membrane of the bacteria. However, this idea has been challenged [23]. The intrinsic resistance of *Pseudomonas aeruginosa* is due to efflux pumps in the bacteria that pump out most antibiotics [24].

Acquired Resistance Resistance to quinolones can develop in three ways, of which the third one is very rare: (1) by developing point mutations in the targets, DNA gyrase or topoisomerase IV enzymes, (2) by decreasing the intracellular concentration of the quinolone and (3) by the acquisition of mobile elements carrying the *qnr* gene which confers resistance to quinolones. Multiple mutations can occur over time, resulting in cumulative increase of resistance (Sect. 3.3.2.12). Since quinolones are synthetic antibiotics, there is no gene available in nature coding for any enzyme that can degrade the antibiotics (Sect. 2.8). Also, since the gene for DNA gyrase is present in the chromosome and not in a plasmid, the resistance cannot be easily transferred to other bacteria (Sect. 2.6). By the same reasoning, bacteria cannot acquire resistance from other bacteria. However, it is theoretically possible to transfer chromosome encoded resistance determinants by natural transformation or transduction.

Resistance Development by Target Modification Numerous bacterial mutants that are resistant to quinolones have been studied, and the mutations have been identified. The region of gyrA and gyrB where most of these mutations are frequently found is called the QRDR (quinolone resistance-determining region). The protein sequence in this region is conserved in most bacterial species suggesting that the region is important for proper functioning of DNA gyrase enzyme. These mutations have been most studied in *E. coli*. All quinolone-resistant *E. coli* clinical isolates were found to have point mutations in *gyrA* but not in *gyrB*. However, mutants obtained in vitro showed equal probability of mutations in *gyrA* and *gyrB* [25]. Results in *S. aureus* also indicated that mutations in both *gyrA* and *gyrB* can be responsible for resistance to quinolones [26]. Fluoroquinolones are broad-spectrum antibiotics that are known to interact with GyrA and GyrB subunits of DNA gyrase and ParC and ParE subunits of DNA topoisomerase IV. All ciprofloxacin-resistant isolates of *Salmonella enterica* in the USA from 1999 to 2003 were analyzed and found to have mutations in the QRDR of *gyrA* but not in *gyrB*, *parC*, or *parE* [27]. Since *Mycobacterium tuberculosis* does not have topoisomerase IV, fluoroquinolones target only DNA gyrase in *M. tuberculosis,* and all known mutations in the gyrase gene have been compiled and reviewed [28]. Out of 1220 resistant isolates of *M. tuberculosis* that were sequenced, 64% had mutations in QRDR of *gyrA* while 3% had mutations in QRDR of *gyrB*. The most common mutations (54%) are in codons 90, 91, or 94 of *gyrA*. High levels of resistance to nalidixic acid can be achieved by single mutations in QRDR, but more than one mutation is usually needed for significant resistance to ciprofloxacin, making it a more effective antibiotic [25]. However, there are also reports of mutations that confer high levels of resistance to ciprofloxacin but not to nalidixic acid in *Stenotrophomonas maltophilia* [29] and *Salmonella enterica typhimurium* [30].

Resistance   Development   by   Decreasing   Intracellular   Concentration of the Antibiotic   This can take place in two ways: (a) by decreasing the permeability of the drugs through the cell membrane and (b) by overexpression of active efflux pumps that nonspecifically pump out drugs from the cytoplasm to outside the cell. Quinolones enter the cell through porins present in the outer membrane of gram-negative bacteria. Transport of quinolones into the cell can be decreased by decreasing the expression of porins. *E. coli* outer membrane contains three types of porins, OmpA, OmpC, and OmpF (omp = outer membrane protein). Mutations in OmpF or decreased expression of OmpF increase the resistance to some quinolones. Since many other drugs are also transported through OmpF porin, these mutations will also confer resistance to all those antibiotics such as β-lactams, tetracyclines, and chloramphenicol. Mutations in the mar (*m*ultiple *a*ntibiotic *r*esistance) operon cause resistance to multiple antibiotics. Mutations in marA, one of the genes in the mar operon decreases the amount of OmpF expressed and thus lowers the uptake of quinolones by the cell.

   In gram-positive bacteria, which do not have any outer membrane, the cellular concentration of quinolone is lowered by active efflux pumps. Note that the term "active" here means "energy dependent." Energy from ATP or other high energy compounds is used to transport drugs against a concentration gradient (the concentration of the antibiotic outside the cell is higher than inside). Similar active efflux systems are also responsible for resistance to tetracyclines.

Resistance Development by Gene Acquisition   As mentioned above, a resistance gene against quinolones is not expected since these are synthetic antibiotics. In 1998 there was a report of a plasmid-borne gene, *qnr* (*q*ui*n*olone *r*esistance), that confers resistance to quinolones in addition to many other antibiotics [31]. The *qnr* gene product, which protects the DNA gyrase from inhibition by quinolones, is part of an integron present in the plasmid [32]. This gene is very rare and was found in only a few clinical isolates out of the numerous ones tested, most of them from the same geographic location in Alabama, USA [33].

Anticancer Drugs That Target Human Topoisomerases   Topoisomerases play an essential role in replication, transcription, recombination, and DNA repair.

Topoisomerase inhibitors prevent these functions and cause cell death. Bacterial type II topoisomerases (gyrase and Topo IV) are the targets of antibacterial antibiotics, while human topoisomerases are targets of anticancer antibiotics (see Sect. 1.1 for justification for the use of the terminology). Camptothecins are anticancer drugs that target type IB topoisomerases, while etoposide, anthracyclines (doxorubicin, daunorubicin), and mitoxantrone are anticancer drugs that inhibit type IIA topoisomerases (Top2) in humans [34]. Anticancer drugs that target human topoisomerase II (Top2) result in increased levels of Top2-DNA covalent complexes, which are called Top2 poisons since they block transcription and replication [35]. Another antibiotic, actinomycin D (dactinomycin), which can inhibit DNA replication and also functions as an anticancer agent by inhibiting RNA polymerase, is discussed later (Sect. 5.7).

5.4   Nitroheterocyclic Aromatic Compounds as Antibiotics

Unlike quinolones and fluoroquinolones, which inhibit DNA synthesis, some antibiotics such as the nitroheterocycles function by cleaving DNA and thereby inhibiting its replication to synthesize new DNA. Two types of antibiotics belonging to this nitroheterocycles category of antibiotics are the nitroimidazoles (e.g., Metronidazole) (Fig. 5.7) and nitrofurans (e.g., Nitrofurantoin) (Fig. 5.8), of which, the mechanism of action of nitrofurans is not yet very clear.

5.4.1   Nitroimidazoles: Antibiotics That Cleave DNA

The first nitroimidazole antibiotic, azomycin (2-nitroimidazole, Fig. 5.7) was discovered in 1953 as an alkaloid produced by *Streptomyces* spp. [36]. However, today the most common ones are derivatives of 5-nitro-imidazoles. There are several

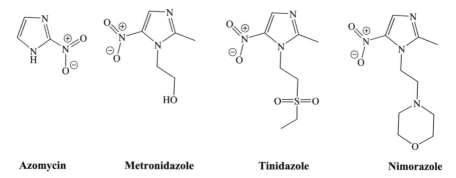

Azomycin   Metronidazole   Tinidazole   Nimorazole

Fig. 5.7 Structures of some nitroimidazoles

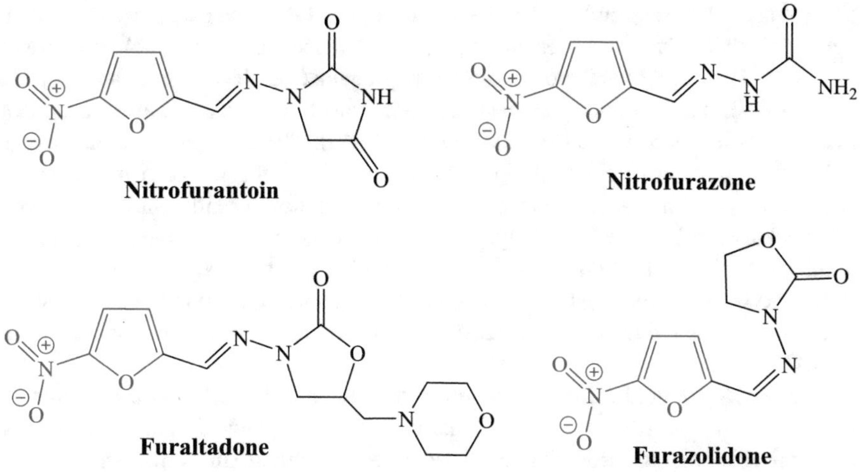

Fig. 5.8 Structures of some commonly used nitrofurans. The 5-nitrofuran group is shown in *red*

examples, of which the most effective antibiotic is "metronidazole" (sold under various brand names, including Flagyl and Metrogil). Metronidazole is a synthetic derivative of azomycin, the first natural product nitroimidazole. It was first sold in France in 1960 and was approved by FDA in 1963. Metronidazole works against anaerobic or microaerophilic microorganisms, including bacteria and parasites. Its antibacterial activity was discovered by accident in 1962 when it was found to cure both trichomonad vaginitis and bacterial gingivitis in a patient [37]. Later, in the 1970s, other uses of metronidazole became widespread such as against *Giardia lamblia* (also known as *Giardia duodenalis or Giardia intestinalis,* and causes giardiasis), *Entamoeba histolytica* (causes dysentery), *Clostridium difficile* (causes colitis, an inflammation of the large intestine). It is also used to treat Crohn's disease and ulcers caused by *Helicobacter pylori* and the dermatological conditions rosacea and acne.

Metronidazole is different in several ways from other antibiotics discussed till now. It was first developed as a drug against parasites but then gained wide acceptance as antibacterial agent. It is a synthetic derivative of a natural product. However, in spite of having a natural product analog and in spite of being a widely used antibiotic, resistance development to the drug is rare. Its mechanism of action is also different from that of most other antibiotics because it is not a competitive or suicide inhibitor of any enzyme but functions by reacting chemically with DNA. Another property of metronidazole that is rare is that it is a prodrug and thus needs to be activated by the pathogen and sometimes by the host for it to be functional.

Mechanism of action of metronidazole is different from that of most other antibiotics since it works by degrading DNA by a chemical reaction that is not catalyzed by an enzyme. As mentioned above, metronidazole is active against a broad spectrum of microorganisms with diverse morphologies. One factor common with all microorganisms that are sensitive to metronidazole is that they can grow in anaerobic or microaerophilic conditions. Cleavage of DNA can take place only when the drug is first reduced by a reaction that can take place only in anaerobic bacteria or protozoa. Thus metronidazole has no activity against aerobic bacteria and, for the same reason, has no toxicity against the human host. The mechanism of action of metronidazole has been studied extensively [38–40]. The nitro group of the drug is reduced by transfer of electron from the reduced form of ferredoxin in a reaction catalyzed by the iron-sulfur enzyme, pyruvate: ferredoxin oxidoreductase (POR). Most eukaryotes lack the POR enzyme and so, fail to activate metronidazole [41]. As a substitute for POR, humans have the enzyme pyruvate dehydrogenase, which decarboxylates and oxidizes pyruvate to produce acetate (as acetyl CoA), but instead of producing the reduced form of ferredoxin, it produces NADH, the reduced form of NAD.

The redox potential for the reduction of metronidazole is about 450 mV. Such a low reduction potential is not found in aerobes, which explains the selective toxicity of nitroimidazoles towards anaerobes alone. Since ferredoxin is also involved in photosystem I, an essential part of photosynthesis, metronidazole can also affect plants as has been demonstrated by inhibition of sugar synthesis in sugarcane leaves [38]. Microbial susceptibility to nitroimidazole agents is determined primarily by the reactivity of the cellular electron donor toward the nitroimidazole. Thus, growth of *T. vaginalis* is inhibited at 0.003–0.01 mM metronidazole while 1 mM concentration is needed to inhibit *Anabaena,* and this observation is consistent with the greater reactivity of *T. vaginalis* ferredoxin toward nitroimidazoles than *Anabaena* ferredoxin [42].

If DNA is added to previously reduced drug, no cleavage of the DNA is observed, indicating that the reduced form of the drug is not responsible for DNA damage, rather a short-lived intermediate formed during the reduction of the drug, is responsible for cleavage of DNA. The short-lived toxic intermediates are nitroso and hydroxylamine radicals, which finally decompose into nontoxic end products. A short-lived radical anion intermediate has been shown to interact with and cleave DNA [40]. The reduction of the drug also helps its activity in another way. Being a small molecule, metronidazole enters the microbial cell by the process of diffusion through the membrane. Rate of entry of the drug by diffusion depends on the concentration difference on the two sides of the membrane. Reduction of the drug decreases the concentration of the unreduced form of the drug inside the cell, thus increasing the concentration gradient and helps the entry of more of the drug into the cells. By the same reasoning, lack of the reduction reaction will decrease the rate of entry of the drugs into aerobic cells, as was first pointed out by Ings et al. [39]. The extent of DNA damage is also proportional to percentage of the nucleotides A and T. This explains why *E. histolytica, T. vaginalis,* and *Bacteroides*, which have much greater than 50% A + T are sensitive to metronidazole, while *Rhodospirillum* and *R. acidophila,* which have much less than 50% A + T are significantly less susceptible to nitroimidazoles.

Metronidazole is also used as a cure for the disease rosacea, a skin condition that causes redness in the face and primarily affects people of northwestern European descent. This effect of metronidazole is due to its anti-inflammatory activity

[43]. Metronidazole acts synergistically with palmitoleic acid, a lipid that is usually present in the skin, to inhibit free radical generation by human neutrophils. This has been proposed as a possible mechanism of action of metronidazole against rosacea and acne [44].

Resistance to Metronidazole Several factors are responsible for the widespread use of the drug. It functions against a wide variety of microorganisms with diverse morphologies, it is relatively inexpensive compared to other antibiotics, and it has favorable pharmacological properties such as low molecular weight, efficient absorption from the intestinal tract, limited binding to serum proteins and easy tissue penetration. However, in spite of the widespread use, resistance development to the drug is rare. For example, metronidazole has been the drug of choice for four decades against *Bacteroides fragilis* which is the most commonly isolated anaerobic pathogens in humans. *B. fragilis* is part of the normal flora of the human colon and is known to exhibit resistance to most antibiotics but not to metronidazole. The first case of metronidazole-resistant strain of *B. fragilis* from India was reported in 2001 [45]. Metronidazole resistance has been reported in helicobacter pylori. The resistance is due to a nonsense mutation in the rdxA gene, which codes for NADPH nitroreductase enzyme [46]. The mutation results in the synthesis of an inactive truncated enzyme that is unable to reduce the metronidazole. Although metronidazole is a semisynthetic drug, it is significantly different from its natural product analog (Fig. 5.7). Because of this and also because it is prodrug, for which the actual drug is a reactive intermediate, there is no known gene for degradation or inactivation of metronidazole (Sect. 2.8).

5.4.2 Nitrofurans: Multiple Possible Mechanisms of Action

Derivatives of 5-nitrofuran (Fig. 5.8) can function as synthetic broad-spectrum antibiotics that are effective against both gram-negative and gram-positive bacteria as well as against anaerobic pathogens. Examples of nitrofurans include furazolidone, nitrofurazone, and nitrofurantoin. The most commonly used nitrofuran is nitrofurantoin (1-[(5-nitrofurfurylidene) amino] hydantoin). Nitrofurantoin is taken orally and is rapidly absorbed; however, its serum concentration does not reach high level because it is quickly excreted through urine. Thus its concentration can reach inhibitory levels only in the urine, and so nitrofurantoin, which is sold under the trade name Macrobid, is most commonly used for urinary tract infections. The development of resistance to other widely used antibiotics, such as ciprofloxacin (Sect. 5.3), and the lack of new antibiotics in the pipeline have renewed our interest in nitrofurans.

Mechanism of Action of Nitrofurans Nitrofurantoin was first approved by the FDA in 1953. However, even after six decades since its approval, its mechanism of action is still not clear. Several different mechanisms of action have been proposed for nitrofurans. At one time, it was believed to function as antibiotic by cleaving DNA [47]. It has been reported to inhibit a number of bacterial enzymes of the carbohydrate metabolic pathways and also inhibit bacterial cell wall synthesis [48]. It is known that in order to be effective, the nitrofurans need to be first activated by being reduced. One of the intermediates formed during the reduction of the nitro group is believed to be responsible for the antibiotic activity of nitrofurans [49]. These intermediates were shown to attack bacterial ribosomal proteins nonspecifically, causing complete inhibition of protein synthesis [50]. The authors also reported a novel mechanism of action for nitrofurantoin which does not require the production of reactive metabolites formed during reduction of nitrofurantoin.

Resistance to Nitrofurans Since nitrofurans are synthetic antibiotics, there is no gene known for degradation or inactivation of the antibiotics (Sect. 2.8). Resistance development by point mutations is also rare but possible. Cumulative increase in resistance can result from sequential development of mutations in the different nitro-reducing enzymes in *E. coli*. These mutations were shown to be mostly concentrated in the nitroreductase genes *nfsA* and *nfsB* [51]. The fact that there are multiple possible mechanisms of action of nitrofurans probably explains why clinically significant resistance development to the drugs is rare. There is also a fitness cost associated with antibiotic resistance (Sect. 2.9). Resistant bacteria usually grow slower than non-resistant bacteria. This is especially important for urinary tract infections because if the resistant bacteria cannot grow fast enough, it will be frequently flushed out and so will be unable to maintain the infection, making resistance development less likely [49].

5.5   RNA Synthesis: Background Biochemistry Information

Genetic information stored in DNA is copied to make messenger RNA (mRNA). The process is called transcription. A gene is defined as the part of DNA that is transcribed to make RNA, including those that do not code for proteins (note that all transcripts are not messenger RNAs). Bacterial genomes have a few thousand genes, while human genome has ~30,000 genes. DNA-directed RNA synthesis, or transcription, is catalyzed by the enzyme RNA polymerase. Bacterial RNA polymerase is a large enzyme complex (MW ~500,000), and contains two α, one β and one β' subunits. The complex assembles on the DNA in a region preceding (upstream of) the gene and called the promoter, where transcription is initiated. The RNA polymerase has no specific affinity for the promoter. Another small protein known as the σ^{70} protein or sigma (σ) factor, binds to the promoter and assists the RNA polymerase to assemble at the promoter by first binding to the σ^{70} protein. During initiation, the template DNA partially unwinds to form the "open promoter complex" and RNA polymerase starts copying the template strand (which is complementary to the coding strand that has the gene). RNA chain elongation is very similar to DNA synthesis. Complementary NTPs bind to the active site, form H-bond to the previous nucleotide, a new phosphodiester bond is formed, and a PPi is released. Hydrolysis of PPi by pyrophosphatase makes the reaction thermodynamically feasible (Sect. 1.7.7). Since the sigma factor is needed only to bind specifically to the promoter site, it is no longer needed after the RNA synthesis has started. In fact, it slows down the synthesis. After about 10 ribonucleotides have joined to form a chain, the sigma factor leaves the complex so that the complex is no longer tightly bound to the promoter site and can move along the DNA and continue RNA synthesis. This process is called promoter clearance. Transcription continues till it reaches a termination site where the process ends, and the RNA is dissociated from the transcription complex.

5.6 Rifamycins

Rifamycins were first isolated in 1957 from the soil bacteria, *Amycolatopsis rifamycinica*. This species of bacteria has undergone several name changes. It was originally known as *Streptomyces mediterranei*. Its name was then changed to *Nocardia mediterranei* and then later to *Amycolatopsis mediterranei,* and then finally in 2004 was renamed as *Amycolatopsis rifamycinica* [52]. For a long time, this was the only species of bacteria known to produce the antibiotic until it was also discovered in *Salinispora* group [53]. Of the several rifamycins produced by bacteria, the antibiotic that is most commonly used is rifampicin (also known as rifampin) which is a synthetically modified version of rifamycin (Fig. 5.9).

Mechanism of Action of Rifamycins There are very few antibiotics known that inhibit RNA synthesis. Rifamycins form a family of antibiotics that inhibit RNA synthesis by binding to RNA polymerase, the enzyme that catalyzes the process of transcription (Sect. 5.5). The antibiotic binds to the β-subunit of RNA polymerase within the DNA/RNA channel. It inhibits RNA synthesis by directly blocking the path of the elongating RNA when the transcript becomes 2 to 3 nucleotides in length [54]. Binding to RNA polymerase is very tight, and so a very small amount of the antibiotic is needed to inhibit RNA synthesis. The binding is specific for RNA polymerase; the activity of DNA polymerase is not affected. Rifamycins are broad-spectrum bactericidal antibiotics that inhibit both gram-negative and gram-positive bacteria, including obligate intracellular bacteria. They can easily cross the bacterial cell wall and membrane and bind to the target. Since the structure of RNA polymerase is very similar to most bacterial species, the antibiotic has a broad spectrum of activity. Eukaryotic RNA polymerase is significantly different such that rifampicin does not bind to it and thus has little toxicity. One advantage of rifamycins as antibiotics is that they can penetrate into mammalian tissues and also into cells. Thus they are effective against intracellular bacteria, a property that many antibiotics do not have since they cannot cross the mammalian cell membrane. It is for this reason that rifampicin is effective in the treatment of mycobacterial diseases, such as tuberculosis and leprosy. Other antibiotics are not effective since mycobacteria are obligate intracellular bacteria and live within the host cells where

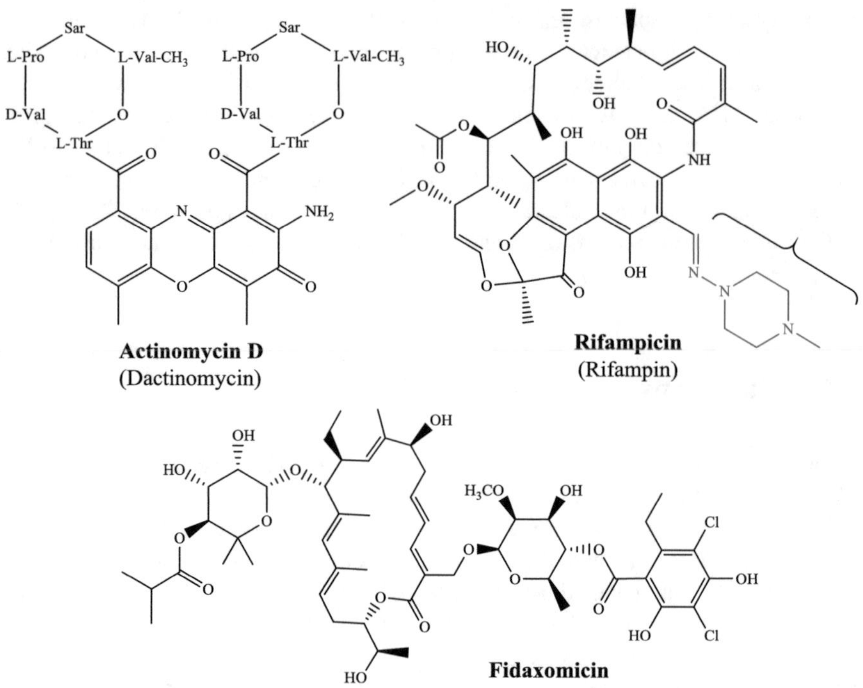

Fig. 5.9 Antibiotics that inhibit RNA synthesis. The methyl group is on the N of valine in Actinomycin D. Sar is sarcosine. The group shown in red is the synthetic modification made in Rifamycin SV to make Rifampicin

other antibiotics cannot penetrate. Although rifampicin has a broad spectrum of activity, it is mostly used for treating tuberculosis, leprosy, and meningitis. Since rifampicin easily enters all tissues, use of the drug results in red or orange-colored sweat, urine, and tears.

Resistance to Rifampicin One drawback of rifampicin is that bacteria develop resistance to the drug very quickly. Resistance development is usually by acquiring point mutations in the RNA polymerase that prevent the binding of the antibiotic. So rifampicin is always used in combination with other antibiotics. It is also used in combination with the antifungal drug amphotericin B (Sect. 8.1.5). When many rifampicin-resistant mutants were sequenced, all mutations were found to be within a small region of the 1342 amino acid long β-subunit. Mutations involving 8 conserved amino acids were identified in 64 of 66 rifampicin-resistant isolates of *Mycobacterium tuberculosis* from diverse geographical sources. All these mutations were clustered within a 23 amino acid region that is involved in binding to the antibiotic [55]. There is cross-resistance among rifamycins, so a clinical isolate of *M. tuberculosis* that is resistant to rifampicin is also resistant to other rifamycins. Almost all rifampicin-resistant strains are also resistant to isoniazid [56].

5.7 Actinomycin D (Dactinomycin)

Actinomycin D (Fig. 5.9) was the first antibacterial antibiotic that was found to have anticancer activity. Its relative, actinomycin A, was the first antibiotic isolated from *Actinomyces antibioticus* (now *Streptomyces antibioticus*) by Waksman and Woodruff in 1940. It functions by binding to DNA at the transcription initiation complex and preventing RNA synthesis by RNA polymerase. It was proposed that actinomycin binds to a premelted DNA conformation, called β-DNA that is found within the transcriptional complex. This acts to immobilize the complex, preventing the elongation of growing RNA chains [57]. Since actinomycin D inhibits RNA synthesis in both bacteria as well as humans, it was used as anticancer drug [58]. However, because of its toxicity, it is no longer used in cancer chemotherapy. Actinomycin D and its fluorescent derivative, 7-amino actinomycin D are used in research as dyes for cells and are used to differentiate between dead and live cells. Since actinomycin D binds to DNA, it also inhibits DNA replication in bacteria (Sect. 5.3).

5.8 Fidaxomicin: A New Antibiotic with a New Target

Fidaxomicin is a macrocyclic antibiotic (Fig. 5.9) obtained from the actinomycete *Dactylosporangium aurantiacum*. It is a narrow-spectrum antibiotic for the treatment of *Clostridium difficile*-associated diarrhea. Both Rifamycins and fidaxomicin bind to RNA polymerase and inhibit RNA synthesis. However, their mechanisms of action are different. Rifamycins prevent the extension of RNA beyond the first 2–3 nucleotides, while fidaxomicin affects an earlier step by blocking the formation of the open promoter complex (Sect. 5.5) in which the two strands of the DNA are separated locally prior to RNA synthesis [59]. The two antibiotics act synergistically and exhibit no cross-resistance. These results suggest that they have different mechanisms of action and act at two separate stages of the transcription process [60].

References

1. Wainwright M (2001) Acridine—a neglected antibacterial chromophore. J Antimicrob Chemother 47:1–13
2. Streisinger G, Okada Y, Emrich J, Newton J, Tsugita A, Terzaghi E, Inouye M (1966) Frameshift mutations and the genetic code. Cold Spring Harb Symp Quant Biol 31:77–84
3. Denny WA, Turner PM, Atwell GJ, Rewcastle GW, Ferguson LR (1990) Structure-activity relationships for the mutagenic activity of tricyclic intercalating agents in *Salmonella typhimurium*. Mutat Res 232:233–241
4. Ferguson LR, Denny WA (2007) Genotoxicity of non-covalent interactions: DNA intercalators. Mutat Res 623:14–23

5. Stevenson P, Sones KR, Gicheru MM, Mwangi EK (1995) Comparison of isometamidium chloride and homidium bromide as prophylactic drugs for trypanosomiasis in cattle at Nguruman, Kenya. Acta Trop 59:77–84
6. Lesher GY, Forelich EJ, Gruett MD, Bailey JH, Brundage RP (1962) 1,8-naphthyridine derivatives, a new class of chemotherapeutic agents. J Med Chem 5:1063–1065
7. King DE, Malone R, Lilley SH (2000) New classification and update on the quinolone antibiotics. Am Fam Physician 61:2741–2748
8. Emmerson AM, Jones AM (2003) The quinolones: decades of development and use. J Antimicrob Chemother 51(Suppl 1):13–20
9. Galante D, Pennucci C, Esposito S, Barba D (1985) Comparative in vitro activity of ciprofloxacin and five other quinoline derivatives against gram-negative isolates. Drugs Exp Clin Res 11:331–334
10. Smith SM, Eng RH (1985) Activity of ciprofloxacin against methicillin-resistant Staphylococcus aureus. Antimicrob Agents Chemother 27:688–691
11. Ridgway GL, Mumtaz G, Gabriel FG, Oriel JD (1984) The activity of ciprofloxacin and other 4-quinolones against Chlamydia trachomatis and Mycoplasmas in vitro. Eur J Clin Microbiol 3:344–346
12. Lawrence LE, Wu P, Fan L, Gouvei KE, Card A, Casperson M, Denbleyker K, Barrett JF (2001) The inhibition and selectivity of bacterial topoisomerases by BMS-284756 and its analogues. J Antimicrob Chemother 48:195–201
13. Collin F, Karkare S, Maxwell A (2011) Exploiting bacterial DNA gyrase as a drug target: current state and perspectives. Appl Microbiol Biotechnol 92:479–497
14. Drlica K, Zhao X (1997) DNA gyrase, topoisomerase IV, and the 4-quinolones. Microbiol Mol Biol Rev 61:377–392
15. Hoshino K, Kitamura A, Morrissey I, Sato K, Kato J, Ikeda H (1994) Comparison of inhibition of Escherichia coli topoisomerase IV by quinolones with DNA gyrase inhibition. Antimicrob Agents Chemother 38:2623–2627
16. Khodursky AB, Zechiedrich EL, Cozzarelli NR (1995) Topoisomerase IV is a target of quinolones in Escherichia coli. Proc Natl Acad Sci U S A 92:11801–11805
17. Hawkey PM (2003) Mechanisms of quinolone action and microbial response. J Antimicrob Chemother 51(SI):29–35
18. Chen CR, Malik M, Snyder M, Drlica K (1996) DNA gyrase and topoisomerase IV on the bacterial chromosome: quinolone-induced DNA cleavage. J Mol Biol 258:627–637
19. Kohanski MA, Dwyer DJ, Hayete B, Lawrence CA, Collins JJ (2007) A common mechanism of cellular death induced by bactericidal antibiotics. Cell 130:797–810
20. Dwyer D, Kohanski M, Hayete B, Collins J (2007) Gyrase inhibitors induce an oxidative damage cellular death pathway in Escherichia coli. Mol Syst Biol 3:91
21. Drlica K, Hiasa H, Kerns R, Malik M, Mustaev A, Zhao X (2009) Quinolones: action and resistance updated. Curr Top Med Chem 9:981–998
22. Walsh TJ, Standiford HC, Reboli AC, John JF, Mulligan ME, Ribner BS, Montgomerie JZ, Goetz MB, Mayhall CG, Rimland D, Stevens DA, Hansen SL, Gerard GC, Ragual RJ (1993) Randomized double-blinded trial of either Rifampin with Novobiocin or trimethoprimsulfamethoxazole against methicillin-resistant Staphylococcus aureus colonization: prevention of antimicrobial resistance and effect of host factors on outcome. Antimicrob Agents Chemother 37:1334–1342
23. Nikaido H (1998) Antibiotic resistance caused by gram negative multidrug efflux pumps. Clin Infect Dis 27(suppl 1):S32–S41
24. Li XZ, Livermore DM, Nikaido H (1994) Role of efflux pump(s) in intrinsic resistance of Pseudomonas aeruginosa: resistance to tetracycline, chloramphenicol, and norfloxacin. Antimicrob Agents Chemother 38:1732–1741
25. Ruiz J (2003) Mechanisms of resistance to quinolones: target alterations, decreased accumulation and DNA gyrase protection. J Antimicrob Chemother 51:1109–1117

26. Ito H, Yoshida H, Bogaki-Shonai M, Niga T, Haytori H, Nakamura S (1994) Quinolone resistance mutations in the DNA Gyrase gyrA and gyrB genes of *Staphylococcus aureus*. Antimicrob Agents Chemother 38:2014–2023

27. Stevenson JE, Gray K, Barrett TJ, Medalla F, Chilller TM, Angulo FJ (2007) Increase in nalidixic acid resistance among non-Typhi *Salmonella enterica* isolates in the United States from 1996 to 2003. Antimicrob Agents Chemother 51:195–197

28. Maruri F, Sterling TR, Kaiga AW, Blackman A, van der Heijden YF, Mayer C, Cambau E, Aubry A (2012) A systematic review of gyrase mutations associated with fluoroquinolone resistant *Mycobacterium tuberculosis* and a proposed gyrase numbering system. J Antimicrob Chemother 67:819–831

29. Valdezate S, Vindel A, Baquero F, Cantón R (1999) Comparative in vitro activity of quinolones against *Stenotrophomonas maltophilia*. Eur J Clin Microbiol Infect Dis 18:908–911

30. Gu Y, Xu X, Lin L, Ren X, Cui X, Hou X, Cui S (2013) Functional characterization of quinolone-resistant mechanisms in a lab-selected *Salmonella enterica typhimurium* mutant. Microb Drug Resist 19:15–20

31. Martinez-Martinez L, Pascual A, Jacoby GA (1998) Quinolone resistance from a transferable plasmid. Lancet 351:797–799

32. Tran JH, Jacoby GA (2002) Mechanism of plasmid mediated quinolone resistance. Proc Natl Acad Sci U S A 99:5638–5642

33. Jacoby GA, Chow N, Waites KB (2003) Prevalence of plasmid-mediated quinolone resistance. Antimicrob Agents Chemother 47:559–562

34. Pommier Y, Leo E, Zhang HL, Marchand C (2010) DNA topoisomerases and their poisoning by anticancer and antibacterial drugs. Chem Biol 17:421–433

35. Nitiss JL (2009) Targeting DNA topoisomerase II in cancer chemotherapy. Nat Rev Cancer 9: 338–350

36. Maeda K, Osato T, Umezawa H (1953) A new antibiotic, azomycin. J Antibiot (Tokyo) 6:182

37. Shinn DLS (1962) Metronidazole in acute ulcerative gingivitis. Lancet 279:1191

38. Edwards DI (1980) Mechanisms of selective toxicity of metronidazole and other nitroimidazole drugs. Br J Vener Dis 56:285–290

39. Ings RMJ, McFadzean JA, Ormerod WE (1974) The mode of action of metronidazole in *Trichomonas vaginalis* and other micro-organisms. Biochem Pharmacol 23:1421–1429

40. Muller M (1983) Mode of action of metronidazole on anaerobic bacteria and protozoa. Surgery 93:165–171

41. Samuelson J (1999) Why metronidazole is active against both bacteria and parasites. Antimicrob Agents Chemother 43:1533–1541

42. Vidakovic M, Crossnoe CR, Neidre C, Kim K, Krause KL, Germanas JP (2003) Reactivity of reduced [2Fe-2S] ferredoxins parallels host susceptibility to nitroimidazoles. Antimicrob Agents Chemother 47:302–308

43. Miyachi Y, Imamura S, Niwa Y (1986) Anti-oxidant action of metronidazole: a possible mechanism of action in rosacea. Br J Dermatol 114:231–234

44. Akamatsu H, Oguchi M, Nishijima S, Asada Y, Takahashi M, Ushijima T, Niwa Y (1990) The inhibition of free radical generation by human neutrophils through the synergistic effects of metronidazole with palmitoleic acid: a possible mechanism of action of metronidazole in rosacea and acne. Arch Dermatol Res 282:449–454

45. Chaudhry R, Mathur P, Dhawan B, Kumar L (2001) Emergence of metronidazole-resistant bacteroides fragilis, India. Emerg Infect Dis 7:485–486

46. Goodwin A, Kersulyte D, Sisson G, Veldhuyzen van Zanten SJ, Berg DE, Hoffman PS (1998) Metronidazole resistance in Helicobacter pylori is due to null mutations in a gene (rdxA) that encodes an oxygen-insensitive NADPH nitroreductase. Mol Microbiol 28:383–393

47. McCalla DR, Reuvers A, Kaiser C (1971) Breakage of bacterial DNA by nitrofuran derivatives. Cancer Res 31:2184–2188

48. Munoz-Davila MJ (2014) Role of old antibiotics in the era of antibiotic resistance. Highlighted nitrofurantoin for the treatment of lower urinary tract infections. Antibiotics 3:39–48

49. Sandegren L, Lindqvist A, Kahlmeter G, Andersson DI (2008) Nitrofurantoin resistance mechanism and fitness cost in Escherichia coli. J Antimicrob Chemother 62:495–503
50. McOsker CC, Fitzpatrick PM (1994) Nitrofurantoin: mechanism of action and implications for resistance development in common uropathogens. J Antimicrob Chemother 33(Suppl A):23–30
51. Whiteway J, Koziarz P, Veall J, Sandhu N, Kumar P, Hoecher B, Lambert IB (1998) Oxygen insensitive nitroreductases: analysis of the roles of nfsA and nfsB in development of resistance to 5-nitrofuran derivatives in Escherichia coli. J Bacteriol 180:5529–5539
52. Bala S, Khanna R, Dadhwal M, Prabagaran SR, Shivaji S, Cullum J, Lal R (2004) Reclassification of *Amycolatopsis mediterranei* DSM 46095 as *Amycolatopsis rifamycinica* sp. nov. Int J Syst Evol Microbiol 54:1145–1149
53. Kim TK, Hewavitharana AK, Shaw PN, Fuerst JA (2006) Discovery of a new source of rifamycin antibiotics in marine sponge actinobacteria by phylogenetic prediction. Appl Environ Microbiol 72:2118–2125
54. Campbell EA, Korzheva N, Mustaev A, Murakami K, Nair S, Goldfarb A, Darst SA (2001) Structural mechanism for rifampicin inhibition of bacterial RNA polymerase. Cell 104:901–912
55. Telenti A, Imboden P, Marchesi F, Lowrie D, Cole S, Colston MJ, Matter L, Schopfer K, Bodmer T (1993) Detection of rifampicin-resistance mutations in *Mycobacterium tuberculosis*. Lancet 341:647–650
56. Chaisson RE (2003) Treatment of chronic infections with rifamycins: is resistance likely to follow? Antimicrob Agents Chemother 47:3037–3039
57. Sobell HM (1985) Actinomycin and DNA transcription. Proc Natl Acad Sci U S A 82:5328–5331
58. Demain AL, Sanchez S (2009) Microbial drug discovery: 80 years of progress. J Antibiot 62:5–16
59. Artsimovitch I, Seddon J, Sears P (2012) Fidaxomicin is an inhibitor of the initiation of bacterial RNA synthesis. Clin Infect Dis 55(S2):S127–S131
60. Babakhani F, Seddon J, Sears P (2014) Comparative microbiological studies of transcription inhibitors fidaxomicin and the rifamycins in *Clostridium difficile*. Antimicrob Agents Chemother 58:2934–2937

Chapter 6
Antibiotics That Inhibit Protein Synthesis

Abstract Antibiotics that inhibit protein synthesis are discussed. Background biochemistry information on translation is provided. Antibiotics presented include puromycin, aminoglycosides, tetracyclines, chloramphenicol, macrolides, lincosamides, streptogramins, oxazolidinones, pleuromutilins, mupirocins, and peptide deformylase inhibitors. Mechanisms of action of the antibiotics and resistance development against them are discussed.

6.1 Protein Synthesis: Background Biochemistry Information

After the sequence of the DNA is copied to make mRNA by the process of transcription, the mRNA is translated to make protein in the last step of biological information flow. In this process, which is known as translation, protein is synthesized based on the sequence of the mRNA. The translational machinery is called a ribosome, which is a large complex made of proteins and RNA, called ribosomal RNA or rRNA. The ribosome moves along the mRNA as it synthesizes the protein by joining amino acids corresponding to the genetic code that is prescribed in the sequence of the mRNA. The genetic code for each amino acid is called the codon for that amino acid. Amino acid corresponding to a codon will be brought to the ribosome and joined to the previous amino acid. The ribosome then moves to the next codon, and protein synthesis continues.

It is widely known that three bases make a codon for each amino acid. Since there are only 4 bases and 20 amino acids to code for, the codon for each amino acid must contain 3 bases. So there are $4^3 = 64$ codons, which is more than enough to code for the 20 amino acids. A table of codons and the corresponding amino acid can be found in any biochemistry textbook and is not presented in this book. Note that it is most sensible for nature to have evolved to have three bases per codon. If two bases made a codon, then there can be only $4^2 = 16$ possible codons, which is not enough to code for the 20 amino acids. If codons had 4 bases, then there would be $4^4 = 256$ codons, which are too many for only 20 amino acids.

Since three bases make a codon, a sequence of mRNA can be translated in three ways depending on which base is considered to be the starting base of codon. So the arbitrary sequence ...AUGCAUGCAUGC... can be read as ...AUG-CAU-GCA-UGC... or ...A-UGC-AUG-CAU-GC... or ....AU-GCA-UGC-AUG-C.... These three ways of reading the sequence are called the three possible reading frames. Since DNA is double stranded, the second strand will also have three reading frames. So for any given length of DNA sequence, there can be six possible reading frames. Of these, there is only one correct reading frame for a particular gene. However, it is sometimes possible that another gene is also present in the same region of DNA and is read from a different reading frame.

Since the mRNA is a long continuous chain, how is it determined by the ribosome which codon is the start of the gene? The beginning of a gene is called the start codon or initiation codon, which is usually an AUG in the mRNA (ATG for the corresponding DNA). In the same way synthesis of a protein will stop when a stop codon is reached. There are three stop codons: UAG, UGA, UAA (TAG, TGA, TAA in DNA). Translation stops when the ribosome reaches a stop codon in the same reading frame. Note that, since there are three reading frames, the stop codon must be in the same reading frame as the start codon. A stop codon in one reading frame will have no effect on translation taking place in another reading frame.

Since there are 64 codons for only 20 amino acids, there can be multiple codons possible for most amino acids. Except for Met and Trp, all other amino acids have more than one codon each. Because of this degeneracy of the genetic code, a single base mutation sometimes results in a codon that still specifies for the same amino acid. These are called silent mutations. So all mutations in the DNA do not result in a changed protein. Point mutations that do change the amino acid sequence are called missense mutations. A single base mutation may convert a codon for a certain amino acid to a stop codon. Such mutations are called nonsense mutations and will result in a truncated protein being synthesized. An insertion or deletion of a few bases (that is not a multiple of three) into a gene may change the reading frame of the mRNA and thus completely change the sequence of the protein starting from the point of insertion or deletion. Such mutations are called frameshift mutations.

Besides the ribosomal and messenger RNAs, there is a third type of RNA called transfer RNA or tRNA, which plays an important role in translation. Each tRNA forms a covalent bond to an amino acid and brings the right amino acid (based on the mRNA sequence) to the ribosome to form peptide bond to the previous amino acid in the sequence. Since there are 20 different amino acids that are present in proteins, every cell must contain at least 20 different tRNA species, each one specific for one amino acid. All tRNAs are small single-stranded RNAs, ~ 70–95 bases long. Although the 20 different tRNAs have different sequences, their three-dimensional structures are similar. Each tRNA is single stranded but forms intramolecular base pairs to form several stem loop structures as shown in Fig. 6.1. This is called the cloverleaf structure. The number of bases in each arm of the tRNA is more or less constant, except in the variable arm, which can have 3–21 bases. The region at the 5' and 3' ends of the tRNA molecule are base paired, forming what is known as the acceptor stem or amino acid stem because an amino acid forms covalent bond at the

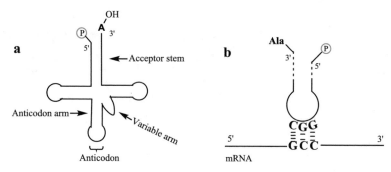

Fig. 6.1 Representation of the structure of tRNA and its interaction with mRNA. (**a**) Cloverleaf structure of tRNA (**b**) Codon–anticodon binding shown for the alanine codon. The direction of the tRNA has been reversed in order to show binding to the mRNA

5' end of the stem. The base at the 3' end of all tRNAs is always an "A." The tRNAs are named after the amino acid they carry, e.g., tRNA^{gly} or tRNA^{phe}. The tRNA and mRNA molecules interact through hydrogen bonding between bases in the antico-dons in the tRNA and codons in mRNA. Codon and anticodon are antiparallel. So the 5' end of the anticodon pairs with 3' end of the codon. The tRNA transcripts undergo several further modifications at the 5' end, 3' end, and the various loops, including the anticodon loop.

In the first step of protein synthesis, each amino acid forms a covalent bond to the nucleotide "A" at the 3' end of the corresponding tRNA specific for that particular amino acid. An acyl (ester) bond is formed between the carboxy group of the amino acid and the hydroxyl group at the 3' position of ribose at the 3' end of the tRNA (Fig. 6.2). Energy for the reaction comes from the hydrolysis of ATP to AMP and PPi. Immediate hydrolysis of PPi by pyrophosphatase makes the reaction thermo-dynamically favorable and moves the reaction forward (Sect. 1.7.4). So two equiv-alents of ATP (actually two phosphoester bonds) are needed for the activation of each amino acid such that the reaction is metabolically irreversible (Sect. 1.7.7 and Fig. 1.13). There is a separate enzyme for catalyzing the acyl bond formation of each amino acid to the corresponding tRNA. The energy of the acyl bond between the amino acid and the tRNA will be used later to form a peptide bond between this amino acid and the previous amino acid in the sequence of the protein. After the aminoacylated tRNAs are synthesized, they bind to an elongation factor called EF-Tu (elongation factor thermos unstable). The complex then binds to the ribosome for protein synthesis.

Ribosome, The Site of Protein Synthesis Protein synthesis is carried out by a complex of proteins and RNA called the ribosome. All ribosomes contain two different subunits. Each subunit is very large. For such a large complex, the size is not measured by the combined molecular weight of the complex but is determined by the rate at which it sediments from a solution when it is centrifuged. Large complexes are assigned an "S" value called the Svedberg unit. *E. coli* ribosome has a size of 70S. Complexes that sediment faster are assigned a lower S value. If the two

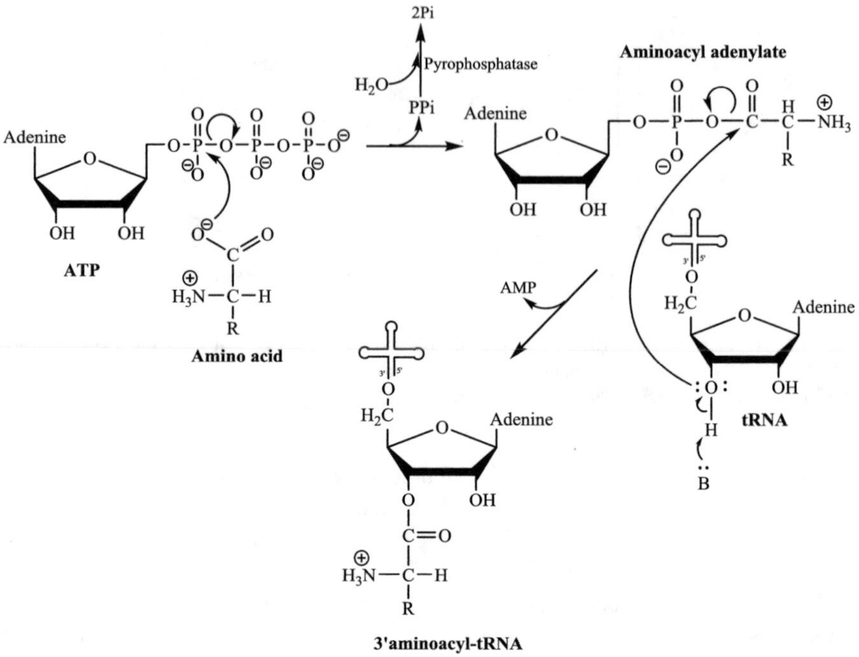

Fig. 6.2 Activation of amino acids and formation of aminoacyl-tRNA. (Structure not drawn to scale)

subunits of the ribosome are separated and centrifuged, they have S values of 30S and 50S. Note that the S values are not additive. So the 30S and 50S together do not make an 80S complex but a 70S complex. Each subunit of the ribosome is made up of proteins and RNA. The 50S subunit consists of two RNAs and 31 proteins. The 30S subunit contains one RNA and 21 proteins.

The 2009 Nobel Prize in Chemistry was awarded to Venkatraman Ramakrishnan, Ada Yonath, and Thomas Steitz, who determined the structures of the 30S and the 50S subunits of the ribosome and studied the mechanism of protein synthesis and its inhibition in bacteria by antibiotics [1–3]. For protein synthesis, the ribosome binds to the mRNA and at the same time binds to three different tRNAs. This explains why a ribosome needs to be so large. It has sufficient space to bind the three tRNAs at three binding sites, which are named as the *A*-site for binding to the incoming *a*minoacyl tRNA, the *P*-site for binding to the *p*eptidyl-tRNA that has been formed in the previous step, and the *E*-site for the *e*xiting tRNA that brought the previous amino acid. The interaction between the anticodon in the tRNA and the codon in the mRNA takes place in the 30S subunit while the 3′ end of the tRNA containing the amino acid is positioned at the active site of the enzyme where peptide bond formation takes place. Thus, the 30S subunit has the decoding center while the 50S subunit contains the catalytic site, called the peptidyl transferase center (PTC) and the protein exit tunnel.

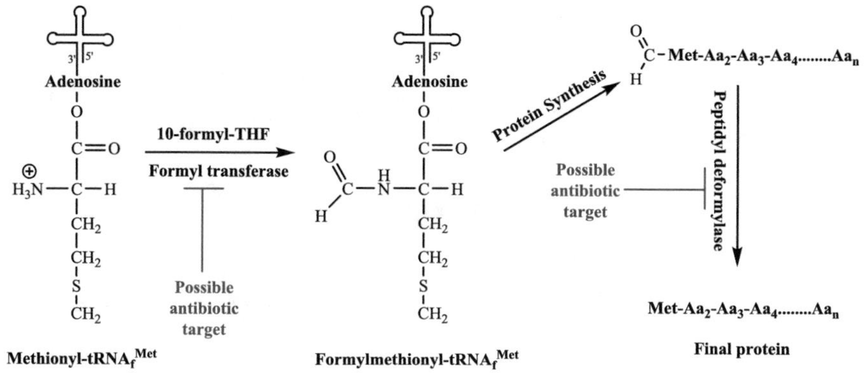

Fig. 6.3 Formylation and deformylation of bacterial proteins and possible antibiotic targets

Initiation of Translation The first codon of most proteins is AUG, and almost all proteins start with the amino acid methionine. Besides the initiation codon, the gene may contain other AUG codons that also code for methionine. How does the ribosome differentiate between AUG initiation codon and AUG in the middle of a gene? In bacteria, the initiation codon is always preceded by certain sequence called the ribosome binding site (rbs), which is also known as the Shine Dalgarno sequence, which is a purine-rich sequence such as AGGAGGT. This site is recognized, and the ribosome assembles at this site. The first AUG after the rbs is the start codon, and protein synthesis starts from here. The tRNAs present in the cell also differentiate between two types of AUG codons. In the infecting bacteria (but not in the human host) the methionine of the initiator tRNA is formylated. Thus all cells contain two types of tRNAs carrying methionine: a N-formylmet-tRNA, which recognizes and binds to the initiator AUG and tRNA^{Met} (without the formyl group), which recognizes and binds to the internal AUG. So all bacterial proteins start with a formyl methionine at the N-terminus. Later the methionine is either deformylated or removed completely (Fig. 6.3). Since this formylation and deformylation reactions take place only in the infecting bacteria and not in the human host, either of these steps can be target for development of new antibiotics. This is discussed further in Sect. 6.2.8.4.

In prokaryotes, initiation of translation takes place by forming an initiation complex by binding of the 30S subunit at the rbs of the mRNA along with the initiation factors IF1, IF2, and IF3, thus automatically positioning the fmet-tRNA^{metf} at the first codon. The 50S subunit then binds to the 30S subunit to form the complete 70S ribosome and the initiation factors are released. This way the fmet-tRNA^{metf} is positioned and bound to the first codon at the P-site of the ribosome. The tetracyclines and aminoglycosides are two families of antibiotics that are effective at this stage of the ribosomal cycle (Sects. 6.2.2 and 6.2.4). A second tRNA containing the amino acid corresponding to the second codon, and EF-TU GTP then bind to the A-site (also known as the decoding center) of the ribosome. Once correct base

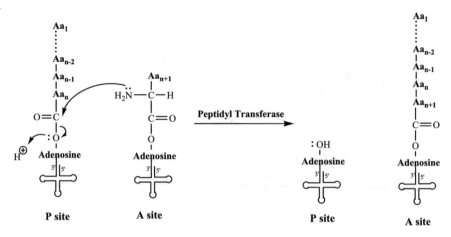

Fig. 6.4 Peptidyl transferase reaction during protein synthesis

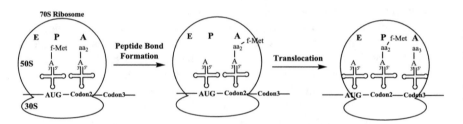

Fig. 6.5 Peptide bond formation followed by translocation

pairing of the codon and anticodon takes place, the GTP bound to EF-TU is hydrolyzed, and the aminoacylated tRNA is released at the A-site of the ribosome.

Chain Elongation The formyl methionine from the P-site is then transferred to the second amino acid at the A-site to form a peptide bond. The reaction is catalyzed by the ribozyme, peptidyl transferase, which is a part of the 50S subunit of the ribosome. Two antibiotics that affect this stage of synthesis are puromycin and chloramphenicol (Sects. 6.2.1 and 6.2.5). The peptidyl transferase reaction takes place by a nucleophilic attack of the amino group of the amino acid at the A-site on the carbonyl group of the amino acid at the P-site, thereby placing the newly formed peptide chain at the A-site and the tRNA without amino acid at the P-site (Fig. 6.4). In the next step, the ribosome moves downstream along the mRNA by one codon, a process known as translocation (Fig. 6.5). The process is catalyzed by an incoming EF-G (elongation factor-G), which also has GTPase activity. Binding of EF-G close to the A-site forces the peptidyl tRNA from the A-site. This positions the tRNA containing the dipeptide at the P-site and the free tRNA for the first codon at the E-site, from where it is eventually ejected. A new tRNA containing amino acid corresponding to the third codon then binds at the A-site. This process is repeated till the whole protein is synthesized.

Chain Termination   The protein synthesis stops when any stop codon in the right reading frame is positioned at the A-site. There are three possible stop codons (UAA, UGA, and UAG) in the mRNA. There is no tRNA that can bind to the stop codons. So protein synthesis stops here, and the synthesized protein is released with the help of some release factors and the ribosome dissociates into its 50S and 30S components, which can then assemble again at the start codon to start the synthesis of another protein molecule.

Error Rate of Protein Synthesis   During protein synthesis, the ribosomes make about one mistake for every 10,000 peptide bonds formed between amino acids. This error rate is much higher than that for DNA synthesis. The reason is obvious. There is only one copy of the chromosomal DNA and so its sequence must be very accurate. So the cells have developed several mechanisms for lowering the error rate of DNA synthesis (Sect. 2.4). However, there are many molecules of each protein made. So if some of them are inactive due to wrong sequence, the loss is compensated by the many other active molecules of the same protein that have the right sequence. For the same reason error rate of transcription is also higher than that of replication.

6.2   Antibiotics That Inhibit Protein Synthesis

For protein synthesis to take place, first, the mRNA has to be made. So, inhibitors of RNA synthesis will also have a secondary effect on protein synthesis. Antibiotics that inhibit RNA synthesis have already been discussed in Sect. 5.6. Besides these, there are also antibiotics that directly inhibit the synthesis of new proteins. One such antibiotic, chloramphenicol, is often used in research to increase the yield of DNA or RNA from bacteria. Chloramphenicol is added to rapidly dividing bacteria (in log phase). This stops protein synthesis within minutes while the cells continue to synthesize DNA and RNA for some time longer. Thus the amount of DNA or RNA calculated per ml of cells or per mg of protein will be higher.

Each of the steps in the process of protein synthesis can be a potential target for antibiotics development, and there are many such antibiotics already known. Only the most commonly used antibiotics will be described here. For an excellent review of all other antibiotics, one can read a review article on the subject [4]. As mentioned above, an error in the sequence of one protein molecule has no deleterious effect because there are multiple other copies of the protein molecule that will have the right sequence. The same reasoning applies to the effectiveness of antibiotics. To be effective, these antibiotics should be able to cause changes in the protein sequence every time the ribosome synthesizes the protein and not just occasionally. This requires tight binding of the antibiotic to the ribosome. Antibiotics can function by inhibiting any of the steps of translation described above (Sect. 6.1) and thus slow down the growth of the bacteria. Most inhibitors of protein synthesis are bacteriostatic, so proper protein synthesis can resume once the antibiotic is removed. Thus the effect of these antibiotics is to prevent the growth of the bacteria. However, aminoglycosides, which are also inhibitors of protein synthesis, are bactericidal. Antibiotics can inhibit protein synthesis by targeting either the 30S or the 50S subunit. Examples of antibiotics that target the 30S subunit include spectinomycin, tetracycline, and the aminoglycosides kanamycin and streptomycin, while example of those that target the 50S subunit, include clindamycin, chloramphenicol, linezolid, and the macrolides erythromycin, clarithromycin, azithromycin, and tylosin.

6.2.1 Puromycin

Puromycin is an antibiotic obtained from *Streptomyces alboniger*. It is a nonselective antibiotic that interferes with protein synthesis in both prokaryotes and eukaryotes and so has no clinical use due to its toxic effect in the host. However, it is used in research, especially cell culture research. Puromycin is an aminonucleoside antibiotic containing a 3′-amino-*N,N*-dimethyladenosine linked by an amide bond to a O-methyltyrosine. Its structure (Fig. 6.6) resembles that of the 3′ end of aminoacyl tRNA with some differences. The differences are shown in red in Fig. 6.6. It does not contain a tRNA and has an unusual amino acid linked to a modified adenosine through an amide bond instead of the usual ester bond. Because of the similarity, puromycin binds to the A-site of ribosome, and undergoes a peptidyl transferase reaction forming a protein ending with puromycin residue. Since it does not have a tRNA it cannot bind tightly to the A-site and also cannot

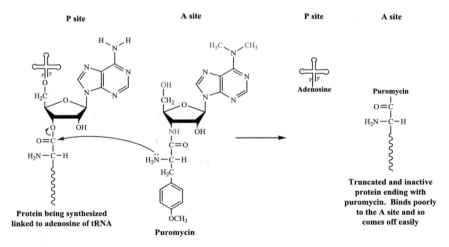

Fig. 6.6 Mechanism of action of puromycin

bind to mRNA since it lacks the anticodon. So truncated proteins of various lengths, each ending with puromycin, come off the ribosome, and further chain extension does not take place.

Puromycin also causes membrane damage. The truncated protein prematurely released by puromycin can be incorporated into the membrane, thus creating channels in it, making the membranes leaky [5]. This mechanism is similar to that of aminoglycosides which also damage the membranes (Sect. 6.2.2).

6.2.2 Aminoglycosides

Aminoglycosides contain three sugar rings linked by glycosidic bonds and contain several amino groups (Fig. 6.7). The first aminoglycoside, streptomycin, was discovered in 1943 by Selman Waksman, Albert Schatz, and Elizabeth Bugie at Rutgers University from the soil bacteria *Streptomyces griseus* (Sects. 1.1 and 1.5). Streptomycin has a streptidine ring, while kanamycin, gentamicin, and neomycin have streptamine rings. Amikacin is a semisynthetic derivative of kanamycin designed to prevent resistance development to the antibiotic. Spectinomycin is not an aminoglycoside, it is an aminocyclitol which is similar to aminoglycosides and so is often included in this group. Spectinomycin, which was discovered in 1961, is produced by *Streptomyces spectabilis* as well as many other organisms, including cyanobacteria and various plant species. *Streptomyces* is the largest antibiotic-producing genus, producing more than two-thirds of the clinically useful natural antibiotics, including spectinomycin, streptomycin, neomycin, tetracycline, chloramphenicol, and many others [6]. The cyanobacteria have also emerged as an important source of many antibiotics, including anticancer, antibacterial, and antiviral drugs [7]. Aminoglycosides are no longer used in humans due to toxicity to the kidneys and ears.

As mentioned above, aminoglycosides contain several amino and guanido groups (Fig. 6.7). At physiological pH, all these are protonated to form $-NH_3^+$ groups. This polycationic property of the aminoglycosides helps them to bind to the negatively charged phosphate groups of the 16S RNA at the A-site of the 30S ribosome subunit. This prevents the binding of the aminoacyl tRNA to the A-site of the ribosome and thus inhibits protein synthesis. Spectinomycin also follows the same mechanism of action. It inhibits protein synthesis by binding to the 30S subunit of the bacterial ribosome. Besides the protonated amino groups, other structural features of the aminoglycosides result in high affinity for specific regions of the RNA, especially bacterial rRNA [8]. However, it has also been reported that there is not much difference between the binding affinities of aminoglycosides to prokaryotic and eukaryotic A-site decoding region (where codon–anticodon binding takes place) of the 16S ribosomal RNA. There have been many studies on the binding of aminoglycosides with specific sequences in the ribosomal RNA, and the consensus sequences for binding to various antibiotics have been determined [9]. The specific binding of aminoglycosides and other ribosomes targeting antibiotics to the

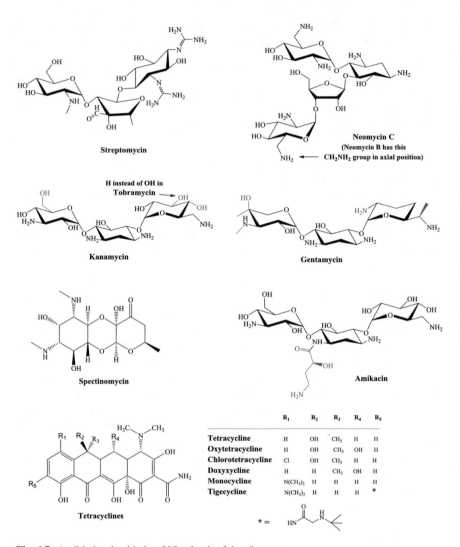

Fig. 6.7 Antibiotics that bind to 30S subunit of the ribosome

ligand-binding sites at the various bulges and internal loops of RNA has been
reviewed [10]. The different aminoglycosides bind to different sites of the ribosomal
RNA. Kanamycin, gentamicin, and neomycin all bind in a similar manner to the
A-site of the ribosome, where the 16S RNA is located within the 30S subunit. The
nucleotides that interact with the antibiotics have been identified. The binding of
these antibiotics to the A-site at the decoding region (the codon–anticodon region)
interferes with the recognition of the right tRNA and thus results in mistranslation
and also interferes with the translocation (Sect. 6.1) of the peptidyl-tRNA from the
A-site to the P-site [8].

Inhibitors of protein synthesis are usually bacteriostatic, but the aminoglycosides are bactericidal. Numerous studies have been done on the mechanism of action of various aminoglycosides. Various mechanisms have been suggested to explain how aminoglycosides have a bactericidal effect. This complex issue has been explained in a review article [11]. Aminoglycosides have many effects on the cell, of which the four main types are: (1) blocking of the ribosome, (2) misreading of the genetic code, (3) membrane damage, and (4) irreversible uptake of the antibiotics. All these four effects are believed to contribute to the bactericidal activity of aminoglycosides. At low concentrations (2 µg/ml), streptomycin binds mostly to the predominant ribosomes that are in the process of elongation during protein synthesis. For large mRNAs, multiple ribosomes, separated from each other by approximately 100 nucleotides, can be seen engaged in translation at the same time. These are known as polysomal ribosomes or polysomes. The antibiotic distorts the structure of ribosomes by binding to them and causes misreading of the codons because wrong aminoacyl tRNAs are able to bind to the A-site without matching the codon present in the mRNA at that position. At higher concentrations (20 mg/ml), streptomycin is able to bind to all ribosomes including those that are engaged in initiation. It binds to the initiation complex tightly and prevents the translocation step (Sect. 6.1) and so the ribosome is unable to move along the mRNA and prevents new initiations from taking place. So protein synthesis completely stops, and the mRNAs can be seen with only one ribosome at the initiation site with the antibiotic bound to it and no other ribosome bound in the rest of the mRNA.

6.2.3 Resistance to Aminoglycosides

There are two major mechanisms for resistance development to aminoglycosides.

By Target Modification Point mutations in the target (the ribosome) may alter its structure so that the antibiotic binds to it less tightly. However, resistance development to aminoglycosides by point mutations in the target is not very common. There can be two possible reasons for this. The ribosomes carry out the very essential function of protein synthesis that is needed in all species. The sequence of rRNA is highly conserved in all species. So any structural change in the ribosome resulting from point mutations is not tolerated. Another reason is that all organisms have multiple copies of the genes that code for rRNA. So to develop resistance to aminoglycosides, mutations will have to be generated in all the copies of these genes, which is highly unlikely. Nevertheless, there are many reports of bacteria being resistant to aminoglycosides due to point mutations in one of the many components of the ribosome. Streptomycin, an aminocyclitol aminoglycoside, interacts with the ribosome accuracy center, so-called because the accuracy of translation in *E. coli* is influenced by three interacting ribosomal proteins, S4, S5, and S12, that are present in this region. The interplay between these three proteins in the accuracy center has been highly conserved for billions of years [12]. Since there is only one copy of the *rpsL* gene, which codes for the ribosomal protein S12, resistance to streptomycin is frequently due to mutations in this gene.

Another component of the 30S ribosome is the 16S rRNA which makes several contacts with the S12 protein. The sequence and structure of the 16S rRNA are highly conserved. Mutations in the 16S rRNA are known to confer resistance to streptomycin in *Mycobacterium smegmatis* since the strain contains only one rRNA operon (*rrn*) [13]. Most of these mutations are in the 530 (nucleotide number) region, which is part of the aminoacyl-tRNA binding site and so is important for correct translation of the mRNA sequence. Mutations in this region disrupt the hydrogen bonding that is necessary for the stem loop structure of the 16S rRNA and has been shown to confer resistance to streptomycin.

By Modification of the Antibiotics Bacteria can become resistant to aminoglycosides by acquiring genes that code for antibiotic modifying enzymes. The most common enzymes are acetyltransferases, phosphotransferases, and nucleotidyltransferases, which catalyze acetylation, phosphorylation, or adenylation respectively of various hydroxyl and amino groups of the aminoglycoside antibiotics. The modified antibiotics confer resistance to the bacteria in two ways. They may be transported through the bacterial membrane less efficiently than the unmodified antibiotics due to charge differences. Molecules that do enter the cells may bind poorly to the ribosome because of changed steric and electrostatic factors resulting from these modifications. The original source of these antibiotic modifying enzymes is probably the microorganisms that produce these antibiotics. They protect themselves from the antibiotics with the help of these modifying enzymes. The gene *aadA*, which encodes an adenylation enzyme that modifies streptomycin and spectinomycin, was first reported in 1985 [14]. The *aadA* genes are the only characterized genes that encode both streptomycin and spectinomycin resistance, and many of these genes are found as gene cassettes in integrons [15].

Amikacin Developed to Prevent Resistance Development The antibiotic amikacin (Fig. 6.7) was made by synthetic modification of kanamycin such that bacteria are slow to develop resistance to it. The amino group of kanamycin at position 1 was acylated with 4-amino-2-hydroxybutyrate to obtain amikacin. The large aminohydroxybutyryl group sterically prevents the acetylation, phosphorylation, and adenylation reactions that normally take place for resistance development to aminoglycosides. In order to prevent resistance development to the drug, the use of amikacin has been regulated and made limited. As explained before, the more frequently an antibiotic is used, the greater is the probability of resistance development (Sect. 2.6). However, in spite of these favorable properties of the antibiotic, resistance to amikacin has been reported. Point mutations that confer resistance to other aminoglycosides can also result in resistance to amikacin. For example, mutations in the *rpsL* gene of the S12 ribosomal protein or in the 530 or 915 region of the *rrs* gene of the 16S rRNA can provide resistance to amikacin. Mutation 1400 from A to G in the *rrs* gene of *Mycobacterium tuberculosis* can provide high-level resistance to kanamycin and amikacin since there is only one copy of the *rrs* gene [16].

6.2.4 Tetracyclines

As discussed before, chlortetracycline was discovered as a yellow antibiotic from *Streptomyces aurofaciens* and was then named aureomycin which was later renamed as chlortetracycline (Sect. 1.5). Later the first semisynthetic antibiotic, tetracycline was made by catalytic hydrogenation of chlortetracycline (Sect. 2.8). As the name suggests, the structure of all tetracyclines contains four rings (Fig. 6.7). Tetracyclines are active against a wide range of infections with minimal side effects. Tetracyclines, except minocycline, are known to bind to food and chelate metal ions such as calcium, magnesium, aluminum, and iron which prevents its absorption from the digestive system and so should not be administered with food. Oxytetracycline is a broad-spectrum antibiotic sold by Pfizer under the brand name Terramycin and has better absorption property than tetracycline. Doxycycline is another derivative of oxytetracycline that is widely used today since it has better pharmacological properties. Tigecycline (trade name Tygacil), approved by the FDA in the USA in 2005, belongs to the glycylcycline class of antibiotics. As shown in Fig. 6.7, it is a 9-*t*-butylglycylamido derivative of minocycline and has increased potency. One of the latest and popular tetracycline analogs, Omadacycline, is an aminomethylcycline. It was approved by FDA in 2018 and has a broad spectrum of activity, including against MRSA. Another analog, Sarecycline (trade name Seysara) was approved by the FDA in 2018 and specifically used for the treatment of acne due to its high selectivity against *Cutinebacterium acnes* [17].

All tetracyclines have the same mechanism of action. They inhibit protein synthesis by binding to the 30S subunit of the ribosome. Similar to the aminoglycosides, tetracyclines inhibit the binding of amino-acyl tRNA to the A-site of the ribosome. The 7S ribosomal protein is part of the binding site. A highly conserved region of 16S rRNA may also be part of the binding site, which explains the broad spectrum of activity of tetracycline [18]. Tigecycline also binds at the A-site of the 30S subunit of the ribosome and inhibits bacterial protein synthesis with potency 3- and 20-fold greater than that of minocycline and tetracycline, respectively [19]. Tetracyclines can bind equally well to both prokaryotic (70S) and eukaryotic (80S) ribosomes. However, tetracycline is still considered to be selective for bacteria because the antibiotic is actively transported (transported against a concentration gradient by using energy for the process, Sect. 1.7.8) through the bacterial plasma membrane resulting in high enough concentration inside the cell. Eukaryotic cells do not accumulate tetracycline in the cell, and so their ribosomes are not affected by the antibiotic at the dosage used. This makes tetracyclines selective against bacteria.

Resistance Development Since the different tetracyclines have the same mechanism of action, resistance usually develops to all tetracyclines simultaneously. As in case of most natural antibiotics, resistance to tetracyclines can be either by acquiring point mutations in the target, which is the 30S subunit such that it binds poorly to the antibiotic, or by acquisition of genes coding for enzymes that confer resistance to antibiotic. While point mutations provide weak resistance (Sect. 2.4), the resistance genes are more common and confer strong resistance to tetracyclines. There are many tetracycline resistance genes reported in many bacteria, and based on their sequence similarities, they can be classified into many types that are named as Tet A, B, C, D, E, K, L, M, O, P, Q, and X. Based on their mechanisms of action these tetracycline resistance genes can be of three types: antibiotic efflux (Tet A-E, K, L, P), target protection (Tet M, O, Q), and antibiotic inactivation (Tet X) [18].

Tetracycline Efflux The most common method of resistance development against tetracycline is by lowering the concentration of the drug in the cell. This is achieved by acquiring a gene coding for a protein that acts as a pump to transport the antibiotic to the outside of the cell [20]. One such tetracycline resistance gene that is most studied is *tetA*, which codes for a 41 kDa transmembrane protein that transports the antibiotic through the membrane to the outside. Since even a small amount of tetracycline is pumped outside where the concentration of the antibiotic is much higher, this classifies as active transport (that is why it is called a pump) and requires energy for such transport. As explained before (Sect. 1.7.8), one possible source of energy for active transport can be an already existing gradient. In this case, it uses energy from a proton gradient that always exists in bacteria. The pH outside is lower (higher H^+ concentration) than inside. The transmembrane TetA protein transports tetracycline outside at the same time H^+ is transported into the cell. Thus the TetA protein is called a tetracycline/H^+ antiporter (Sect. 1.7.8). Similar to TetA, other tetracycline resistance proteins have also been reported, such as TetB, TetC, TetD, TetE, TetG, TetK, and TetL. All these proteins have similar structure, which consists of 12 transmembrane regions and five loops in the cytoplasm and five loops in the periplasm.

Ribosome Protection A less studied method of tetracycline resistance is by proteins that bind to the ribosome and thereby protect it from binding to tetracycline [21]. Examples of such proteins are TetM, TetO, and TetQ, which are found in a variety of bacteria. These proteins have sequence homology to the elongation factors, EF-G and EF-Tu, and like them, also have GTPase activity, suggesting that the ribosome protection proteins have probably evolved from bacterial elongation factors [18]. TetM and TetO have been shown to be able to dislodge tetracycline that is bound to ribosome, a process that requires GTP. Binding of tRNA that is normally inhibited by tetracycline is protected in the presence of TetM [22]. The TetM or TetO protein specifically binds to tetracycline-blocked ribosomes but does not bind during normal protein synthesis [23].

Inactivation of Tetracycline Genes that code for enzymes that inactivate tetracycline by modifying it have been reported but are not very common. The enzyme, TetX modifies the antibiotic in the presence of oxygen and NADPH [24]. TetX enzyme binds stoichiometric amount of flavin adenine dinucleotide (FAD). Mechanism of inactivation of tetracycline by this flavin-dependent monooxygenase in the presence of oxygen and NADPH has been reported [25].

6.2.5   *Chloramphenicol*

Chloramphenicol (Fig. 6.8), which was previously known as chloromycetin was the first broad-spectrum antibiotic developed and was isolated from *Streptomyces venezuelae,* a soil bacterium (Sect. 1.5). Chloramphenicol has several advantages as an antibiotic. It can be easily synthesized and is available at a very low cost, which is one of the main advantages of the antibiotic. It has a very broad range of activity against most bacteria, including anaerobes. One big advantage of the drug is that it can easily penetrate all tissues, including the cerebrospinal fluid (CSF), and so can be used to treat meningitis. It is also one of the very few antibiotics that can enter human cells and so can be used against intracellular bacteria. However, in spite of the many advantages of the drug, chloramphenicol is not used in the USA due to its side effects, including aplastic anemia, which can sometimes be fatal. Limited use of chloramphenicol in the developed countries has had one advantage: many clinical isolates of pathogenic bacteria are still sensitive to chloramphenicol [26]. Recently there has been renewed interest in the antibiotic particularly because of the growing problem of resistance development to other commonly used antibiotics.

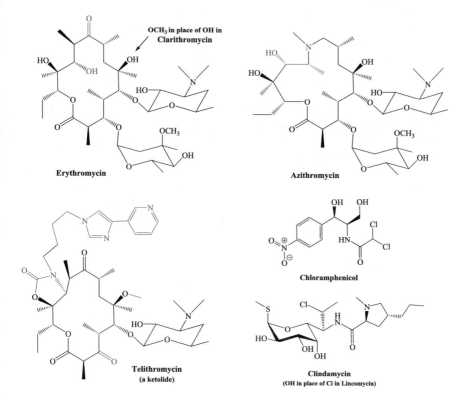

Fig. 6.8 Antibiotics that bind to 50S subunit of the ribosome. Regions of difference are highlighted in red

Mechanism of Action of Chloramphenicol   Chloramphenicol functions by binding to the 50S subunit of the ribosome and inhibiting protein synthesis. Note that this way, it is different from the aminoglycosides and tetracyclines, which bind to the 30S subunit of the ribosome. Since its binding is reversible, chloramphenicol is a bacteriostatic antibiotic (Sect. 1.6) and stops protein synthesis immediately without affecting DNA or RNA synthesis. So in the research laboratory, it is often used to increase the yield of DNA since it increases the DNA to protein ratio in a bacterial culture. Chloramphenicol specifically binds to bacterial ribosome but not to the human ribosome. However, it does inhibit protein synthesis in the mitochondria, which are believed to be derived from bacteria and so have ribosomes of size 70S, similar to that of bacteria, as opposed to the 80S ribosomes present in the cytoplasm. A possible consequence of this can be suppression of bone marrow function by inhibition of mitochondrial protein synthesis in bone marrow stem cells. Chloramphenicol functions by binding to the A-site of the 50S subunit of the ribosome and affects proper binding of the aminoacyl tRNA to the A-site and thus inhibits the peptidyl transferase activity.

Resistance to Chloramphenicol   Resistance to chloramphenicol can be by several mechanisms: by inactivation of the antibiotic, by target modification or by lowering the intracellular concentration of the drug by either decreased permeability into the cell or by efflux pumps that transport the antibiotic to the outside. The most common method of resistance development to chloramphenicol is by modification of the drug. This is carried out by the enzyme chloramphenicol acetyl transferase (CAT), the gene for which is usually on a plasmid and so can be easily spread to other bacteria (Sect. 2.6). The CAT enzyme is found in numerous species of bacteria and depending on their sequences, can be classified into several types and subtypes [23]. The enzyme catalyzes the transfer of an acetyl group from acetyl CoA to either of the two hydroxyl groups of the drug [27]. The acetylated chloramphenicol is unable to function as an antibiotic. As discussed in Sect. 1.7.6, acetyl CoA is a thioester and is a commonly used high energy compound in the cell for various reactions. Florfenicol is a semisynthetic derivative of chloramphenicol in which the hydroxyl group at C-3 is replaced by a fluorine atom, and the nitro group is also replaced with a sulfomethyl group. Replacement of the OH group prevents acetylation of the antibiotic, and so florfenicol is resistant to inactivation by CAT enzyme. Currently, florfenicol is approved for use in veterinary medicine only.

Another method of resistance to chloramphenicol is a specific efflux protein, CmlA, that is found in several species of bacteria and provides resistance to both chloramphenicol and florfenicol. Other less common methods for resistance include inactivation of the antibiotic by phosphotransferases, mutations of the target site, and permeability barriers [28].

6.2.6 The MLS Group of Antibiotics

6.2.6.1 Macrolides

Macrolides are large lactones that are bonded to one or more deoxy sugars. By definition, a lactone is a cyclic ester, and in macrolides, they are usually 14–16 carbon rings. Those with 14 atom rings, such as clarithromycin and erythromycin, are metabolized differently from those with 16 atom rings, such as spiramycin and leucomycin A3 (josamycin). Four macrolide antibiotics are currently available for use in the USA: erythromycin, clarithromycin, azithromycin, and telithromycin, of which erythromycin is the most commonly used and the first to be discovered. There are several other macrolides that are not used much. Erythromycin was obtained in 1952 from *Streptomyces erythreus,* while the others are semisynthetic derivatives of erythromycin. Azithromycin was made later in 1989. Although the ribosomes of both gram-positive and gram-negative organisms are susceptible to macrolides, these antibiotics are mainly used against gram-positive bacteria since they are unable to enter the porins of gram-negative bacteria [25]. Gram-positive bacteria can accumulate 100 times more of the antibiotic than the gram-negatives. As can be seen in Fig. 6.8, erythromycin contains an amino group, which exists in the protonated cationic form at neutral pH. This ionized form is less permeable to cells than the neutral form, explaining why the drugs are more active at alkaline pH. Most macrolides are destroyed by stomach acid and so are administered intravenously or with enteric coating (that is stable at the acidic pH of the stomach but dissolves at the basic pH of the small intestine). Clarithromycin and azithromycin are more active than erythromycin against several gram-negative bacteria as well as mycobacteria. Mycobacteria have an unusually thick and hydrophobic cell wall which acts as a permeability barrier for many antibiotics. The MIC of clarithromycin was 32 and 64 times lower than that of erythromycin for *M. smegmatis* and *M. avium,* respectively [29]. Azithromycin, but not erythromycin, is highly effective against the periodontal pathogen *Aggregatibacter actinomycetemcomitans,* and *Porphyromonas gingivalis,* which are difficult to eliminate by conventional therapy [30]. Since macrolides are easily transported into most cells, azithromycin can enter the gingival crevicular fluid in tooth pockets to treat periodontitis.

Mechanism of Action of Macrolides The mechanism of action of macrolides has been a matter of controversy for some time. Macrolides as well as lincosamides and streptogramins (known as the MLS group) inhibit bacterial growth by inhibiting protein synthesis. They bind to the 23S rRNA of the 50S subunit of the ribosome. This blocks the path by which growing peptides exit the ribosome resulting in the dissociation of peptidyl-tRNA from ribosomes during translocation. A study of macrolides, lincosamides, and streptogramins demonstrated that all of these MLS drugs cause dissociation of peptidyl-tRNA from the ribosome, however, they may cause dissociation of peptidyl-tRNA containing between two to ten amino acid residues depending on how closely they bind to the peptidyl transfer center, and thus on the space available between the peptidyl transferase center and the drug. The number of amino acids in the released peptide varies from two to four for lincosamides, six to eight for erythromycin and pristinamycin, and nine to ten for telithromycin [31]. Selectivity of macrolides is due to the fact that they do not bind to human ribosomes, which have a 40S and a 60S subunit, while bacterial ribosomes have a 30S and a 50S subunit. Binding of the macrolides to the 50S subunit is reversible, which explains the bacteriostatic nature of the antibiotics.

Resistance to Macrolides The most common resistance to macrolides is due to a methylase enzyme that methylates an adenine residue in a highly conserved region of the 23S ribosomal RNA, which is a peptidyl transferase ribozyme. The methylated ribozyme does not bind the macrolide, and so the peptidyl transferase activity is not inhibited. The same methylation can give rise to resistance to macrolides, lincosamides, and streptogramins type B (the MLS group) [32]. The gene for the methylase enzyme called *erm* (*e*rythromycin *r*esistance *m*ethylase) is also present in macrolide producer bacteria and provides them resistance to the drug. The methylase gene (*erm*) is present on a plasmid or transposon and thus, can be transferred to other bacteria. So resistance can spread easily. Other means of resistance, such as cell impermeability or drug inactivation, have been reported but are less common. The macrolides can be inactivated by erythromycin esterases which are encoded by *ereA* and *ereB* (*e*rythromycin *r*esistance *e*sterase) [33].

Telithromycin Telithromycin, the first ketolide to be approved for clinical use, is a semisynthetic derivative of erythromycin made by removing the neutral sugar, L-cladinose from C3 position of the macrolide ring and by oxidation of the 3-hydroxyl to a 3-keto functional group [34]. Telithromycin has better acid stability than erythromycin. It is acid stable and so can be taken orally. Because of its excellent penetration through the skin, it is used for a variety of skin infections. Telithromycin is active against gram-positive cocci that are resistant to erythromycin due to the presence of the methylase gene, *erm* [35]. Binding of telithromycin to the 50S ribosome is very tight because it binds at two sites compared to only one site for erythromycin [36]. Telithromycin was approved by the European Commission in 2001 and by the US FDA in 2004. However, in 2007 the FDA curtailed the use of telithromycin because of controversy regarding safety.

6.2.6.2 Lincosamides

The three types of antibiotics, macrolides, lincosamides, and streptogramins, are often grouped together (MLS group) because they have a similar mechanism of action and inhibit protein synthesis by binding to the 50S ribosome. The lincosamides include the antibiotics lincomycin and clindamycin. Lincomycin, which was isolated from *Streptomyces lincolnensis*, obtained from the soil in Lincoln, Nebraska, USA, was the first lincosamide to be discovered. Clindamycin which has a chlorine atom replacing a OH group of lincomycin (Fig. 6.8), shows better antibiotic properties. Resistance to lincosamide is conferred by the methylase enzyme (Erm), which gives cross-resistance to macrolides. Resistance to lincosamides is also provided by modification of the drugs by phosphorylation or adenylation (by LinB enzyme).

6.2.6.3   Streptogramins

Streptogramins can belong to either of two unrelated groups, A and B (Fig. 6.9) and are named after the soil bacteria, *Streptomyces graminofaciens*, from which they were first isolated in 1953, but it was not until 1998 that they were first used clinically. The group A streptogramins are polyunsaturated macrolactones, and the group B streptogramins are cyclic hexa- or hepta- depsipeptides, which are peptides in which one or more amide bonds are replaced by ester bonds (for another example of depsipeptide see Sect. 3.3.3.6). The two groups of streptogramins are made by the producer strains in a 70:30 ratio.

Mechanism of action of the two streptogramins is the same as that of lincosamides and macrolides. They bind to the 50S subunit of the bacterial ribosome and inhibit protein synthesis. Each of the two streptogramins, A or B, has only moderate activity and is bacteriostatic in nature. However, the combination of the two has a synergistic effect and shows 100-fold higher activity that results in a bactericidal effect

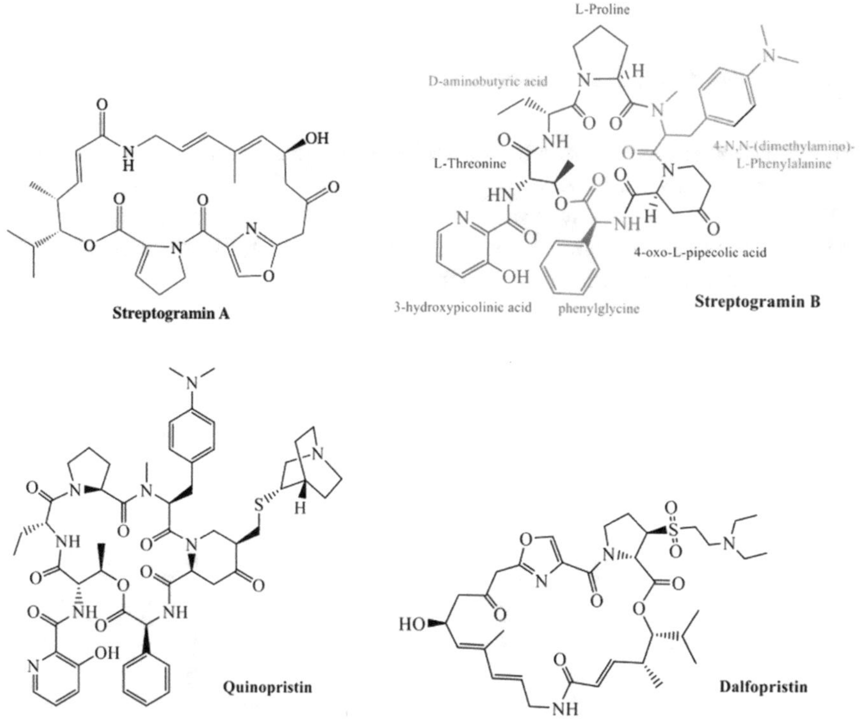

Fig. 6.9 Streptogramins

[37]. Two examples of streptogramins are pristinamycin and virginiamycin, of which, the latter is no longer used as an antibiotic but is used subtherapeutically in animals (Sect. 2.10.2). A popular streptogramin antibiotic that is used therapeutically is Synercid, which is a combination of quinupristin and dalfopristin (Fig. 6.9), two semisynthetic antibiotics derived from the natural streptogramin, pristinamycin. These derivatives were made to increase their solubility in water since the natural streptogramins are poorly soluble in water, which limits their clinical use. This combination also has a synergistic activity and a broader spectrum of activity than either antibiotic alone. Individually, the two streptogramins are bacteriostatic, while the combination is bactericidal. The combination Synercid is also effective against MRSA and VRE.

Resistance development against streptogramins is similar to that of macrolides. As mentioned above, post-transcriptional methylation of 23S ribosomal rRNA confers resistance to *m*acrolide, *l*incosamide and *s*treptogramin B-type antibiotics (MLS). One mechanism of resistance to streptogramin is the enzyme streptogramin B lyase (aka. Virginiamycin B lyase), which catalyzes the linearization of the cyclic antibiotic to make it inactive [38]. As has been discussed before (Sect. 2.10.2), resistance against Synercid had already developed even before it was introduced into the market. This was because of the rampant subtherapeutic use of the related antibiotic Virginiamycin.

6.2.7   Two Truly New Antibiotics

Very few new antibiotics have been discovered in the last few decades; most new antibiotics that have been developed have been derivatives of existing ones. Two truly new antibiotics that have been approved for human use in the last two decades are oxazolidinone and pleuromutilins. Structures of Linezolid, an oxazolidinone and Lefamulin, a pleuromutilin, are shown in Fig. 6.10.

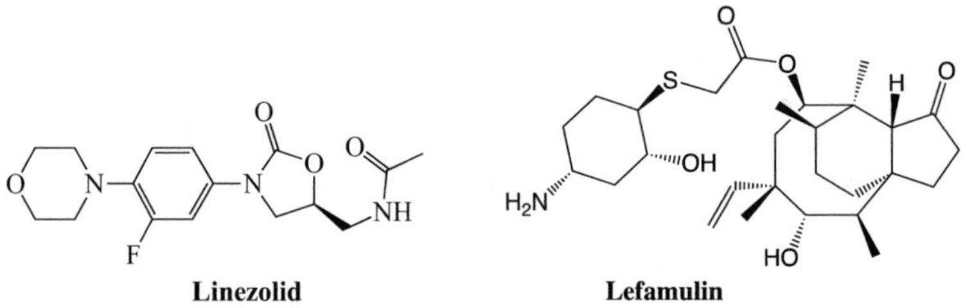

Linezolid   **Lefamulin**

Fig. 6.10 Linezolid and Lefamulin

6.2.7.1 Oxazolidinones

A novel agent to be introduced as antibiotic in the last 20 years is the oxazolidinone linezolid, which was discovered in the 1990s and approved for clinical use in 2000. Oxazolidinones have been known since the 1950s as monoamine oxidase inhibitors. Their antibiotic property with a new mechanism of action was first reported in 1987 [39]. Although other oxazolidinones have been studied, linezolid is the most clinically useful oxazolidinone antibiotic [40]. Linezolid is currently a very expensive but very effective antibiotic and is used as an antibiotic of last resort when others have failed. Although it is an expensive drug, one great advantage of linezolid is that it can be administered orally and is easily absorbed into the blood. It is used mostly against gram-positive bacteria. It can penetrate into the cerebrospinal fluid and so is effective against MRSA infection of the cerebrospinal fluid [41].

Mechanism of Action Linezolid functions by binding to the 23S RNA of the 50S subunit of the ribosome and preventing the initiation of protein synthesis. Although linezolid binds at a site that is close to the binding site of chloramphenicol and macrolides, its mechanism of action is different because it interferes with the formation of the initiation complex [42, 43].

Resistance Development Since linezolid is a synthetic antibiotic, resistance development to it is rare (Sect. 2.8). Moreover, since the mechanism of action of linezolid is different from that of other protein synthesis inhibitors, including those that bind close to it in the 50S subunit of ribosome, there is no cross-resistance between linezolid and other antibiotics such as chloramphenicol and lincosamides. Gram-negative bacteria are intrinsically resistant to linezolid because of the presence of efflux pumps that pump out the drug [43]. Resistance development in gram-negative bacteria is due to point mutations in the 23S RNA, which inhibit the binding of the linezolid [44].

6.2.7.2 Pleuromutilins

Pleuromutilins are antibiotics produced by the fungus *Pleurotus mutilus*. Antibiotic activity of pleuromutilin against *S. aureus* was first reported in 1951. Tiamulin was the first semisynthetic derivative of pleuromutilin that was approved in 1979 for use in veterinary medicine. The next pleuromutilin to be approved for use in animals was valnemulin in 1999. In 2007, retapamulin was the first pleuromutilin to be approved for topical use in humans. Toxicity of pleuromutilins prevented their approval for systemic use in humans. However, the present crisis resulting from increased antimicrobial resistance (AMR) against most antibiotics has renewed the interest in pleuromutilins. This led to the synthesis of lefamulin in 2006 [45]. Lefamulin (Fig. 6.10), a semisynthetic derivative, was the first pleuromutilin that was successfully tested in clinical trial for systemic use in patients [46]. In 2019, lefamulin was approved by FDA for both oral and intravenous use in humans to treat community-acquired bacterial pneumonia (CABP).

Mechanism of Action   Lefamulin shows both bacteriostatic and bactericidal activity against both gram-positive and gram-negative bacteria. It shows activity against both typical and atypical CABP pathogens, including some anaerobes. It has a very unique mechanism of action. It binds to the peptidyl transferase center (PTC) of the bacterial 50S ribosome, thereby preventing the peptidyl transfer reaction between the tRNAs at the P-site and A-site, thus stopping bacterial protein synthesis.

Resistance Development   Although tiamulin and valnemulin have been in use for several decades in the treatment of animals, neither was approved for subtherapeutic use (Sect. 2.10.2) in animals. This explains why no widespread resistance has developed against the antibiotic, and cross-resistance to lefamulin, the new drug for humans, has also not been observed. Pleuromutilins have the unique mechanism of action of inhibiting peptidyl transfer reaction, which no other commercial antibiotic has. This accounts for the lack of cross-resistance to other antibiotics that function by inhibiting protein synthesis, as described above.

6.2.8   Protein Synthesis Antibiotics with Unusual Mechanisms of Action

6.2.8.1   Thermorubin

The antibiotic thermorubin isolated from thermophilic actinomycete, *Thermoactinomyces antibioticus*, inhibits protein synthesis in both gram-positive and gram-negative bacteria but is inactive against fungi and higher eukaryotes. Its mechanism of action is different from other protein synthesis inhibitors in its binding to the ribosome [47]. Its structure is similar to that of tetracycline but does not contain any chiral center (Fig. 6.11). It binds weakly to both the 30S and 50S subunit individually but binds to the full 70S ribosome about 100 times more tightly, suggesting that the high-affinity binding site is formed only when the two subunits are combined. After binding, it results in a unique conformational change in the ribosomal RNA structure and thus inhibits the initiation stage of protein synthesis.

6.2.8.2   Fusidic Acid

Fusidic acid is a bacteriostatic natural antibiotic obtained from the fungus *Fusidium coccineum* and is effective against gram-positive but not gram-negative bacteria. The drug is approved for use in many countries but not in the USA, where it is used only topically in creams and eyedrops [48]. In 2015, the US FDA has granted a qualified infectious disease product (QIDP) designation to fusidic acid. The QIDP designation was created by the Generating Antibiotic Incentives Now (GAIN) Act of 2012,

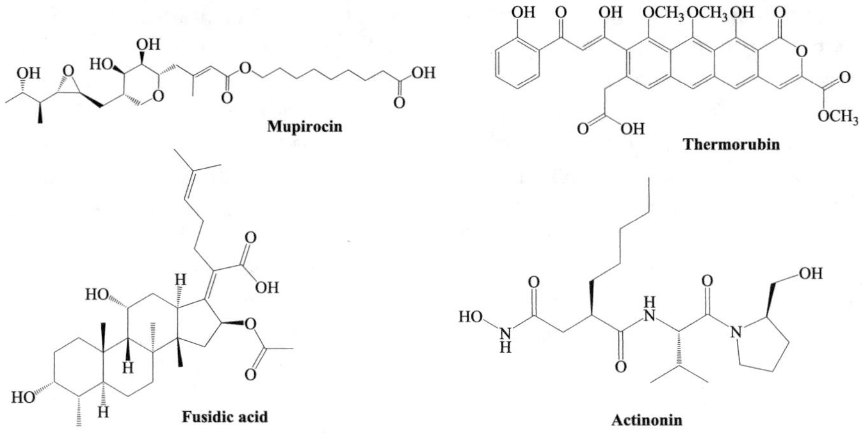

Fig. 6.11 Some protein synthesis antibiotics with unusual mechanisms

which provides certain incentives for the development of new anti-infectives. One advantage of fusidic acid is that it is effective against MRSA (Sect. 3.3.2.10). Fusidic acid is an unusual antibiotic in two ways. Firstly, it has a steroid-like structure, and secondly, its mechanism of action is different from all other antibiotics that inhibit protein synthesis. It does not bind to the ribosome or inhibit peptide bond formation or translocation. It inhibits protein synthesis by interfering with the function of the elongation factor EF-G. The elongation factor plays an essential role in transloca- tion, a process by which the ribosome moves along the mRNA during protein synthesis (Sect. 6.1, Fig. 6.5). EF-G is a GTPase and hydrolyzes GTP to provide energy for the translocation of the peptidyl-tRNA from the A-site to the P-site of the ribosome. In the presence of fusidic acid, EF-G is unable to leave the ribosome after GTP hydrolysis and thus blocks the next cycle in the protein synthesis [49].

Resistance to fusidic acid is mainly due to the development of point mutations in *fusA*, the gene coding for EF-G protein. Because of the mutations, the elongation factor is unable to bind to the antibiotic [50]. Since the drug has not been used much in the USA during the several decades of use in other countries, cases of bacteria resistant to the drug are much less in the USA.

6.2.8.3 Mupirocin

Mupirocin (previously called pseudomonic acid A) is a naturally occurring antibiotic that was first isolated from *Pseudomonas fluorescens* (Fig. 6.11). It was first approved by FDA for topical use as a cream in 1987. It is only used topically because of its rapid and extensive metabolism [51]. The drug is effective against skin infection by gram-positive bacteria, especially against *streptococci* and *staphylo- cocci*, including MRSA. An intranasal form of the drug was approved in 1995. Mupirocin is less effective against gram-negative bacteria, probably because of poor transport through the outer membrane of the bacteria.

Mupirocin has an unusual mechanism of action. It inhibits protein synthesis not by binding to the ribosome but by binding to the bacterial enzyme isoleucyl-tRNA synthetase. This enzyme catalyzes the formation of a covalent bond between the activated isoleucine amino acid and the respective tRNA for isoleucine (tRNA^{Ile}) (Fig. 6.2). Since no tRNA carrying isoleucine will be available, protein synthesis stops at a codon for isoleucine. The drug also has antifungal activity. It was shown to inhibit isoleucyl-tRNA synthetase from *Candida albicans* and thus has a mechanism of action similar to that in bacteria [52].

Since mupirocin has a unique mechanism of action that is not shared by any other antibiotic, there is no cross-resistance observed with any antibiotic. So there is less concern about developing resistance to mupirocin because of exposure to other antibiotics. Low-level resistance development to mupirocin can be due to mutations in the isoleucyl-tRNA synthetase enzyme [53]. High-level resistance development can be due to the acquisition of a gene for a different isoleucyl-tRNA synthetase enzyme that is not inhibited by the antibiotic [54].

6.2.8.4   Peptide Deformylase Inhibitors: Actinonin

Another reaction that takes place in bacteria but not in the host is the use of formylated methionine as the first amino acid in the sequence of nearly all proteins. The formyl methionine serves as a degradation signal at the N-termini of bacterial proteins [55]. After synthesis, proteins are further processed by removing the formyl group by the enzyme peptide deformylase (PDF), and in many cases, this is followed by the removal of the first methionine at the N-terminal end by the enzyme methionine aminopeptidase (Fig. 6.12). Since the aminopeptidase cannot hydrolyze the terminal methionine if it is *N*-blocked, deformylation is also a prerequisite for protein maturation. Both these enzymes are essential for bacterial growth. Apfel et al. have identified several inhibitors of peptidyl deformylase. However, they found that those compounds were only moderately active as antibacterials, although they were potent inhibitors of isolated enzymes. They also found that bacteria quickly developed resistance to the inhibitors [56]. The gene *def*, which codes for peptidyl deformylase, is present in all bacteria but has no mammalian counterpart. Thus, this reaction is unique for the bacteria and inhibiting the enzyme can be an effective strategy for curing all bacterial infections. Peptidyl deformylase is a metalloprotease. The metal ion present in the enzyme is iron, and the PDF inhibitors may function by chelating iron. Derivatives of hydroxamic acid and *N*-formylhydroxylamine with metal chelating activity have been tested for PDF inhibitor activity. Actinonin (Fig. 6.11), a naturally occurring hydroxamic acid pseudopeptide was found to be a potent inhibitor of PDF. The antibiotic binds extremely tightly to PDF with a dissociation constant in the less than nanomolar range [57]. Extensive screening of libraries of compounds has identified other potential antibiotics that inhibit the PDF enzyme. Unfortunately, one such derivative of hydroxamic acid was found to be a potent PDF inhibitor with an IC50 also in the low nanomolar range but did not exhibit antibacterial activity. Possible explanations for the lack of antibacterial

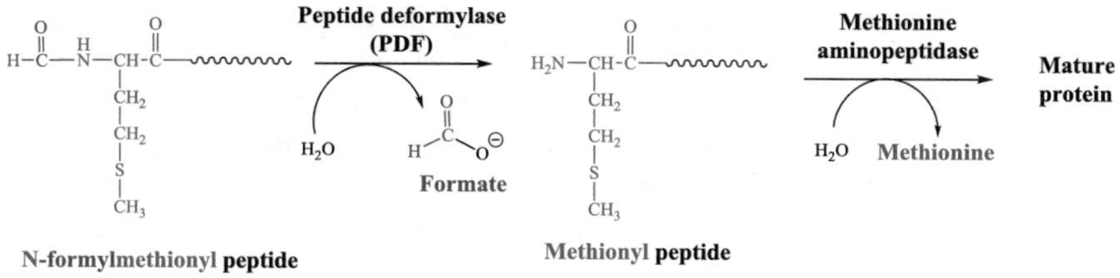

Fig. 6.12 Potential targets for the development of new antibiotics

activity may be poor membrane permeability into the bacterial cell or active efflux of the drug from the cell [58]. Similar screening of a library of compounds has identified an *N*-formylhydroxylamine derivative, BB-3497 [59]. Both actinonin and BB-3497 show activity against gram-positive bacteria including MRSA and VRE and also against gram-negative bacteria. Both antibiotics are bacteriostatic.

Since peptidyl deformylase is a unique target, it holds great promise for developing effective antibiotics. Since there is no such antibiotic already in use, it can be expected that there is no resistance gene already prevalent in nature. Also, since there is no mammalian counterpart of the enzyme, any antibiotic targeting the enzyme will be highly selective. However, in spite of all these advantages, no PDF inhibitor has been approved yet for clinical use.

6.2.8.5 Methionine Aminopeptidase Inhibitors

As mentioned above (Sect. 6.2.8.4), after deformylation of the *N*-terminal methionine of nascent proteins synthesized in bacteria, the methionine is then removed by the enzyme methionine aminopeptidase (abbreviated as Map or MetAP). The deformylated methionine is specifically removed if the next amino acid in the sequence is small and uncharged. All prokaryotes contain at least one copy of the *map* gene that is essential for the survival of the bacteria. *M. tuberculosis* contains two genes *mapA* and *mapB*. The methionine residue is hydrolyzed in 50–70% of proteins in all bacteria. Such activity is required for proper sub-cellular localization of the proteins as well as other essential functions [60]. Thus inhibitors of the enzyme can serve as potential antibiotics. The MetAP enzyme requires a divalent metal ion for activity. The identity of the metal ion in vivo is not known, however, there is indication that Fe(II) is likely the metal. Some catechol-containing compounds that inhibited MetAP were discovered by high throughput screening of a library of compounds [61]. The inhibitor probably functions by chelating the metal ion at the active site.

Methionine aminopeptidase is present not just in bacteria but in all life forms. In fact eukaryotes, including humans, express two MetAP enzymes, MetAP1 and MetAP2, which are required for cell proliferation, tissue repair, and protein degradation. Cells can still survive if one of these genes is deleted. Since the single gene present in prokaryotes is essential for their survival, this can be an effective target for new antibiotic development. In humans, MetAP2 has been shown to be involved in endothelial cell proliferation and thus is a potential target for anticancer drug development. MetAP-2 was identified as the target of the antiangiogenic natural product fumagillin and its drug candidate analog, TNP-470 [62]. Although originally developed as anticancer agents, methionine aminopeptidase (MetAP2) inhibitors were also found to cause weight reduction [63]. One such inhibitor, Beloranib, had progressed up to the third phase of clinical trials when it encountered some safety issues.

References

1. Ban N, Nissen P, Hansen J, Moore PB, Steitz TA (2000) The complete atomic structure of the large ribosomal subunit at 2.4 Å resolution. Science 289:905–920
2. Wimberly BT, Brodersen DE, Clemons WM Jr, Morgan Warren RJ, Carter AP, Vonrhein C, Hartsch T, Ramakrishnan V (2000) Structure of the 30S ribosomal subunit. Nature 407:327–339
3. Yonath A (2005) Antibiotics targeting ribosomes: resistance, selectivity, synergism, and cellular regulation. Annu Rev Biochem 74:649–679. https://doi.org/10.1146/annurev.biochem.74.082803.133130
4. Chopra S, Reader J (2015) tRNAs as antibiotic targets. Int J Mol Sci 16:321–349
5. Davis BD, Chen L, Tai PC (1986) Misread protein creates membrane channels: an essential step in the bactericidal action of aminoglycosides. Proc Natl Acad Sci U S A 83:6164–6168
6. Watve M, Tickoo R, Jog M, Bhole B (2001) How many antibiotics are produced by the genus streptomyces? Arch Microbiol 176:386–390
7. Singh RK, Tiwari SP, Rai AK, Mohapatra TM (2011) Cyanobacteria: an emerging source for drug discovery. J Antibiot 64:401–412
8. Kotra LP, Haddad J, Mobashery S (2000) Aminoglycosides: perspectives on mechanisms of action and resistance and strategies to counter resistance. Antimicrob Agents Chemother 44:3249–3256
9. Wang Y, Rando RR (1995) Specific binding of aminoglycoside antibiotics to RNA. Chem Biol 2:281–290
10. Hong W, Zeng J, Xie J (2014) Antibiotic drugs targeting bacterial RNAs. Acta Pharm Sin B 4:258–265
11. Davis BD (1987) Mechanism of bactericidal action of aminoglycosides. Microbiol Rev 51:341–350
12. Alksne LE, Anthonyt RA, Liebmant SW, Warner JR (1993) An accuracy center in the ribosome conserved over 2 billion years. Proc Natl Acad Sci U S A 90:9538–9541
13. Springer B, Kidan YG, Prammananan T, Ellrott K, Bottger EC, Sander P (2001) Mechanisms of streptomycin resistance: selection of mutations in the 16S rRNA gene conferring resistance. Antimicrob Agents Chemother 45:2877–2884
14. Hollingshead S, Vapnek D (1985) Nucleotide sequence analysis of a gene encoding a streptomycin/spectinomycin adenyltransferase. Plasmid 13:17–30
15. Sandvang D (1999) Novel streptomycin and spectinomycin resistance gene as a gene cassette within a class 1 integron isolated from Escherichia coli. Antimicrob Agents Chemother 43:3036–3038
16. Alangaden GJ, Kreiswirth BN, Aouad A, Khetarpal M, Igno FR, Moghazeh SL, Manavathu EK, Lerner SA (1998) Mechanism of resistance to amikacin and kanamycin in Mycobacterium Tuberculosis. Antimicrob Agents Chemother 42:1295–1297

17. Rusu A, Buta EL (2021) The development of third-generation tetracycline antibiotics and new perspectives. Pharmaceutics 13:2085. https://doi.org/10.3390/pharmaceutics13122085
18. Speer BS, Shoemaker NB, Salyers AA (1992) Bacterial resistance to tetracycline: mechanisms, transfer, and clinical significance. Clin Microbiol Rev 5:387–399
19. Olson MW, Ruzin A, Feyfant E, Rush TS III, O'Connell J, Bradford PA (2006) Functional, biophysical, and structural bases for antibacterial activity of tigecycline. Antimicrob Agents Chemother 50:2156–2166
20. McMurry L, Petrucci RE, Levy SB (1980) Active efflux of tetracycline encoded by four genetically different tetracycline resistance determinants in Escherichia coli. Proc Natl Acad Sci U S A 77:3974–3977
21. Burdett V (1991) Purification and characterization of Tet(M), a protein that renders ribosomes resistant to tetracycline. J Biol Chem 266:2872–2877
22. Burdett V (1996) Tet(M)-promoted release of tetracycline from ribosomes is GTP dependent. J Bacteriol 178:3246–3251
23. Connell SR, Tracz DM, Nierhaus KH, Taylor DE (2003) Ribosomal protection proteins and their mechanism of tetracycline resistance. Antimicrob Agents Chemother 47:3675–3681
24. Speer BS, Salyers AA (1989) Novel aerobic tetracycline resistance gene that chemically modifies tetracycline. J Bacteriol 171:148–153
25. Yang W, Moore IF, Koteva KP, Bareich DC, Hughes DW, Wright GD (2004) TetX is a flavindependent monooxygenase conferring resistance to tetracycline antibiotics. J Biol Chem 279:52346–52352
26. Fernández M, Conde S, de la Torre J, Molina-Santiago C, Ramos JL, Duque E (2012) Mechanisms of resistance to chloramphenicol in Pseudomonas putida KT2440. Antimicrob Agents Chemother 56:1001–1009
27. Murray IA, Shaw WV (1997) O-acetyltransferases for chloramphenicol and other natural products. Antimicrob Agents Chemother 41:1–6
28. Schwarz S, Kehrenberg C, Doublet B, Cloeckaert A (2004) Molecular basis of bacterial resistance to chloramphenicol and florfenicol. FEMS Microbiol Rev 28:519–542
29. Doucet PF, Capobianco JO, Jarlier V, Goldman RC (1998) Molecular basis of clarithromycin activity against Mycobacterium avium and Mycobacterium smegmatis. J Antimicrob Chemother 41:179–187
30. Lai PC, Walters JD (2013) Azithromycin kills invasive Aggregatibacter actinomycetemcomitans in gingival epithelial cells. Antimicrob Agents Chemother 57:1347–1351
31. Tenson T, Lovmar M, Ehrenberg M (2003) The mechanism of a of macrolides, Lincosamides and Streptogramin B reveals the nascent peptide exit path in the ribosome. J Mol Biol 330:1005–1014
32. Brisson-Noel A, Trieu-Cuot P, Courvalin P (1988) Mechanism of action of spiramycin and other macrolides. J Antimicrob Chemother 22(Suppl B):13–23
33. Arthur M, Brisson-Noel A, Courvalin P (1987) Origin and evolution of genes specifying resistance to macrolide, lincosamide and streptogramin antibiotics: data and hypotheses. J Antimicrob Chemother 20:783–802
34. Scheinfeld N (2004) Telithromycin: a brief review of a new ketolide antibiotic. J Drugs Dermatol 3:409–413
35. Ackermann G, Rodloff AC (2003) Drugs of the 21st century: telithromycin (HMR 3647)—the first Ketolide. J Antimicrob Chemother 51:497–511
36. Kostopoulou ON, Petropoulos AD, Dinos GP, Choli-Papadopoulou T, Kalpaxis DL (2012) Investigating the entire course of telithromycin binding to Escherichia coli ribosomes. Nucleic Acids Res 40:5078–5087
37. Mast Y, Wohlleben W (2014) Streptogramins—two are better than one! Int J Med Microbiol 304:44–50

38. Korczynska M, Mukhtar TA, Wright GD, Berghuis AM (2007) Structural basis for streptogramin B resistance in *Staphylococcus aureus* by virginiamycin B lyase. Proc Natl Acad Sci U S A 104:10388–10393

39. Slee AM, Wuonola MA, McRipley RJ, Zajac I, Zawada MJ, Bartholomew PT, Gregory WA, Forbes M (1987) Oxazolidinones, a new class of synthetic antibacterial agents: in vitro and in vivo activities of DuP 105 and DuP 721. Antimicrob Agents Chemother 31:1791–1797

40. Livermore DM, Mushtaq S, Warner M, Woodford N (2009) Activity of oxazolidinone TR-700 against linezolid-susceptible and -resistant staphylococci and enterococci. J Antimicrob Chemother 63:713–715

41. Naesens R, Ronsyn M, Druwé P, Denis O, Ieven M, Jeurissen A (2009) Central nervous system invasion by community-acquired methicillin-resistant Staphylococcus aureus: case report and review of the literature. J Med Microbiol 58:1247–1251

42. Shinabarger DL, Marotti KR, Murray RW, Lin AH, Melchior EP, Swaney SM, Dunyak DS, Demyan WF, Buysse JM (1997) Mechanism of action of oxazolidinones: effects of linezolid and eperezolid on translation reactions. Antimicrob Agents Chemother 41:2132–2136

43. Schumacher A, Trittler R, Bohnert JA, Kummerer K, Pages J-M, Kern WV (2007) Intracellular accumulation of linezolid in *Escherichia coli, Citrobacter freundii* and *Enterobacter aerogenes*: role of enhanced efflux pump activity and inactivation. J Antimicrob Chemother 59:1261–1264

44. Besier S, Ludwig A, Zander J, Brade V, Wichelhaus TA (2008) Linezolid resistance in Staphylococcus aureus: gene dosage effect, stability, fitness costs, and cross-resistances. Antimicrob Agents Chemother 52:1570–1572

45. Watkins RR, File TM (2020) Lefamulin: a novel semisynthetic pleuromutilin antibiotic for community-acquired bacterial pneumonia. Clin Infect Dis 71:2757–2762. https://doi.org/10.1093/cid/ciaa336

46. Novak R (2011) Are pleuromutilin antibiotics finally fit for human use? Ann N Y Acad Sci 1241:71–81I

47. Bulkley D, Johnson F, Steitz TA (2012) The antibiotic thermorubin inhibits protein synthesis by binding to inter-subunit bridge B2a of the ribosome. J Mol Biol 416:571–578

48. Fernandes P, Pereira D (2011) Efforts to support the development of fusidic acid in the United States. Clin Infect Dis 52(suppl 7):S542–S546

49. Martemyanov KA, Liljas A, Yarunin AS, Gudkov AT (2001) Mutations in the G-domain of elongation factor G from *Thermus thermophilus* affect both its interaction with GTP and fusidic acid. J Biol Chem 276:28774–28778

50. Besier S, Ludwig A, Brade V, Wichelhaus TA (2003) Molecular analysis of fusidic acid resistance in Staphylococcus aureus. Mol Microbiol 47:463–469

51. Parenti MA, Hatfield SM, Leyden JJ (1987) Mupirocin: a topical antibiotic with a unique structure and mechanism of action. Clin Pharm 6:761–770

52. Nicholas RO, Berry V, Hunter PA, Kelly JA (1999) The antifungal activity of mupirocin. J Antimicrob Chemother 43:579–582

53. Yanagisawa T, Lee JT, Wu HC, Kawakami M (1994) Relationship of protein structure of isoleucyl-tRNA synthetase with pseudomonic acid resistance of *Escherichia coli*. A proposed mode of action of pseudomonic acid as an inhibitor of isoleucyl-tRNA synthetase. J Biol Chem 269:24304–24309

54. Piatkov KI, Vu TTM, Hwang CS, Varshavsky A (2015) Formyl-methionine as a degradation signal at the N-termini of bacterial proteins. Microbial Cell 2:10

55. Apfel CM, Locher H, Evers S, Takacs B, Hubschwerlen C, Pirson W, Page MGP, Keck W (2001) Peptide deformylase as an antibacterial drug target: target validation and resistance development. Antimicrob Agents Chemother 45:1058–1064. https://doi.org/10.1128/AAC.45.4.1058-1064.2001

56. Gilbart J, Perry CR, Slocombe B (1993) High-level mupirocin resistance in Staphylococcus aureus: evidence for two distinct isoleucyl-tRNA synthetases. Antimicrob Agents Chemother 37:32–38

57. Chen DZ, Patel DV, Hackbarth CJ, Wang W, Dreyer G, Young DC, Margolis PS, Wu C, Ni ZJ, Trias J, White RJ, Yuan ZY (2000) Actinonin, a naturally occurring antibacterial agent, is a potent deformylase inhibitor. Biochemistry 39:1256–1262

58. Thorarensen A, Douglas MR Jr, Rohrer DC, Vosters AF, Yem AW, Marshall VD, Lynn JC, Bohanon MJ, Tomich PK, Zurenko GE, Sweeney MT, Jensen RM, Nielsen JW, Seest EP, Dolak LA (2001) Identification of novel potent hydroxamic acid inhibitors of peptidyl deformylase and the importance of the hydroxamic acid functionality on inhibition. Bioorg Med Chem Lett 11:1355–1358

59. Clements JM, Beckett RP, Brown A, Catlin G, Lobell M, Palan S, Thomas W, Whittaker M, Wood S, Salama S, Baker PJ, Rodgers HF, Barynin V, Rice DW, Hunter MG (2001) Antibiotic activity and characterization of BB-3497, a novel peptide deformylase inhibitor. Antimicrob Agents Chemother 45:563–570

60. Kanudia P, Mittal M, Kumaran S, Chakraborti PK (2011) Amino-terminal extension present in the methionine aminopeptidase type 1c of *Mycobacterium tuberculosis* is indispensable for its activity. BMC Biochem 12:35. https://doi.org/10.1186/1471-2091-12-35

61. Wang WL, Chai SC, Huang M, He HZ, Hurley TD, Ye QZ (2008) Discovery of inhibitors of Escherichia coli methionine aminopeptidase with the Fe(II)-form selectivity and antibacterial activity. J Med Chem 51:6110–6120

62. Zhang Y, Yeh JR, Mara A, Ju R, Hines JF, Cirone P, Griesbach HL, Schneider I, Slusarski DC, Holley SA, Crews CM (2006) A chemical and genetic approach to the mode of action of fumagillin. Chem Biol 13:1001–1009

63. Joharapurkar AA, Dhanesha NA, Jain MR (2014) Inhibition of the methionine aminopeptidase 2 enzyme for the treatment of obesity. Diabetes Metab Syndr Obes 7:73–84

Chapter 7
Antibiotics That Affect the Membrane and Other Structural Targets

Abstract Antibiotics as well as antiseptics and disinfectants that affect the bacterial cell membrane and other structural targets are discussed. Background biochemistry information on lipids and membranes is provided. The antibiotics include gramicidin, tyrocidine, polymyxin, daptomycin, bacteriocins, lantibiotics, triclosan, isoniazid, magainin, and defensin. Inhibitors targeting the cell division protein, FtsZ is also included. Mechanisms of action of the antibiotics and resistance development against the antibiotics are discussed.

7.1 Background Biochemistry Information

7.1.1 Function of Biological Membranes

All biological cells are surrounded by at least one membrane. In addition, within eukaryotic cells there are compartments (known as organelles) that are also surrounded by membranes. Thus, a membrane is an essential part of all living species. The cytoplasm of all species is enclosed by a cell membrane, known as the cytoplasmic membrane. Bacteria can be classified into two broad classes depending on the staining property of their cell walls: gram-positive and gram-negative. Gram-positive bacteria contain one cell membrane that is surrounding by a thick cell wall, while Gram-negative bacteria contain two membranes; an inner cell membrane and an outer membrane with a thin cell wall present between the two membranes (Fig. 1.3). Membranes have many functions. One main function is to define the boundary of the cells and of organelles within eukaryotic cells. However, the membrane does not merely act as a diffusion barrier for all molecules between the outside and inside of the cell. It selectively transports molecules from outside to inside (for example, nutrients) and from inside to outside (for example, waste products). Membranes also contain protein receptors that bind to extracellular signals known as first messenger (for example, hormones) and create a response signal known a second messenger inside the cell. Another function of the membrane is to maintain a concentration gradient of various molecules and ions between the two sides of the membrane. This is achieved by pumping the molecule or ion through the membrane from one side to the other. These gradients serve as energy that is used to selectively transport various molecules or ions through the membrane. One important ion gradient maintained by all bacteria is a proton gradient. Using energy from respiration (oxidation of carbohydrates, proteins, and lipids), protons are pumped through the membrane to the outside, thus creating a proton gradient that is always maintained. A gradient is equivalent to energy. The potential difference between the two sides of the cell membrane is called protonmotive force, which drives the synthesis of ATP. As the H$^+$ ions move back into the cell through an enzyme complex called ATP synthase, the energy of the gradient is used to synthesize ATP.

7.1.2 Composition of the Membrane

The two major components of cell membranes are phospholipids and proteins. Other minor components include carbohydrates (usually attached to the proteins) and steroids. Structure of phospholipids contains three important parts: fatty acids, glycerol, and phosphates. Fatty acid is defined as a carboxylic acid with a long hydrophobic chain (Fig. 7.1). Fatty acids are called saturated if the hydrophobic chain contains no carbon-carbon double bonds and are called unsaturated if they contain at least one carbon-carbon double bond. Because of the double bonds, unsaturated fatty acids have two potential stereoisomers, *cis* and *trans* at each

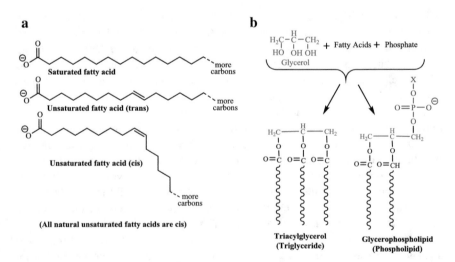

Fig. 7.1 Structures of some lipids. (**a**) Saturated and unsaturated (*cis* and *trans*) fatty acids. (**b**) Formation of triglycerides and phospholipids by condensation of glycerol, fatty acids, and various phosphates. X can represent any one of the following: hydrogen, choline, ethanolamine, serine, glycerol, phosphatidyl glycerol or myo-inositol, some of which can result in extra positive or negative charges

carbon-carbon double bond. Of these, only the cis isomers are found in nature. Because of the cis configuration unsaturated fatty acid chains contain a bend at each double bond (Fig. 7.1). This prevents close packing of the fatty acid molecules resulting in a lower melting point for the *cis* unsaturated fatty acids compared to either the saturated or the *trans* unsaturated fatty acids, which can pack efficiently and so have higher melting points. More the degree of unsaturation (number of double bonds), lower will be the melting point of the fatty acid.

At cellular pH (usually close to neutral), the carboxyl groups of fatty acids are ionized. Thus at body pH fatty acids will contain an anionic carboxylate group attached to a long hydrophobic chain, which is the typical structure of detergent molecules. Since detergents can disrupt biological membranes, fatty acids present as carboxylates are toxic. So free fatty acids are rarely present in cells. Majority of the lipids stored in the body are not as fatty acids but as triglycerides, which contain ester linkages between the three hydroxyl groups of glycerol and carboxyl groups of three fatty acid molecules. Triglycerides do not function as detergent and are more hydrophobic than fatty acids, a property that is important for efficient storage. Another class of lipid includes the phospholipids, which contain two fatty acid chains forming ester linkages to the two hydroxyl groups at C1 and C2 of glycerol, while the hydroxyl group at C3 of glycerol forms an ester linkage with phosphate. So phospholipids can be viewed as a molecule containing two long hydrophobic chains attached to a small ionic region. This structure is commonly referred to as a polar head with two nonpolar tails. The phosphate may form a second ester link to one of several other groups such as choline, serine, and ethanolamine (Fig. 7.1). Some of these have an extra positive charge making the polar head more polar. This resembles the structure of a typical detergent molecule such as sodium dodecyl sulfate, which has a polar head and one nonpolar tail. At low concentrations, detergent molecules are soluble in water as a result of the polar head. However, at higher concentrations, they form spherical clusters called micelles in order to keep the hydrophobic tails out of contact with water, a phenomenon known as the hydrophobic effect. Detergent molecules, because of a polar head and one nonpolar tail, have a conical shape. When many conical molecules cluster together, they form a spherical micellar structure in which the polar heads face the water and the hydrophobic tails interact with each other in the interior of the sphere. Phospholipids, with a polar head and two nonpolar tails, have a rectangular shape. When many rectangular molecules cluster together, they form a flat layer with all polar heads facing one side and all nonpolar tails facing the other side. In order to not have to interact with water, the hydrophobic side of the monolayer interacts with another monolayer of phospholipid to form a bilayer (Fig. 7.2). This is the structure of the cell membrane, which is a bilayer of phospholipid with the polar head facing aqueous environments on both sides: the cytoplasm inside the cell and the extracellular medium outside.

Besides phospholipids, other major components of the cell membrane are proteins and glycoproteins, which are necessary for transport of various chemicals through the membrane, catalyzing various metabolic reactions and cell to cell communication, especially in eukaryotes. Lipopolysaccharides (LPS), also known as endotoxins, constitute an essential and major component of the outer leaflet of the

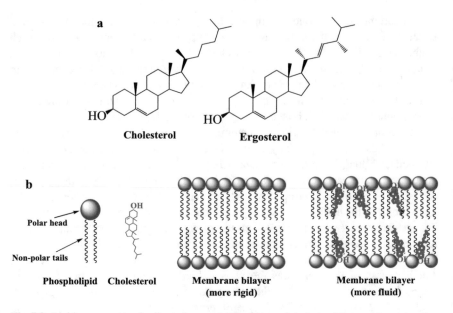

Fig. 7.2 Lipid components of cell membranes. (**a**) Structures of cholesterol (present in mammalian membranes) and ergosterol (present in fungal membranes) and (**b**) Effect of cholesterol on the fluidity of the mammalian cell membrane (ergosterol in fungal cell membrane will have a similar effect)

outer membrane of gram-negative bacteria. The function of the LPS is to stabilize the membrane structure and protect the bacteria from toxic compounds in the environments, including antibiotics. LPS of the infecting bacteria can elicit a strong immune response in the mammalian host and is the cause of septic shock (septicemia) in humans. One property of the membrane that is essential for carrying out its functions is the fluidity of the bilayer. For example, transport of molecules cannot take place through a rigid membrane. Also enzymes and other proteins present in the membrane constantly move horizontally along the membrane. Thus, fluidity of the membrane is essential for its proper functioning. Bacterial cells need to maintain the fluidity of their membranes as their surrounding environment changes from one temperature to another. They do this by adjusting the proportion of unsaturated phospholipids in their membranes. As the cells move from higher to lower surrounding temperature, they replace some of the phospholipids in the membrane with more unsaturated ones. As explained above, *cis* unsaturated fatty acids have lower melting points than saturated ones and thus are able to maintain membrane fluidity at low temperatures. Since humans maintain a fixed body temperature, it is not necessary to adjust the proportion of unsaturated phospholipids in the membrane. However, human cells need to maintain fluidity appropriate for the 37 °C body temperature. This is done by inserting cholesterol molecules between phospholipid molecules in the cell membrane (Fig. 7.2). Cholesterol is a steroid, which is defined as a molecule having a particular arrangement of four hydrophobic aliphatic rings, as shown in

Fig. 7.2. Cholesterol also has a polar head in the form of a hydroxyl group at one end of the molecule. Insertion of the cholesterol makes the phospholipid molecules more spread out, and so the fluidity of the membrane is increased. However, the function of cholesterol in the membrane is not as simple as mentioned above. Because of the four fused rings, the cholesterol molecule is actually more rigid than the phospholipid molecules and helps to make the membrane stronger. Cholesterol can be said to maintain the balance between fluidity and rigidity of the membrane. A detailed discussion of the effect of cholesterol is beyond the scope of this book. One can refer to several papers on this topic for more details [1–3]. Fungi maintain their membrane fluidity by a similar method in which steroid molecules are inserted between phospholipid molecules, however, the steroid is not cholesterol but ergosterol. Ergosterol has a similar structure as cholesterol, but the two are significantly different such that ergosterol is an effective target for development of antifungal drugs (Sect. 8.1.1).

7.2 Inhibition of Bacterial Membrane Function

Maintenance of the proper structure of the cell membrane is vital for survival of microorganisms. Any chemical agent or process that disrupts the membrane structure will be toxic for the cell and so can be used to kill the microorganisms. A complete disruption of the membrane is not needed to kill the cells. Even minor disruptions may make the membrane more permeable and thereby dissipate the various concentration gradients through the "holes" made in the membrane. Each of these gradients across the membrane serves some essential function; especially important is the proton gradient (Sect. 7.1.1). If the proton gradient is dissipated, the cells will be unable to make ATP, and thus it will have a bactericidal effect. Based on how these chemical agents are used, they can be classified either as antiseptics and disinfectants or as antibiotics.

7.2.1 Antiseptics and Disinfectants That Disrupt Microbial Cell Membrane

Both antiseptics and disinfectants are used to kill microorganisms. Disinfectants kill microorganisms on inanimate objects. The difference between an antiseptic and an antibiotic is matter of debate. Antiseptics can be defined as chemicals that are used to kill or inhibit the growth of microorganisms present on tissues, especially on the surface of the skin, while antibiotics are used to kill or inhibit growth of microorganisms in systemic infections as well as topical infections (Sect. 1.1). Some disinfectants and antiseptics can have mechanisms of action other than disruption of the cell membrane; however, those agents are not discussed here. A more detailed

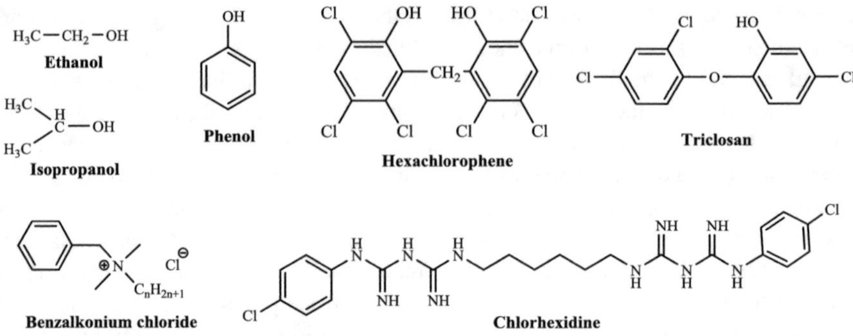

Fig. 7.3 Disinfectants that disrupt bacterial cell membrane

discussion of all antiseptics and disinfectants can be found in a review on the topic [4].

Alcohols, including ethanol and isopropanol, and n-propanol, are used both as disinfectants and as antiseptics and comprise the main active ingredients in hand sanitizers, a terminology that has now become a household name throughout the world as a result of the current Covid-19 pandemic. Alcohols function by disrupting hydrophobic interactions. Proper functioning of the membrane is highly dependent on hydrophobic interaction, which is an essential factor for maintaining the bilayer structure of the phospholipids, the native structure of the proteins in the membrane as well as the interactions of these proteins with the phospholipids. Since alcohols work by disrupting the membrane and proteins, they have broad spectrum of activity against all microorganisms such as bacteria (including mycobacteria), fungi as well as viruses. The antimicrobial activity of alcohols is best in the concentration range of 60–90%.

Phenol (earlier name carbolic acid) is one of the oldest known disinfectant and antiseptic. Today it is used mostly as a disinfectant. Phenol and other phenolic compounds function by denaturing membrane proteins by disrupting hydrophobic interaction and thus making the membrane leaky. Loss of membrane integrity kills the cells since the cytoplasmic contents are released. Phenolics also have antifungal and antiviral activities. Bis-phenols (two phenolic aromatic rings connected by a bridge) also have antimicrobial activity. One example is hexachlorophene (Fig. 7.3), which has bacteriostatic action against gram-positive bacteria and is used in medicated soaps. It inhibits respiration by inhibiting the electron transport chain in the membrane. At higher concentrations, it causes cell leakage. It has a broad spectrum of activity but is not used much due to its toxicity, which is a result of disruption of mammalian cell membrane. Another example of bis-phenol is triclosan, which functions by disrupting the cytoplasmic membrane. Triclosan has antibacterial activity, mostly against gram-positive bacteria. However, its activity against gram-negative bacteria as well as yeast can be enhanced in the presence of EDTA. The combination of triclosan and EDTA has been shown to increase outer membrane permeability [5]. Triclosan is widely used in surgical scrubs and is found in many health care products such as shampoos, deodorants, toothpastes, and mouthwashes. However, the concentrations of triclosan present in these products are not high enough to cause disruption of hydrophobic interaction in the membrane phospholipids and proteins. Since triclosan's mechanism of action at this low concentration involves inhibition of a specific metabolic step, this is discussed further along with other specific antibiotics (Sect. 7.3.1).

The most widely used antiseptic in health care products, particularly handwashes and mouthwashes, is chlorhexidine which is a biguanide (Fig. 7.3). It has a broad spectrum of activity and is bactericidal. Chlorhexidine kills cells quickly. It penetrates the outer membrane by passive diffusion and then disrupts the inner membrane causing leakage of cell components. After entering the cytoplasm, chlorhexidine can cause the cytoplasmic proteins to coagulate and stop the cell leakage; however, the cells will no longer be viable. The drug also shows activity against yeast but not against mycobacteria. It has activity against lipid-enveloped viruses but not against non-enveloped viruses [4].

Another class of compounds that functions as disinfectants and antiseptics includes detergents (Sect. 7.1.2). One such disinfectant is benzalkonium chloride (Fig. 7.3). It contains a quaternary ammonium ion and a long hydrophobic tail and thus can be described as a cationic detergent (aka surfactant). These detergents first penetrate the cell wall and then disrupt the cytoplasmic membrane causing leakage of cellular material and cell lysis [6]. In Gram-negative bacteria, the detergent first damages the outer membrane and then the inner membrane. Similar to chlorhexidine, benzalkonium ion also shows activity against lipid-enveloped viruses but not against non-enveloped ones.

Resistance to antiseptics and disinfectants is mostly intrinsic. Most bacterial spores are intrinsically resistant because the spore coats are impermeable to the antiseptics and disinfectants. Mycobacteria are intrinsically resistant to many antiseptics and disinfectants because their cell walls are impermeable to the chemical agents. Their cell walls are made of arabinogalactan esterified to mycolic acid, which makes the structure much different from the bacterial cell wall. Gram-negative bacteria are more resistant than gram-positive bacteria because the outer membrane serves as a permeability barrier to some of these chemical agents. Acquired resistance to antiseptics and disinfectants is rare, but several cases have been reported and have been reviewed extensively [4].

Although chlorhexidine is considered as an antiseptic, bacteria can develop resistance against it and so, it must not be used too frequently. Resistance happens due to the presence of a plasmid-borne *qac* (quaternary ammonium compound) gene family comprising of *qacA*, *qacB,* and *qacC* genes for a multidrug efflux pump [7]. Transcription of the genes is upregulated by chlorhexidine. The efflux pump may also work for other antiseptics and antibiotics and thus provide resistance against them. Moreover, since the genes are on a plasmid, there may be other resistance genes on the plasmid that will provide resistance of other antibiotics as well.

7.2.2   Antibiotics that Function by Disrupting Microbial Cell Membrane

Two major types of antibiotics that function by disrupting the cell membrane are the antimicrobial peptides and the polyenes. By definition, antibiotics are more selective against microorganisms compared to mammalian cells. However, since the membranes of all species have significant similarities, these antibiotics have some toxicity since mammalian cell membranes may also be affected. However, in spite of their poor selectivity, many membrane-acting antibiotics have been approved for therapeutic use.

7.2.2.1   Antimicrobial Peptides (AMPs)

There are several examples of small peptides functioning as antibiotics. These are known as antimicrobial peptides (AMPs), and their target is usually the bacterial membrane. One example of peptide antibiotic, bacitracin, which inhibits bacterial cell wall synthesis, has already been discussed before (Sect. 3.3.3.1). The AMPs (not to be confused with adenosine monophosphate) are 10–50 amino acid long, membrane-acting natural antibiotics that are produced by a variety of species, including mammals, arthropods, plants, and function as a defense mechanism against infecting pathogens. More than 1800 AMPs have been isolated from a wide range of organisms [8]. They are usually short, cationic peptides that function as broad-spectrum antibiotics and can even be effective against multidrug-resistant bacteria. In addition to antimicrobial activity, AMPs have other functions in the producing organism, such as chemokine production, angiogenesis, wound healing, and apoptosis. Facing a crisis-like situation due to the rapidly emerging antibiotic resistance, scientists have shown increased interest in AMPs. Development of resistance against these antibiotics is rare because their target site is the membrane and structure or composition of the membrane cannot be easily changed by simply acquiring mutations. The properties that make AMPs better antibiotics than others are (1) broad spectrum of activity, (2) rapid bactericidal activity, (3) extremely rare development of resistance, and (4) lack of cross-resistance with other antibiotics [9].

Specificity is an essential requirement of an effective antibiotic. To be an effective membrane-acting antibiotic, the AMPs must have selective effect on the membrane of the infecting microorganism and not on the mammalian host. Although membranes of all species contain phospholipids as the major component, there are fundamental differences between membranes of infecting microorganisms and the mammalian host, explaining the selectivity of the AMPs. The main differences are in the membrane composition and the transmembrane potential. The membrane is made of mostly phospholipids and proteins. As described above, one main difference between bacterial, fungal, and mammalian membranes is the presence or absence of sterols. Fungal membranes contain ergosterol, mammalian membranes contain cholesterol, while bacterial cells do not contain any sterol. Thus biosynthesis of ergosterol is a target for the development of specific antifungal antibiotics (Sect. 8.1.1). There can be variation in the charge of the polar heads of the cell membranes. Besides the negative charge on the oxygen of the phosphate, the polar head of some phospholipids such as cardiolipin and phosphatidyl serine have an extra negative charge. The lipopolysaccharide (LPS) part of the outer membrane of gram-negative bacteria and the teichoic or teichuronic acids in the cell wall of gram-positive bacteria also add to the negative charge of the outer membrane. On the other hand, mammalian membranes contain mostly neutral lipids. Phosphatidyl-choline and sphingomyelin present on the extracellular side of the bilayer are neutral phospholipids.

The difference in the total charges on the two sides of the cell membrane, resulting mainly from the extent of proton flux across the membrane, creates an electrochemical gradient, also known as a transmembrane potential, in which there are more negative charges inside the cell than outside. The transmembrane potential is different for microbial and mammalian cell membranes depending on the cell type. This difference is one of the reasons for the selective binding of the cationic AMPs to the microbial and not the host cell membranes.

Mechanism of Action of AMPs Mechanism of action of AMPs can include two or more of the following four steps [10]. (1) The AMP should first bind to the membrane. (2) The binding can lead to perturbation of the membrane structure and its function. (3) Binding of some AMPs to the membrane can disrupt membrane-associated events such as cell wall biosynthesis. (4) The AMP can be transported across the membrane to interact with its target in the cytoplasm and thereby inhibit some essential metabolic pathway in the microorganism. Structures of some AMPs have hydrophobic as well as hydrophilic regions. They are positively charged (cationic) at physiological pH and thus are able to bind to the polar head groups of the outer surface of the cytoplasmic membranes of microorganisms which are negatively charged as explained above (Sect. 7.1.2). Interaction of an AMP with the membrane depends on a combination of factors such as size, secondary structure, charge, and hydrophobicity of the AMP. These factors allow the AMP to bind preferentially to bacterial membranes that contain a large proportion of anionic lipids, lack cholesterol, and maintain a high electrochemical gradient across the membrane. Membranes of animal cells contain cholesterol, have lipids with no net charge, and have a weak transmembrane potential.

The cationic AMPs can be divided into three categories [11]. The first type, which includes magainin, is linear before interacting with the cell membrane, and then undergoes a conformational change to adopt an amphipathic α-helical secondary structure after binding to the membrane. The AMPs belonging to the second class are mostly linear and are enriched in proline and arginine. The third category includes cationic peptides such as defensins, which contain disulfide bonds and stable β-sheets. There are also some examples of non-cationic AMPs, such as anionic peptides and aromatic peptides; however, they usually have weak antimicrobial properties. The inherent and/or dynamic conformations of antimicrobial peptides also contribute to their selective toxicity. Furthermore, the AMP may undergo conformational change only when bound to the pathogen membrane but not to the host cell membrane, and the conformational change may be essential for their antimicrobial activity [12].

The mechanism by which AMPs cause cell death has been studied intensively. One way in which the AMPs can function is by disrupting the structure and composition of the membrane. Another mechanism of action is by forming pores in the membrane, which results in leakage of metabolites, depolarization, and eventual cell death. Originally it was believed that such increase of cell membrane permeability is the only mechanism by which AMPs work. But recent data suggest that in some cases, there can be cell death without any increase of membrane permeability, while in some other cases, microorganisms can survive for long period of time after membrane permeabilization, suggesting that the eventual cell death is due to reasons other than membrane disruption [12]. These other reasons may be disruption of intracellular processes such as peptidoglycan or other biopolymer syntheses [13]. Perturbation of the cell membrane can also have an effect on peptidoglycan precursor synthesis and translocation.

It has been proposed that membrane permeation by linear amphipathic alphahelical AMPs such as magainins, cecropins, and dermaseptins can take place by either of two mechanisms: (a) "barrel-stave" mechanism or (b) "carpet-like" mechanism [14, 15]. According to the "barrel-stave model" for the formation of pores, the AMP molecule first binds to the membrane surface as a monomer. This is then followed by joining more monomer units which then together form the pores in the membrane. More monomer units can be further recruited to increase the pore size [11]. The pores allow leakage of cytoplasmic content resulting in cell death. The structures of the pores are such that the hydrophobic amino acids interact with the membrane lipids while the hydrophilic amino acids face the inside of the pores, which allows for leakage of the ionic and polar contents of the cytoplasm [13]. According to the "carpet-like" model, the AMP binds to the polar heads of the membrane either as monomers or oligomers, which are then dissociated into monomers and cover the membrane in a carpet-like manner. The amphipathic peptides act as detergents and break up the membrane into micelle-like structures that are surrounded by the AMPs on the surface. For gram-negative bacteria, after the outer membrane is permeabilized this way, the AMP monomers can easily move to the inner membrane and disrupt it in the same manner. Although there are more than 1800 antimicrobial peptides, only some of the most commonly used AMPs are discussed further.

7.2.2.2 Gramicidins

As discussed before (Sect. 1.5), gramicidin, discovered in 1939, was the first clinically tested antibiotic [16, 17]. It was named gramicidin because it killed only gram-positive bacteria. As most other AMPs, gramicidin is only used topically because of toxicity caused by systemic use. A combination of gramicidin, neomycin, and polymyxin B is commonly used to treat eye infections. Naturally occurring

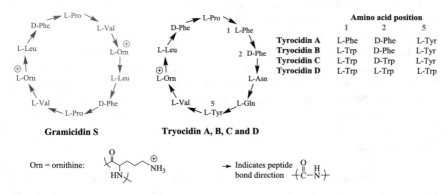

Gramicidin A, B and C
The eleventh amino acid is Trp or Phe or Tyr in Gramicidine A, B and C respectively.

	Amino acid position		
	1	2	5
Tyrocidin A	L-Phe	D-Phe	L-Tyr
Tryocidin B	L-Trp	D-Phe	L-Tyr
Tryocidin C	L-Trp	D-Trp	L-Tyr
Tryocidin D	L-Trp	L-Trp	L-Trp

Gramicidin S   **Tryocidin A, B, C and D**

Orn = ornithine: → Indicates peptide bond direction

Fig. 7.4   Structures of gramicidins and tyrocidins

gramicidin, produced by the soil bacterium, *Bacillus brevis* is a mixture of gramicidin A, B, and C, of which about 80% is gramicidin A. Gramicidins, which are biosynthesized by non-ribosomal enzymes, are linear pentadecapeptide antibiotics containing alternating L- and D-amino acids (Fig. 7.4). Gramicidins B and C differ from gramicidin A at position 11, where the Trp is replaced by Phe and Tyr, respectively. The primary structure of gramicidin A is: HCO-L-Val-Gly-L-Ala-D-Leu-L-Ala-D-Val-L-Val-D-Val-L-Trp-D-Leu-L-Trp-D-Leu-L-Trp-NHCH$_2$CH$_2$OH. Note that glycine is not chiral and so is neither D nor L. All other amino acids in gramicidin have hydrophobic side chains. It is one of the most hydrophobic peptides known, which agrees well with its membrane binding property. Note that both ends of the antibiotic are modified, which protects it from degradation by proteases that may be made by the host. The alternating D and L amino acids make all the peptide bonds in the molecule unusual peptide bonds that cannot be degraded by host proteases.

Mechanism of Action   The antibiotic activity of gramicidins is due to the formation of cation-selective ion channels in the membrane. After binding to the membrane, gramicidin forms a hydrogen-bonded head-to-head dimer and acquires a helical structure to form a channel that spans the membrane bilayer. Since this is the oldest known membrane-acting AMP, gramicidin has been used often in research and has helped us understand how ion channels function in membranes. The channel formed by gramicidin is specific for univalent cations only. The transport of ions and water throughout the length of the channel is by a single file process, which means cations and water molecules cannot pass each other within the channel [18]. Gramicidin A has four Trp residues, which are important for the channel formation due to the amphipathic character of Trp. The alternating L and D arrangement allows hydrophobic side chains of all amino acids to project outwards from the ion channel toward the hydrophobic region of the membrane while the polar peptide backbone faces the inside of the channel. The outer hydrophobic surface of the gramicidin dimer helps to interact with the hydrophobic phospholipids, while the polar inner surface allows monovalent cations to leak out. The structure function relationship of gramicidin has been explained in a review article [19]. The dimerization of gramicidin is reversible, resulting in opening and closing (also called gating) of the channel. The channel is open only when the molecule dimerizes and closed when it separates into monomers since the length of a monomer is not of sufficient length to form a transmembrane channel. Another antibiotic, gramicidin S in spite of the similar name, has a different structure and is discussed later (Sect. 7.2.2.4).

7.2.2.3   Tyrocidine

As discussed above (Sect. 7.2.2.2), the peptide antibiotic gramicidin was discovered in 1939 from the soil bacterium *Bacillus brevis*. Actually, later it was discovered that the bacterial extract contained a mixture of antibiotics. The mixture was then named as Tyrothricin, from which a second antibiotic was purified and was first named as Graminic acid [20]. After it was further purified and characterized, it was renamed as Tyrocidine [21]. The mixture tyrothricin contains 10–20% gramicidin and 40–60% tyrocidine [22]. There are four variations of tyrocidine, which have only minor differences from each other (Fig. 7.4). Like gramicidin, tyrocidine is also a peptide antibiotic, but unlike gramicidin, tyrocidine is a cyclic decapeptide. Also unlike gramicidin, which has only hydrophobic amino acids, tyrocidine contains nine hydrophobic and one cationic amino acid, ornithine. Thus, tyrocidine behaves as a detergent and functions by disrupting bacterial membranes. Because of this property, tyrocidine has a broader spectrum of activity and works against both gram-positive and gram-negative bacteria, while gramicidin functions only against gram-positive bacteria [22]. As is true for all AMPs, tyrocidine also shows some toxicity against mammalian cells, especially red blood cells, and so is used only topically. The cyclic structure of tyrocidine is necessary for the antibiotic activity as a synthetic linear tyrocidine-A decapeptide was shown to have no antibiotic activity [23]. The function of tyrocidines and gramicidins in the producing bacteria, *B. brevis* is believed to be to regulate sporulation. Tyrocidine inhibits RNA synthesis and so is proposed to be involved in the control of gene expression as a repressor-like compound. It has DNA unwinding activity and was shown to relax supercoiled plasmid DNA [24].

7.2.2.4 Gramicidin S

Soon after the discovery of the first gramicidin, another peptide antibiotic, gramicidin S (the S stands for Soviet), produced by *Bacillus brevis* was first reported in 1944 by Georgyi Gause and Maria Brazhnikova from the then Soviet Union [25]. It has a broader spectrum of activity than gramicidin A, B, or C since it has activity against gram-positive and gram-negative bacteria and also against fungi. However, because of its hemolytic activity, it is not used systemically but only as topical medication. Unlike gramicidin A, which is linear, gramicidin S is cyclic decapeptide consisting of two identical pentapeptides with the sequence [D-Phe-L-Pro-L-Val-L-Orn-L-Leu]$_2$. Cyclic structure is formed by peptide bond formation between first and last amino acids in the sequence (Fig. 7.4). Other features of the antibiotic include the unusual D- stereoisomer of Phe, and the unusual amino acid L-ornithine. Ornithine structure has one carbon less than the amino acid lysine and one carbon more than diaminobutyric acid which is present in polymyxin B (Sect. 7.2.2.5). The two side chain amino groups of the two L-ornithine residues are essential for the antibiotic activity of Gramicidin S since modifications of these amino groups result in loss of antibiotic activity even though the hemolytic activity was not affected much [26]. Synthetic peptides containing some or all of the five amino acids present in gramicidin S in the right sequence were tested for activity. Of these, only the linear decapeptide displayed some antibiotic activity [27]. The antibiotic functions by binding to the membrane and increasing the permeability of the membrane. The orientation of the hydrophobic and hydrophilic amino acids gives it a detergent-like property, which explains its action on the membrane. This also explains its toxicity by rapid lysis of red blood cells.

7.2.2.5 Polymyxins and Colistins

Polymyxin belongs to the cationic cyclic lipopeptide family of antibiotics and was discovered in 1947, came into clinical use in 1949 and was then discontinued in the 1970s due to associated neurotoxicity and nephrotoxicity. However, because of the lack of new antibiotics in the pipeline and the rise of antibiotic resistance, particularly multidrug resistance (MDR), there has been a revival of interest in polymyxin since the 1990s. The special structural features of polymyxins include a heptapeptide linked to a tripeptide side chain that is linked by an acyl bond to a branched seven or eight carbon fatty acid chain (Fig. 7.5). All the peptide bonds in the antibiotic are unusual peptide bonds either because of the D-amino acids or because of the unusual amino acid diaminobutyric acid (DAB). This protects the antibiotic from degradation by proteases. For other examples of similar strategy see Sects. 3.1.2, 3.3.3.1, and 7.2.2.2. Another important structural feature of the antibiotic is the presence of multiple positive charges due to the amino group of diaminobutyric acid (DAB) at physiological pH and the long hydrophobic chain that gives the antibiotics detergent-like property. There are five different polymyxins, named polymyxin A-E,

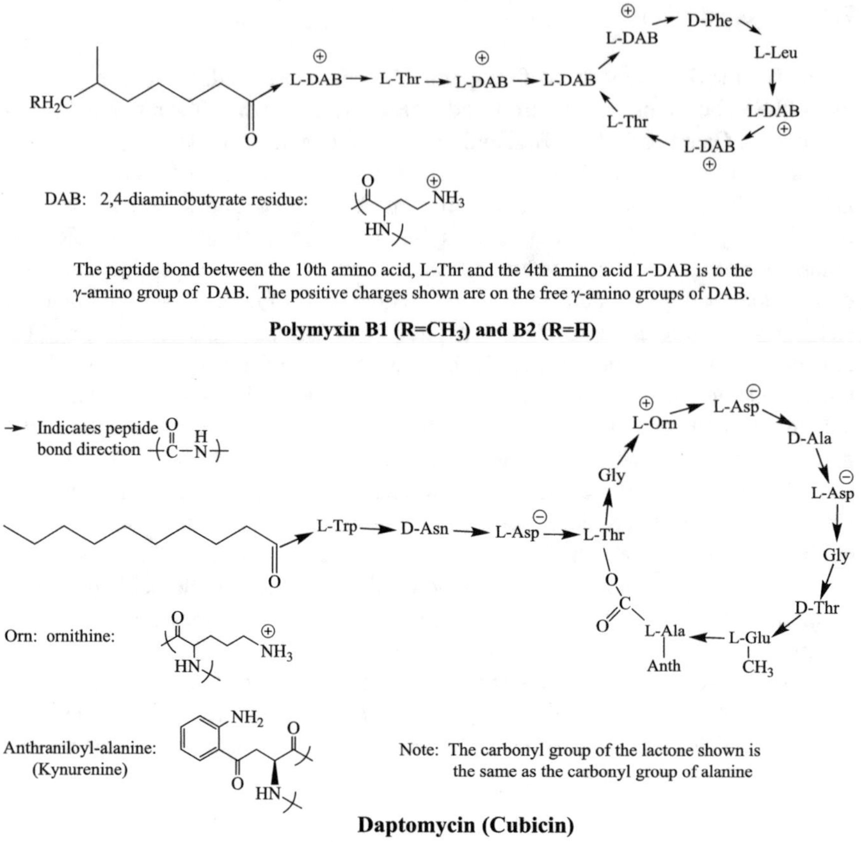
RH2C
O
L-DAB
L-Thr
L-DAB
L-DAB
L-DAB
D-Phe
L-Leu
L-DAB
L-DAB
L-Thr
DAB: 2,4-diaminobutyrate residue:
NH3
HN
The peptide bond between the 10th amino acid, L-Thr and the 4th amino acid L-DAB is to the γ-amino group of DAB. The positive charges shown are on the free γ-amino groups of DAB.
Polymyxin B1 (R=CH3) and B2 (R=H)
Indicates peptide bond direction
L-Trp
D-Asn
L-Asp
L-Thr
Gly
L-Orn
L-Asp
D-Ala
L-Asp
Gly
D-Thr
L-Glu
CH3
L-Ala
Anth
Orn: ornithine:
NH3
HN
Anthraniloyl-alanine: (Kynurenine)
NH2
HN
Note: The carbonyl group of the lactone shown is the same as the carbonyl group of alanine
Daptomycin (Cubicin)

Fig. 7.5 Structures of lipopeptide polymyxins and daptomycin

of which polymyxin A and polymyxin E, also known as colistin (not to be confused with colicin, which is discussed in Sect. 7.2.2.7), have been in clinical use. Colistin and polymyxin B are produced non-ribosomally by *Bacillus polymyxa* subspecies *colistinus* [28]. Polymyxin B and colistin have almost identical primary sequence except at position 6 where D-Phe in polymyxin B is replaced by D-Leu in colistin [29].

Mechanism of Action Polymyxin and colistin are amphiphilic (means the same as amphipathic) molecules due to the presence of five cationic amino groups and a hydrophobic fatty acid chain. These make lipopeptide antibiotics function as detergents. Unlike the gramicidins and tyrocidines, these two amphiphilic antibiotics specifically act against gram-negative bacteria. They damage both the outer and inner membranes of the bacteria. The polymyxins first bind to the lipopolysaccharide (LPS, also known as endotoxin) molecules which are present on the outer leaflet of the outer membrane. The positively charged DAB (diaminobutyric acid) residues interact with the negatively charged phosphates of LipidA, which is the lipid component of LPS in the bacterial outer membrane. Normally the phosphate group of LPS is bonded to the divalent cations Mg^{2+} and Ca^{2+}. The polymyxins displace the divalent cations and bind to the phosphate of LPS. This is followed by insertion of the fatty acyl chain into the outer membrane, thereby destabilizing the outer membrane. This allows the polymyxins to cross the outer membrane and then reach the inner membrane. The drug then binds and inserts into the inner membrane, increases its permeability and leakage of cytoplasmic contents and results in cell lysis and death.

Besides functioning as an antibiotic, it also has potential as an anti-endotoxin. The LPS of the outer membrane of gram-negative bacteria is responsible for endotoxemia (sepsis). Since polymyxin B binds to LPS, its anti-endotoxin activity has been tested in a feline model with positive results [30]. Polymyxin B immobilized on fibers can be used in an extracorporeal hemadsorption cartridge to eliminate endotoxin from peripheral blood circulation [31].

Resistance Development Several mechanisms of resistance development have been reported. Since polymyxins and colistin have similar structure and function, there can be cross-resistance between the two. Since the initial and essential step in the mechanism of action of polymyxins is binding to LPS, most mechanisms of resistance to polymyxins are by modifications in LPS, which stop or reduce this initial interaction [32]. A common modification is by decreasing the net negative charge of LPS, thus preventing the binding of positively charged polymyxins. There are several other mechanisms of resistance development that have been reviewed [33]. One such resistance mechanism is by secretion of proteases that degrade the polymyxin.

7.2.2.6 Daptomycin

Daptomycin, a lipopeptide antibiotic, was discovered as a natural product from *Streptomyces roseosporus* in the late 1980s at Eli Lilly, who did not develop it beyond Phase II clinical trials due to its side effects. The drug was later developed and marketed by Cubist Pharmaceuticals in 2003 under the trade name Cubicin. It is approved in the USA for the treatment of skin infection and for endocarditis. Discovery of daptomycin and its approval for clinical use is very significant for two important reasons. It is the first new class of natural antibiotic to be approved for clinical use in many years. Secondly, daptomycin is effective against a variety of pathogens that are resistant to all current treatments, including vancomycin, which is often used as an antibiotic of last resort. Similar to gramicidin, daptomycin inhibits only gram-positive bacteria because it is unable to cross the outer membrane of gram-negative bacteria. Against gram-positive bacteria, it is more rapidly bactericidal than vancomycin and is also effective against methicillin-resistant *Staphylococcus aureus* (MRSA), vancomycin-resistant *S. aureus* (VRSA), vancomycin-intermediate *S. aureus* (VISA), glycopeptide-intermediate *S. aureus* (GISA) and vancomycin-resistant enterococci (VRE). The structure of daptomycin

(Fig. 7.5) consists of a 13 amino acid peptide linked to a 10-carbon hydrophobic fatty acid chain, and so is described as a lipopeptide antibiotic. Ten of the 13 amino acids form a ring by the formation of a lactone linkage between the carboxyl group of the modified alanine at the C-terminal end of the peptide to the hydroxyl group of the fourth amino acid, threonine. Unlike polymyxin, which contains five cationic amino acids, daptomycin contains one cationic and three anionic amino acids (Fig. 7.5). Daptomycin contains two unusual amino acids, ornithine and 3-anthraniloyl-L-alanine (also known as kynurenine) both of which have one extra amino group in the side chain. However, the pKa of the amino group in the anthraniloyl side chain is less than seven, and so it will not be protonated at physiological pH, whereas the ornithine amino group will be. Bacteria in the stationary growth phase, as occurs in endocarditis, may be better inhibited by daptomycin compared to vancomycin. Daptomycin is soluble in water because of the charged amino acids but is amphipathic because of the lipid chain and some hydrophobic amino acids. Daptomycin shows low occurrence of side effects comparable to other standard antibiotics and can be used systemically.

Mechanism of action of daptomycin involves binding to the membrane and then forming a transmembrane channel resulting in leakage of intracellular ions leading to depolarization of the cell membrane. This loss of membrane potential results in inhibition of synthesis of DNA, RNA, and proteins and eventual cell death. Understanding the mechanism of action of daptomycin has evolved over time. Initially, it was thought that daptomycin inhibits the formation of precursors of cell wall. Then it was proposed that the cell surface molecule, lipoteichoic acid is the target of the drug. These have been proven to be untrue. The current theory is that the target of daptomycin is the cell membrane [34]. As explained above, daptomycin has a net negative charge at physiological pH. However, it was shown that targeting the cytoplasmic membrane in *S. aureus* absolutely requires calcium ion for bactericidal activity. When bound to positively charged calcium ion, daptomycin becomes a de facto cationic peptide which helps it bind to the negatively charged phosphates of the cell membrane. This is followed by insertion of the hydrophobic part of the antibiotic into the membrane, which leads to depolarization and permeabilization of the cell membrane [35]. Conformational change of daptomycin when it binds to calcium ion and the deformation of the membrane structure when the daptomycin calcium complex binds to the membrane has been demonstrated using CD, fluorescence, and NMR spectroscopy [36].

Resistance development against daptomycin is very rare. There is no gene known to confer resistance to the drug, and so cross-resistance to other drugs has also not been reported. However, recently there have been many reports of resistance to daptomycin, mostly occurring by spontaneous mutations, and have been associated with prolonged use. The genes affected by these mutations and the resulting resistance to daptomycin have been reviewed [34, 35].

7.2.2.7 Other AMPs: Defensins, Magainins, Bacteriocins

AMPs are made by not just bacteria but by a variety of species. Two such AMPs that have been studied in detail are magainin which is secreted from the skin of amphibians (frogs and toads), and defensins that is secreted by mammalian polymorphonuclear (PMN) leukocytes. As discussed before (Sect. 7.2.2.1) AMPs can be of three types: those that are linear but become helical after binding to the membrane (e.g., magainin), those that function as linear peptides and those that contain disulfide bonds and β-sheets (e.g., defensin). These two AMPs are discussed here briefly.

Defensins PMNs destroy invading microorganisms by two mechanisms. One mechanism is by the production of reactive oxygen intermediates (ROI) by stimulated phagocytes. A second mechanism is by secretion of several cationic peptides with antimicrobial activity. The peptides are small (32–34 amino acid long), rich in arginine (4–10 residues/molecule), and uniformly rich in cysteine (6 residues/molecule, forming three pairs of intramolecular disulfide bonds), with activity against bacteria, fungi, and certain enveloped viruses [37]. The mixture of these AMPs is collectively known as defensins. Their mechanism of action against bacteria and fungi is by disruption of the cell membrane. It is to be noted that defensins are not the only antibiotics produced by PMNs. Other antibiotics present in PMN granules are lysozyme, which was the first antibiotic discovered by Fleming in 1922 (Sect. 1.5), and lactoferrin which is known to kill many bacterial species. Human lactoferrin was shown to protect lactoferrin-deficient mice during *Aggregatibacter actinomycetemcomitans*-induced bacteremia [38].

Magainins The magainins, which are antimicrobial peptides secreted from the skin of amphibians (frogs and toads), were discovered by Zasloff in 1987 [39]. He reported that the extract from the skin of the African clawed frog *Xenopus laevis* contained two antimicrobial peptides, which he named Magainins. The two peptides, each 23 amino acid long differ from each other at two positions in the sequence. The magainins are amphiphilic, water soluble, and nonhemolytic and inhibited the growth of numerous species of gram-positive and gram-negative bacteria as well as fungi and induced osmotic lysis of protozoa. The 23 amino acid peptide and shorter versions of it were made synthetically and tested for activity. Removal of the first three N-terminal amino acids did not affect the antibiotic activity, but the removal of the fourth amino acid resulted in loss of most activity [40]. The dramatic loss of activity when the peptide length was shortened to less than 19 suggests that magainin functions by forming a transmembrane α-helical structure which is known to require 20 amino acids. In solutions, magainins have a random flexible conformation but transform into an α-helical structure when bound to the membrane surface. The following mechanism of action has been proposed for magainin [41]. The peptide binds to the membrane and forms a helix that lies parallel to the membrane surface, imposing a positive curvature strain on the membrane. To relieve the stress, several magainin helices along with the surrounding lipid molecules form a transmembrane pore allowing cellular ions to leak out. It has been reported that the combined use of the two AMPs from frog skin, magainin, and PGLa results in synergism in bacteria as well as tumor cells. The combination is also hemolytic even though magainin and PGLa used separately are not [42]. It is proposed that the two form a heterodimer of parallel helices with strong membrane permeabilizing activity [41]. Magainin used alone is not very toxic for mammalian cells and so holds promise for systemic use. Currently, it is used topically against skin infections.

Synthetic AMPs Several synthetic AMPs have gone to clinical trials, but many are not found to be successful antibiotics. The main obstacles are their proteolytic degradation and their toxicity. Recently it was reported that NK-18, an 18-amino acid region of a mammalian protein NK-lysin has antibacterial and anticancer activity and functions by interacting with the membrane [8]. Because of the short amino acid sequence, NK-18 can be chemically synthesized at low cost and has the potential to be used clinically. The authors also reported that the peptide functions as antibiotic not just by disrupting the cell membrane but also by binding to DNA in the cytoplasm of the bacteria. Because of the double target of the antibiotic, development of resistance to the antibiotic is difficult.

Bacteriocins are protein toxins produced by bacteria to inhibit the growth of similar or closely related bacterial species. The purpose of killing is usually to have a competitive advantage during nutrient limitation. Thus they can be described as narrow-spectrum antibiotics, although some people object to classifying them as antibiotics. Bacteriocins are named based on the species producing them. For example, the first bacteriocin was discovered by Andre Gratia in 1925 from *E. coli*. Since it could kill *E. coli*, it was named colicin. Thus, the discovery of colicin predates the discovery of penicillin. Colicins, which are large proteins (>500 amino acids) are secreted by the producing cells to the environment. They kill other cells by first binding to a receptor on the outer membrane and are then translocated through the membrane to the periplasm. They then bind to and form a channel in the inner membrane resulting in cell death.

Since bacteriocins are proteins, they cannot be administered orally as they will be digested and intravenous administration will cause an immune response. So the use of bacteriocins as antibiotics has not been very successful. For more details on colicin, one can consult reviews on the subject [43]. One bacteriocin worth mentioning is nisin, which is used in food preservation. Since these proteins are destroyed by digestive enzymes, they are not toxic when ingested. Nisin is a lantibiotic (Sect. 7.2.2.8) that belongs to Class I bacteriocins.

7.2.2.8 Lantibiotics

Lantibiotics, which are defined as lanthionine-containing antibiotics, are peptide antibiotics that contain the cyclic thioether amino acid, lanthionine, or methyl lanthionine as well as some dehydrated amino acids in its sequence (Fig. 7.6). Lanthionine (the name is not related to the element lanthanum) can be described as a thiodialanine or as a monosulfide analog of cysteine (which is formed by

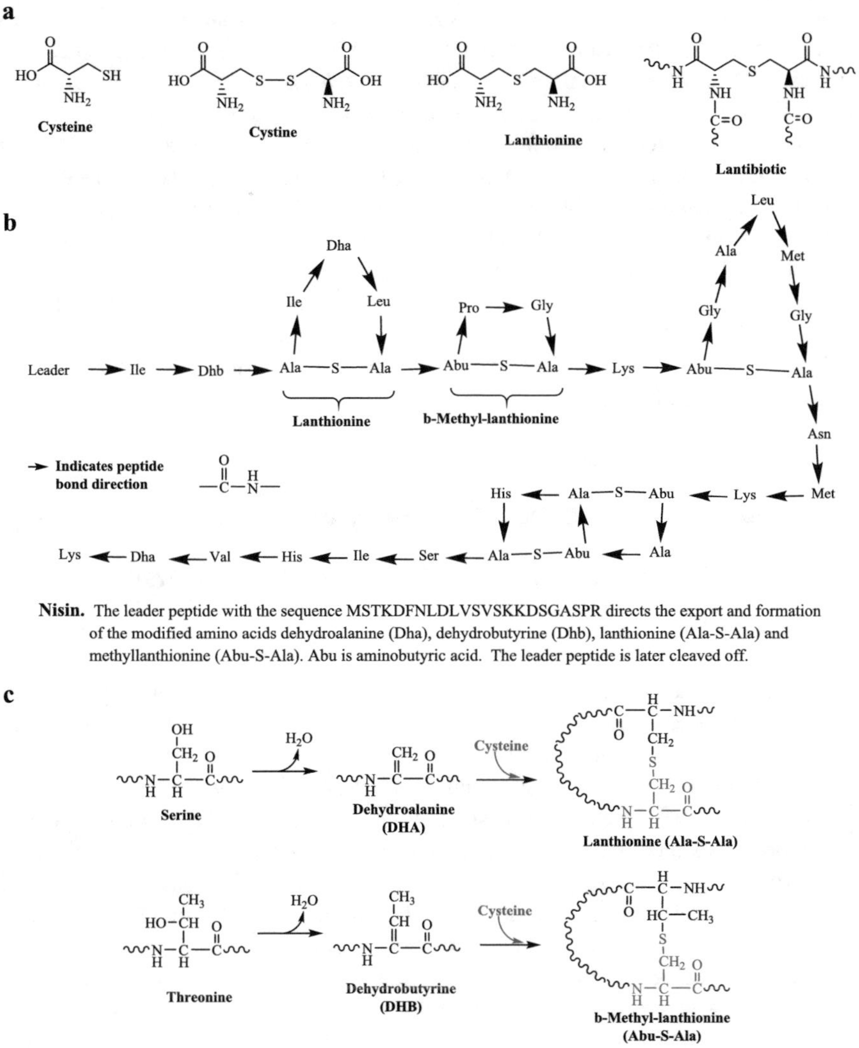
a
Cysteine
Cystine
Lanthionine
Lantibiotic
b
Leader
Lanthionine
b-Methyl-lanthionine
Indicates peptide bond direction
Nisin. The leader peptide with the sequence MSTKDFNLDLVSVSKKDSGASPR directs the export and formation of the modified amino acids dehydroalanine (Dha), dehydrobutyrine (Dhb), lanthionine (Ala-S-Ala) and methyllanthionine (Abu-S-Ala). Abu is aminobutyric acid. The leader peptide is later cleaved off.
c
Serine
Dehydroalanine (DHA)
Cysteine
Lanthionine (Ala-S-Ala)
Threonine
Dehydrobutyrine (DHB)
b-Methyl-lanthionine (Abu-S-Ala)

Fig. 7.6 Lantibiotics. (**a**) Structures of lantibiotic and relevant molecules. (**b**) Structure of Nisin. (**c**) Reactions forming lanthionine

oxidative crosslinking of two cysteines). Lantibiotics are gene-encoded peptides that are ribosomally synthesized in the same way as all other proteins and then are post-translationally modified to include unusual amino acids such as lanthionine Fig. 7.6.

Lantibiotics are produced by gram-positive bacteria such as *Streptococcus* and *Streptomyces* for the purpose of killing other competing gram-positive bacteria. So lantibiotics belong to the broader class of bacterial toxins called the bacteriocins and are classified as Class I bacteriocins. Lantibiotics can be of two types: Type A lantibiotics are long and flexible molecules (e.g., nisin), while Type B lantibiotics are globular (e.g., mersacidin). Since mersacidin inhibits cell wall biosynthesis, it has been discussed in an earlier chapter (Sect. 3.3.3.3). The lantibiotic nisin has been successfully used for more than half a century in the food industry as a preservative. Proteins that are secreted (e.g., nisin) usually contain a leader sequence at the N-terminus that directs it through the membrane (Fig. 7.6). The leader sequence is cleaved before the protein is secreted. Lantibiotics such as nisin kill other bacteria by a dual mode of action: inhibition of cell wall formation and formation of pores in the cell membrane. First nisin binds to Lipid II (C55-PP), whose function is to transport the peptidoglycan monomer from the cytoplasm on one side of the cell membrane to the cell wall on the other side (Sect. 3.2.2). Thus it inhibits cell wall biosynthesis. Many lantibiotics have the ability to bind to Lipid II, but only some of them can use lipid II as an anchor and then insert itself into the cell membrane to form a pore that leads to cell death [44, 45].

7.3   Antibiotics Affecting Other Structural Targets

7.3.1   Triclosan, The Antibiotic That Inhibits Fatty Acid Synthesis

Triclosan has been discussed before as an antiseptic (Sect. 7.2.1). At high concentrations, its antiseptic activity is due to its membrane disrupting property. Triclosan is widely used in surgical scrubs and is found in many health care products such as shampoos, deodorants, toothpastes, and mouthwashes. Such widespread use of the antiseptic is not of much concern since no protein is involved and so resistance development to membrane-acting agents is rare. However, the concentrations of triclosan present in these products are not high enough to cause disruption of hydrophobic interaction in the membrane phospholipids and proteins. At the low concentrations that are used in these products, it still functions as an antibiotic by inhibiting a specific metabolic step in the fatty acid synthesis pathway. Since the target of triclosan is a certain gene, it is probable that widespread use of the antibiotic in health care products will select for strains of bacteria that are resistant to the antibiotic. So it may be reasonable to stop adding triclosan to all health care products. However, there is no unanimous agreement among scientists on this subject. There have been reports published both for and against the idea.

Mechanism of Action of Triclosan Triclosan functions by targeting a fatty acid synthesis pathway which is a spiral pathway involving four enzymes in sequence. The synthesis starts with a condensation of malonyl-ACP with an existing fatty acid chain. The resulting β-ketoacyl-ACP is reduced by an NADPH-dependent β-ketoacyl-ACP reductase. This is followed by a dehydration step to form an alkene. This is then reduced in the final step by enoyl-ACP reductase also known as FabI. The specific enzyme inhibited by triclosan is the enoyl-acyl carrier protein reductase (FabI). It does so by binding to the FabI enzyme and increasing the enzymes affinity for NAD^+. Thus a stable complex of triclosan, FabI, and NAD^+ is formed, which is unable to catalyze fatty acid synthesis. Note that this mechanism of action is very similar to that of isoniazid, which also inhibits the same enzyme in *Mycobacterium tuberculosis* (Sect. 7.3.2). The mechanism is confirmed by the fact that mutations in FabI can confer resistance against triclosan [46]. Another mechanism of resistance development is by overexpressing FabI or by actively pumping out the triclosan from the cells.

7.3.2 Isoniazid: Antibiotic Against Tuberculosis

Isoniazid (Isonicotinylhydrazide or INH, Fig. 7.7) is an effective and inexpensive antibiotic used for the treatment of tuberculosis, which is caused by the bacteria *Mycobacterium tuberculosis*. It was first synthesized in 1912, but its anti-tuberculosis activity was first reported from several laboratories in 1952. Since then, it has been widely used to treat tuberculosis. Although it has been used for a long time, its mechanism of action is still not very clear. It is known to inhibit the synthesis of mycolic acids, which are essential components of mycobacterial cell walls. Many INH-resistant mycobacterial strains were found to have mutations in the catalase peroxidase (*katG*) gene suggesting a role of the catalase gene in the mechanism of action of INH. However, not all resistant mutants were catalase negative, suggesting that other factors may be responsible for resistance. Later, it was shown that a mutation in the mycobacterial gene *inhA* was able to confer resistance to the antibiotic INH [47]. The gene *inhA* codes for the enzyme enoyl-acyl carrier protein reductase, also known as InhA, which is one of the enzymes required in the mycolic acid synthesis pathway. Note that this mechanism of action is similar to that of triclosan which also inhibits the same enzyme in other bacteria (Sect. 7.3.1). Mycolic acids are long-chain fatty acids present in the cell wall of mycobacteria. Structure of mycolic acids is significantly different from that of fatty

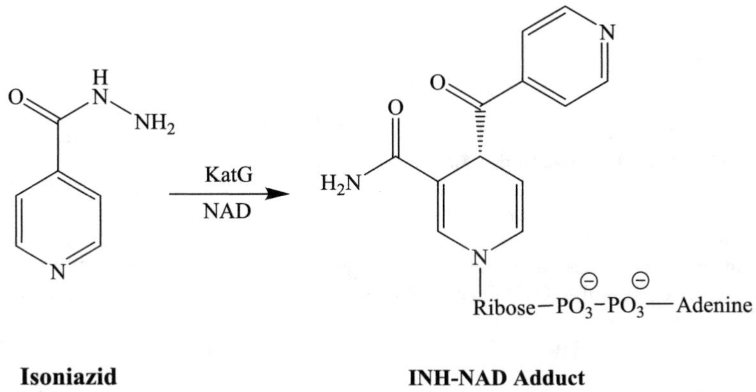

Isoniazid   INH-NAD Adduct

Fig. 7.7 Formation of isoniazid-NAD adduct

acids in other organisms. The special structure of these fatty acids makes mycobacteria resistant to most medical treatments and even allows them to grow inside macrophages and thus be resistant to the host's immune system.

The mechanism of action of isoniazid is complex and has been reviewed [48]. INH is a prodrug and is activated by peroxidation catalyzed by the mycobacterial catalase peroxidase enzyme KatG to generate reactive free radical species that form adducts with NAD^+ and $NADP^+$ (Fig. 7.7). The hydrazide part of the molecule is important for the formation of the free radicals. The covalent INH-NADH adducts are powerful inhibitors of InhA protein, and thus the biosynthesis of mycolic acid is prevented and leads to mycobacterial cell death.

7.3.3   Antibiotics Targeting FtsZ, The Cell Division Protein

An ideal antibiotic is one that targets a process that takes place only in the infecting bacteria but not in the host. One such process is the bacterial cell division process, for which the protein FtsZ is essential and the major protein in the bacterial cell division machinery. Since the protein is present in all bacteria, any antibiotic targeting this protein will have a broad spectrum of activity. Mammalian cell division is a more complex process and does not involve any protein homologous to FtsZ. Thus compounds binding to this protein can be expected to serve as effective antibiotics. Since no such antibiotic already exists in nature, there will be no preexisting resistance gene against any new antibiotic synthesized against the FtsZ target. One such promising antibiotic is a quinuclidine-based FtsZ inhibitor, which also showed synergistic activity with β-lactam antibiotics [49]. The drug is active against multiple antibiotic-resistant bacterial strains, including methicillin-resistant *Staphylococcus aureus* and vancomycin-resistant *Enterococcus faecium*. There are several other promising compounds that are being studied as potential antibiotics targeting the cell division system.

References

1. Rog T, Pasenkiewicz-Gierula M, Vattulainen I, Karttunen M (2009) Ordering effects of cholesterol and its analogues. Biochim Biophys Acta 1788:97–121
2. Eeman M, Deleu M (2010) From biological membranes to biomimetic model membranes. Biotechnol Agron Soc Environ 14:719–736
3. Redondo-Morata L, Giannotti MI, Sanz F (2012) Influence of cholesterol on the phase transition of lipid bilayers: a temperature-controlled force spectroscopy study. Langmuir 28:12851–12860
4. McDonnell G, Russell AD (1999) Antiseptics and disinfectants: activity, action, and resistance. Clin Microbiol Rev 12:147–179
5. Leive L (1974) The barrier function of the Gram-negative envelope. Ann N Y Acad Sci 235:109–129
6. Salton MRJ (1968) Lytic agents, cell permeability and monolayer penetrability. J Gen Physiol 52:277S–252S

7. Cieplik F, Jakubovics NS, Buchalla W, Maisch T, Hellwig E, Al-Ahmad A (2019) Resistance toward chlorhexidine in oral bacteria—is there cause for concern? Front Microbiol 10:587. https://doi.org/10.3389/fmicb.2019.00587

8. Yan J, Wang K, Dang W, Chen R, Xie J, Zhang B, Song J, Wang R (2013) Two hits are better than one: membrane-active and DNA binding-related double-action mechanism of NK-18, a novel antimicrobial peptide derived from mammalian NK-Lysin. Antimicrob Agents Chemother 57:220–228

9. Vooturi SK, Firestine SM (2010) Synthetic membrane-targeted antibiotics. Curr Med Chem 17: 2292–2300

10. Fjell CD, Hiss JA, Hancock RE, Schneider G (2011) Designing antimicrobial peptides: form follows function. Nat Rev Drug Discov 11:37–51

11. Guilhelmelli F, Vilela N, Albuquerque P, Derengowski LS, Silva-Pereira I, Kyaw CM (2013) Antibiotic development challenges: the various mechanisms of action of antimicrobial peptides and of bacterial resistance. Front Microbiol 4:353–364

12. Yeaman MR, Yount NY (2003) Mechanisms of antimicrobial peptide action and resistance. Pharmacol Rev 55:27–55

13. Brogden KA (2005) Antimicrobial peptides: pore formers or metabolic inhibitors in bacteria? Nat Rev Microbiol 3:238–250

14. Oren Z, Shai Y (1998) Mode of action of linear amphipathic alpha-helical antimicrobial peptides. Biopolymers 47:451–463

15. Oren Z, Lerman JC, Gudmundsson GH, Agerberth B, Shai Y (1999) Structure and organization of the human antimicrobial peptide LL-37 in phospholipid membranes: relevance to the molecular basis for its non-cell-selective activity. Biochem J 341:501–513

16. Dubos RJ (1939) Studies on a bactericidal agent extracted from a soil Bacillus. I. Preparation of the agent Its activity in vitro. J Exp Med 70:1–10

17. Dubos RJ (1939) Studies on a bactericidal agent extracted from a soil Bacillus. II. Protective effect of the bactericidal agent against experimental Pneumococcus infections in mice. J Exp Med 70:11–17

18. Finkelstein A, Andersen OS (1981) The gramicidin A channel: a review of its permeability characteristics with special reference to the single-file aspect of transport. J Membr Biol 59: 155–171

19. Kelkar DA, Chattopadhyay A (2007) The gramicidin ion channel: a model membrane protein. Biochim Biophys Acta 1768:2011–2025

20. Hotchkiss RD, Dubos RJ (1940) Fractionation of the bactericidal agent from cultures of a soil *Bacillus*. J Biol Chem 132:791–792

21. Hotchkiss RD, Dubos RJ (1940) Bactericidal fractions from an aerobic sporulating *Bacillus*. J Biol Chem 136:803–804

22. Dubos RJ, Hotchkiss RD, Coburn AF (1942) The effect of gramicidin and tyrocidine on bacterial metabolism. J Biol Chem 146:421–426

23. Kohli RM, Walsh CT, Burkart MD (2002) Biomimetic synthesis and optimization of cyclic peptide antibiotics. Nature 418:658–661

24. Hansen J, Pschorn W, Ristow H (1982) Functions of the peptide antibiotics tyrocidine and gramicidin induction of conformational and structural changes of superhelical DNA. Eur J Biochem 126:279–284

25. Gause GF, Brazhnikova MG (1944) Gramicidin S and its use in the treatment of infected wounds. Nature 154:703–703

26. Nagamurthi G, Rambhav S (1985) Gramicidin-S: structure-activity relationship. J Biosci 7: 323–329

27. Katchalski E, Berger A, Bichowsky-Slomnicki L, Kurtz J (1955) Antibiotically active aminoacid copolymers related to gramicidin *S*. Nature 176:118–119

28. Gupta S, Govil D, Kakar PN, Prakash O, Arora D, Das S, Govil P, Malhotra A (2009) Colistin and polymyxin B: a re-emergence. Indian J Crit Care Med 13:49–53

29. Yu Z, Qin W, Lin J, Fang S, Qiu J (2015) Antibacterial mechanisms of polymyxin and bacterial resistance:679109. https://doi.org/10.1155/2015/679109
30. Sharp CR, DeClue AE, Haak CE, Honaker AR, Reinero CR (2010) Evaluation of the antiendotoxin effects of polymyxin B in a feline model of endotoxemia. J Feline Med Surg 12:278–285
31. Shoji H, Tani T, Hanasawa K, Kodama M (1998) Extracorporeal endotoxin removal by polymyxin B immobilized fiber cartridge: designing and antiendotoxin efficacy in the clinical application. Ther Apher 2:3–12
32. Zavascki AP, Goldani LZ, Li J, Nation RL (2007) Polymyxin B for the treatment of multidrug resistant pathogens: a critical review. J Antimicrob Chemother 60:1206–1215
33. Loutet SA, Valvano MA (2011) Extreme antimicrobial peptide and polymyxin B resistance in the genus *Burkholderia*. Front Microbiol 2:159–166
34. Vilhena C, Bettencourt A (2012) Daptomycin: a review of properties, clinical use, drug delivery and resistance. Mini Rev Med Chem 12:202–209
35. Bayer AS, Schneider T, Sahl HG (2013) Mechanisms of daptomycin resistance in *Staphylococcus aureus*: role of the cell membrane and cell wall. Ann N Y Acad Sci 1277:139–158
36. Jung D, Rozek A, Okon M, Hancock REW (2004) Structural transitions as determinants of the action of the calcium-dependent antibiotic daptomycin. Chem Biol 11:949–957
37. Ganz T, Selsted ME, Szklarek D, Harwig SSL, Daher K, Bainton DF, Lehrer RI (1985) Defensins: natural peptide antibiotics of human neutrophils. J Clin Invest 76:1427–1435
38. Velusamy SK, Poojary R, Ardeshna R, Alabdulmohsen W, Fine DH, Velliyagounder K (2014) Protective effects of human lactoferrin during *Aggregatibacter actinomycetemcomitans* induced bacteremia in lactoferrin-deficient mice. Antimicrob Agents Chemother 58:397–404
39. Zasloff M (1987) Magainins, a class of antimicrobial peptides from *Xenopus* skin: isolation, characterization of two active forms, and partial cDNA sequence of a precursor. Proc Natl Acad Sci U S A 84:5449–5453
40. Zasloff M, Martin B, Chen H-C (1988) Antimicrobial activity of synthetic magainin peptides and several analogues. Proc Natl Acad Sci U S A 85:910–913
41. Nishida M, Imura Y, Yamamoto M, Kobayashi S, Yano Y, Matsuzaki K (2007) Interaction of a Magainin-PGLa hybrid peptide with membranes: insight into the mechanism of synergism. Biochemistry 46:14284–14290
42. Zerweck J, Strandberg E, Kukharenko O, Reichert J, Burck J, Wadhwani P, Ulrich AS (2017) Molecular mechanism of synergy between the antimicrobial peptides PGLa and magainin 2. Sci Rep 7:13153. https://doi.org/10.1038/s41598-017-12599-7
43. Cascales E, Buchanan SK, Duché D, Kleanthous C, Lloubés R, Postle K, Riley M, Slatin S, Cavard D (2007) Colicin biology. Microbiol Mol Biol Rev 71:158–229
44. Cotter PD, Colin Hill C, Ross RP (2005) Bacteriocins: developing innate immunity for food. Nat Rev Microbiol 3:777–788
45. Bierbaum G, Sahl HG (2009) Lantibiotics: mode of action, biosynthesis and bioengineering. Curr Pharm Biotechnol 10:2–18
46. Heath RJ, Rubin JR, Holland DR, Zhang E, Snow ME, Rock CO (1999) Mechanism of triclosan inhibition of bacterial fatty acid synthesis. J Biol Chem 274:11110–11114
47. Banerjee A, Dubnau E, Quemard A, Balasubramanian V, Um KS, Wilson T, Collins D, de Lisle G, Jacobs WR Jr (1994) *inhA*. A gene encoding a target for isoniazid and ethionamide in Mycobacterium tuberculosis. Science 263:227–230
48. Timmins GS, Deretic V (2006) Mechanisms of action of isoniazid. Mol Microbiol 62:1220–1227
49. Chan F-Y, Sun N, Leung Y-C, Wong K-Y (2015) Antimicrobial activity of a quinuclidine based FtsZ inhibitor and its synergistic potential with β-lactam antibiotics. J Antibiot 68:253–258

Chapter 8
Antifungals, Antimalarials, and Antivirals

Abstract This chapter includes a discussion of antifungal, antimalarial, and antiviral drugs. Similarities and differences of these with antibacterial antibiotics are discussed. Drugs presented include amphotericin B, nystatin, azoles, allylamines, morpholines, fluoropyrimidines, DDT, quinine, chloroquine, pyrimethamine, artemisinin, proguanil, amantadine, acyclovir, lamivudine, zidovudine, and HIV protease and neuraminidase inhibitors. Mechanisms of action of the drugs and resistance development against them are discussed. PCR method for the detection of point mutations is also discussed.

8.1 Antifungal Drugs: Antibiotics That Inhibit Growth of Fungi

One main difference between bacterial and eukaryotic membranes is the absence of sterols in bacterial membranes. In mammalian cell membranes, the sterol is cholesterol, while in fungi, it is ergosterol. This difference between fungi and humans makes ergosterol an attractive target for the development of antifungal antibiotics. The different classes of antifungal drugs are polyenes which bind to ergosterol, the azoles, allylamines, and morpholines which inhibit various steps in the synthesis pathway of ergosterol, the pyrimidines, which inhibit the synthesis of DNA and RNA, and lipopeptide antibiotics that inhibit cell wall synthesis in fungi.

8.1.1 Antibiotics That Bind to Ergosterol: Polyenes

These antibiotics contain a long chain of conjugated double bonds. Two examples are amphotericin B and nystatin. The first broad-spectrum antifungal antibiotic (see Sect. 1.1 for justification of the terminology) was amphotericin B, which was discovered in 1953 and was first approved in 1958. Amphotericin A was also discovered simultaneously, but it had much weaker antifungal activity [1]. Earlier, another polyene, Nystatin (Fig. 8.1), was discovered in 1950. However, the drug is

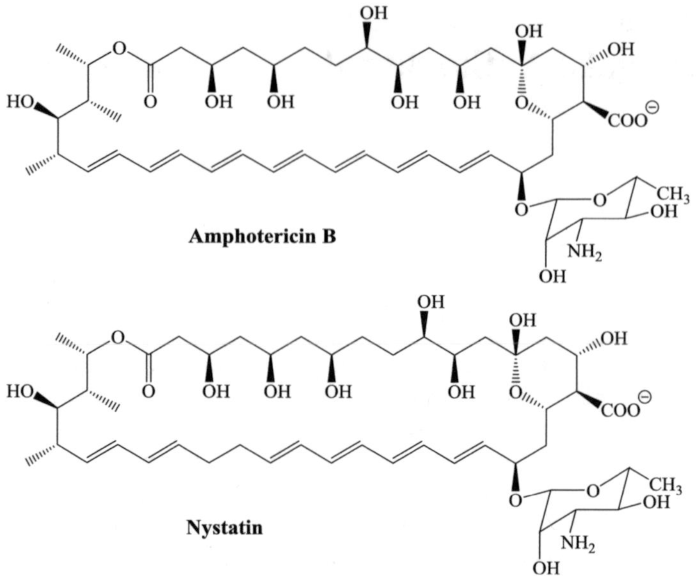

Fig. 8.1 Polyene antifungal drugs: amphotericin B and nystatin

not absorbed through the intestinal walls and was found to be toxic when administered intravenously. For more than two decades since its discovery, amphotericin B
had been the only antifungal drug for systemic use. Since systemic fungal infections
were not very common and most infections were not life threatening, there was little
interest in developing new antifungal drugs. Today the situation is different. There
has been a big increase in the number of fungal infections, and that has resulted in
renewed interest in the study of antifungal drugs. One reason for the drastic increase
in the number of fungal infections is that AIDS patients, who are immunocompro-
mised, are susceptible to infection by opportunistic fungal pathogens that usually do
not infect healthy people. Hematological malignancies and immunosuppression in
transplant recipient also contribute to the increase in number of fungal infections.
Unlike bacteria, fungi are eukaryotes and so have biochemistry that is similar to that
of the host (human). So developing antibiotics that specifically work against fungi
but do not affect the host is very challenging.

As mentioned above, nystatin is too toxic for systemic use and is also not well
absorbed from the intestinal tract and so it is used only for topical application against
fungal infections. The most commonly used polyene antifungal drug is amphotericin
B. It functions by physically binding to ergosterol in the membrane. As discussed
before (Sect. 7.1.2), the function of ergosterol is to maintain fluidity of cell membranes. Ergosterol is present in fungal cell membranes but is absent in animal cell
membranes, which contain cholesterol instead. Because of this difference, ergosterol
serves as a useful target for antifungal drugs. They are not effective against bacteria,
which do not contain sterols but maintain membrane fluidity by adjusting the
proportion of saturated and unsaturated fatty acids (Sect. 7.1.2). Since ergosterol is

also present in cell membranes of trypanosomes, amphotericin B is also effective against sleeping sickness disease. It is believed that the binding of polyenes to ergosterol is simply by hydrophobic interaction. Note that ergosterol has one extra carbon and two extra double bonds compared to cholesterol (Fig. 7.2). Since polyenes also contain a series of double bonds, it is likely that the two interact with each other through these hydrophobic regions. Binding of the polyene to ergosterol possibly increases the fluidity of the membrane that results in leakage of the cell contents, including monovalent ions (K^+, Na^+, H^+, and Cl^-) and small organic molecules. The increase of fluidity can also cause disruption of the essential enzymes present in the membrane. All these factors together may be responsible for fungal cell death.

Toxicity Although amphotericin B is specific for fungal and not human cell membranes, the specificity is not absolute. Since cholesterol and ergosterol have similar structures, the drug does bind to cholesterol to some extent, which accounts for the toxicity of the drug. Dose of the drug used in some patients is limited by its nephrotoxicity especially when used in combination with other antibiotics that are also nephrotoxic, such as aminoglycosides [2]. However, in spite of the high toxicity, amphotericin B is still used clinically because, of the hundreds of polyenes known, amphotericin B has the lowest toxicity for intravenous administration [3].

Resistance Even after five decades of clinical use, resistance development to amphotericin B is rare. Although infrequent, resistant mutants can arise by synthesizing an alternative sterol that is different from ergosterol and binds to the drug less tightly. Some mutants have been shown to produce less amount of ergosterol which reduces the effect of the polyene, while some mutants actually increase the ergosterol content [2]. Overall, it is toxicity and not resistance development that is the major concern in choosing amphotericin B for treatment of patients [3].

8.1.2 Antibiotics That Inhibit Biosynthesis of Ergosterol

Biosynthesis of ergosterol in fungi involves multiple steps. Some of the steps relevant to antibiotic action are shown in Fig. 8.2. Three classes of antibiotics inhibit synthesis of ergosterols (two of which are shown in the figure). Allylamines inhibit an epoxidation reaction of squalene with oxygen, azoles inhibit a demethylation reaction with lanosterol, and morpholines inhibit steps after lanosterol.

8.1.2.1 Azoles

Azoles are five-membered aromatic rings containing either two or three nitrogens (Fig. 8.3). Those that have two nitrogens are called imidazoles (examples include miconazole, ketoconazole, and clotrimazole), while those that have three nitrogens are called triazoles (examples include itraconazole and fluconazole). Miconazole, the

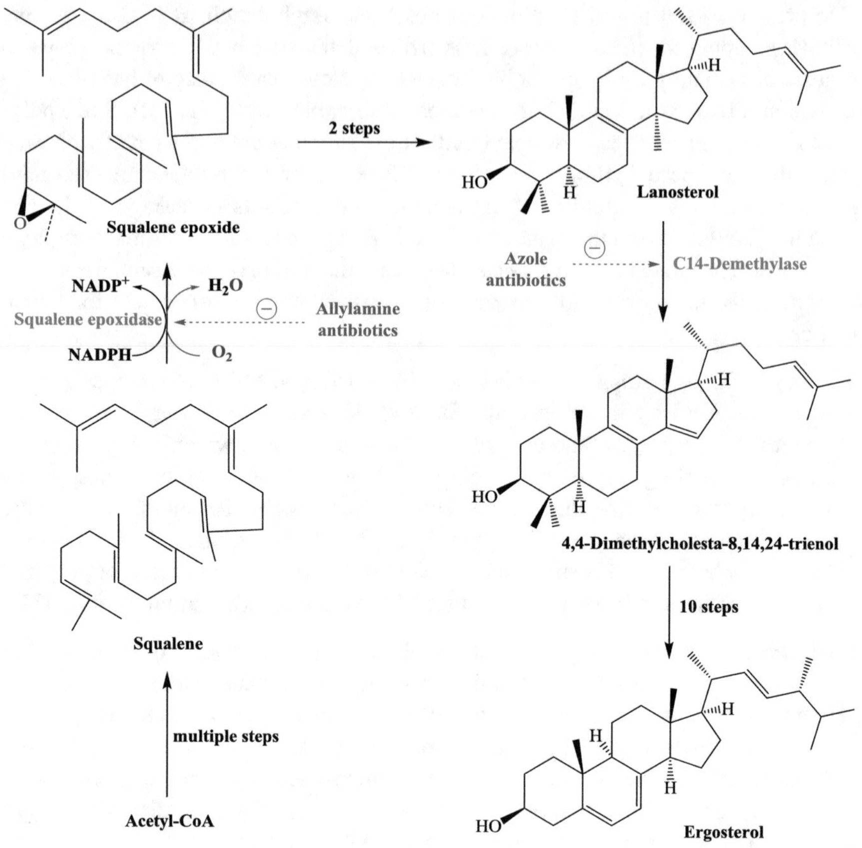

Fig. 8.2 Biosynthesis of ergosterol in fungi

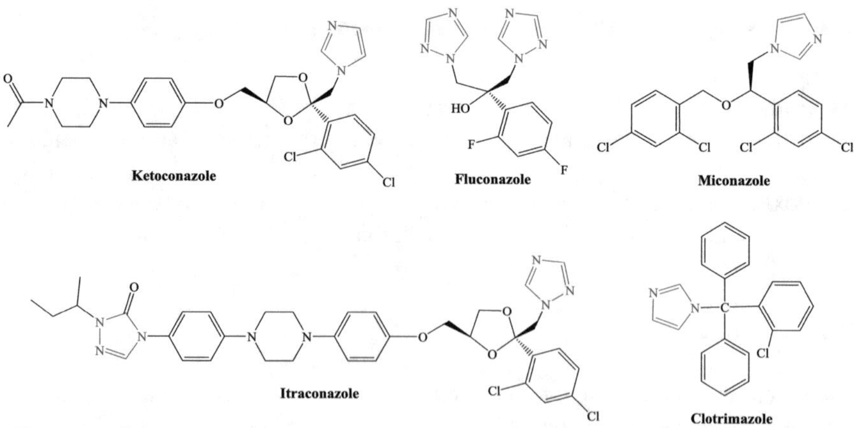

Fig. 8.3 Structures of some azoles used as antifungal drugs. Azole portions are highlighted in *red*

first azole drug to be approved, was later withdrawn from the market due to its toxicity. The other four azoles, ketoconazole, clotrimazole, itraconazole, and fluconazole are frequently used. As can be seen in Fig. 8.3, all the azoles are highly hydrophobic and so are poorly soluble in water. Of these, fluconazole is better than the others not just for its higher water solubility but also for its high bioavailability, less protein binding, wide distribution into body tissues and fluids, long half-life, and lower toxicity [4]. All azoles function by inhibiting the C14 demethylase enzyme, which is a cytochrome P_{450}-dependent enzyme and is encoded by the gene ERG11. This prevents the synthesis of ergosterol resulting in membrane damage and affects the proper functioning of membrane-bound enzymes. It also affects mitochondrial respiration.

Cytochromes P450 (abbreviated as CYPs) are present in most species including humans and carry out a variety of reactions involving electron transfer (oxidation/reduction reactions). The 450 refers to the absorbance at 450 nm, which is due to a heme cofactor that is present in CYPs. In humans, the CYPs are present in the mitochondria in most tissues and play important role in respiration and in metabolism of hormones, cholesterol, vitamin D, and toxic compounds, including various drugs. This raises the possibility of toxicity of azoles depending on how the drugs affect the enzymes in humans. Inhibition of ergosterol biosynthesis is due to binding of the nitrogens of the imidazole or triazole moiety of azoles to the heme iron and to the apoprotein of a cytochrome P-450. The selectivity of azoles is determined by the nitrogen heterocycle and the hydrophobic N-1 substituent of the azole antifungals [5]. The hydrophobic substituent has a greater impact on the selectivity of azoles for inhibition of fungal ergosterol synthesis compared to inhibition of the human enzyme in the pathway for cholesterol synthesis. Of the azoles, ketoconazole is the least selective and so has the highest toxicity.

Development of resistance to azoles can be possible by any one or more of the following methods: (1) increased expression of the target enzyme, (2) increased expression of genes for drug efflux proteins, (3) alterations in sterol synthesis pathway, and (4) decreased affinity of azoles for the target enzyme [6]. Of these, the most common mechanism is by increasing the levels of efflux pump proteins. The intrinsic resistance of *C. krusei* to fluconazole was found to be due to both decreased binding of the drug to the target enzyme and also due to active efflux system [7].

Although it is widely accepted that the mechanisms of action of amphotericin B and azoles involve either binding to ergosterol or inhibiting its biosynthesis, an alternative mechanism involving reactive oxygen species (ROS) and peroxynitrite has also been proposed [8]. It was shown that the azole drug itraconazole, but not fluconazole, led to the formation of ROS, which in turn leads to lipid peroxidation in the fungal pathogen *Cryptococcus gattii*. Amphotericin B also caused lipid peroxidation due to production of oxidative and nitrosative radicals. This effect could be antagonized by peroxynitrite scavengers.

8.1.2.2   Allylamines and Morpholines

An allylic position is the carbon attached to a carbon-carbon double bond. Allylamines contain an amino group on the carbon next to the double bond. Structures of two allylamine antibiotics, terbinafine and naftifine, are shown in Fig. 8.4. These two allylamines affect ergosterol synthesis by inhibiting the enzyme squalene epoxidase, which catalyzes the epoxidation of squalene to form squalene epoxide. This step precedes the demethylase step that is the target of azoles (Fig. 8.3). Because of the inhibition, there is an accumulation of squalene in the fungal cell, and cell death is primarily due to accumulation of squalene rather than due to ergosterol deficiency. High levels of squalene may increase membrane permeability, leading to disruption of cellular organization [7]. Resistance development to allylamines is rare. However, resistance is expected to increase because of the increased use of the antibiotics. Since allylamines, azoles, and polyenes have targets at different steps of the same pathway, it can be expected that mutations in the genes corresponding to the pathway can result in cross-resistance to more than one of these antibiotics.

Amorolfine, which belongs to another class of antibiotics, the morpholines inhibit the same pathway at a later step. Fungicidal properties of morpholines were first reported in 1965. These antibiotics are primarily used against plant pathogenic fungi. The drug is absorbed through the roots and leaves. There are several morpholines available, of which, amorolfine can be used as a topical antifungal in human. It is the active ingredient in nail lacquer that is used for treatment of fungal infection of toes and fingernails. As shown in Fig. 8.4, amorolfine has a chiral center, and the S-enantiomer is much more active than the R enantiomer. Morpholines function by inhibiting two enzymes, D14 reductase and D7-D8 isomerase, in the post-lanosterol step of the ergosterol biosynthesis pathway [9].

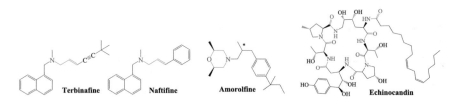

Fig. 8.4 Allylamine, morpholine and echinocandin antifungal drugs. The allyl groups in the allylamines (terbinafine and naftifine) and the morpholine portion of amorolfine, and the peptide bonds in echinocandin are highlighted in *red*. The star in amorolfine indicates chiral center at which the S stereoisomer has more antifungal activity than the R

8.1.3 Antibiotics That Inhibit Biosynthesis of Fungal Cell Wall

The structure of the fungal cell wall, which contains mannan, chitin, and α- and β-glucans, is unique to the fungal kingdom. So the fungal cell wall is a potential target for development of antifungal antibiotics. One such compound is echinocandin, which is a lipopeptide antibiotic (Fig. 8.4) that inhibits the enzyme β-(1 $\rightarrow$ 3) glucan synthetase, which is essential for fungal cell wall synthesis. Inhibition of the enzyme results in structural changes such as growth of pseudohyphae, thickened cell wall, and buds failing to separate from mother cells. Cells also become osmotically sensitive [7]. Resistance development to the antibiotic is by acquiring mutations in the glucan synthetase protein such that it binds poorly to echinocandin. Since the lipopeptide antibiotic cannot cross the fungal membrane, resistance development by preventing entry or by using efflux pumps are not possible mechanisms.

8.1.4 Flucytosine: Antimetabolite Antibiotic That Inhibits Fungal DNA and Protein Syntheses

One of the oldest antifungal (antimycotic) agents is flucytosine (5-fluorocytosine), a synthetic fluorinated analog of cytosine. It was first synthesized in 1957 as a potential antitumor agent; however, it was not found to be effective against tumors. Later it was found to have antifungal activity. Since its antifungal activity is not very strong and also because of high frequency of resistance development, it is used in combination with other antifungals, such as amphotericin B and fluconazole.

Flucytosine itself has no antifungal activity. It is taken up by the fungal cells by the enzyme cytosine permease, which is also the transport system for adenine and hypoxanthine. Once inside the cell, the 5-fluorocytosine is converted to 5-fluorouracil (Fig. 8.5). It is 5-fluorouracil that is responsible for the antifungal activity. However, 5-fluorouracil cannot be prescribed as antifungal drug because it is not taken up significantly by fungal cells and also, as explained before (Sect. 4.3.7), it is highly toxic to mammalian cells. Selectivity of flucytosine for fungal cells is a result of the absence of cytosine deaminase in mammalian cells [10]. Also, flucytosine is not absorbed well by mammalian cells. Some toxicity seen with flucytosine is due to its conversion to fluorouracil by microbes in the intestines and subsequent absorption into mammalian cells.

Mechanism of Action of Flucytosine In general, fluorinated analogs of substrates function in one or both of two possible ways. (1) Because fluorine and hydrogen are of similar sizes and each can form only one bond and thus not have a formal charge, the fluorinated substrate is able to bind to the active site of the enzyme. Thereafter it may act as an inhibitor in that reaction or it can be recognized as a substrate and get

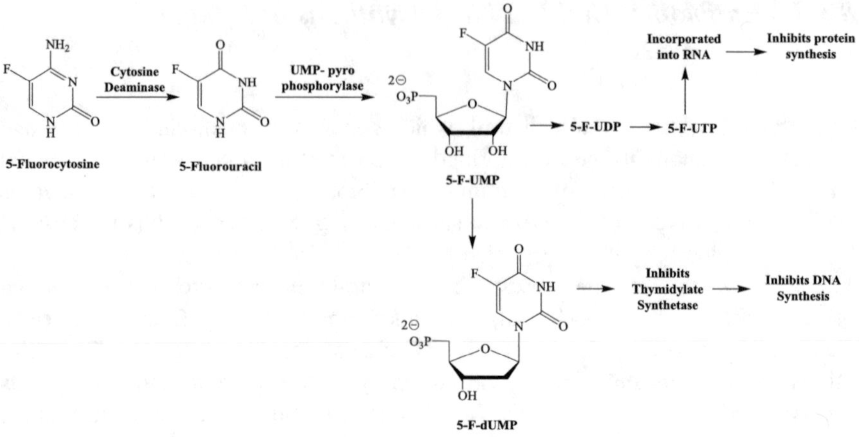

Fig. 8.5 Mechanism of antifungal activity of flucytosine

incorporated into the product, which can then function as an inhibitor in the next reactions in the metabolic pathway. (2) Although fluorine and hydrogen are of similar sizes, they have a large difference in electronegativities. If the usual mechanism of action of the reaction involves removal of the hydrogen as H^+, that step cannot take place with the fluorinated analog because fluorine cannot be removed as a F^+, and thus, the analog will function as a suicide inhibitor (Sect. 1.7.3).

Antifungal activity of 5-fluorocytosine takes place in two ways (Fig. 8.5). After 5-fluorocytosine is taken in by fungal cells, it is first deaminated by the enzyme cytosine deaminase to form 5-fluorouracil. Next the 5-fluorouracil is converted by the enzyme UMP pyrophosphorylase to 5-fluorouridylate (5F-UMP), which is further phosphorylated to form 5F-UDP and then 5F-UTP, which is recognized as a substrate for RNA biosynthesis and is incorporated into fungal RNA. This affects the aminoacylation of tRNA (Sect. 6.1) and thus affects the biosynthesis of proteins [11]. Another way by which 5-fluorouracil inhibits fungal growth is by forming 5-fluorodeoxyuridine monophosphate, which functions as a suicide inhibitor of thymidylate synthase as explained before (Sect. 4.3.7). This affects the synthesis of thymidine nucleotide, which is essential for DNA synthesis. The question arises whether inhibition of only one of these two pathways is important while the other is only incidental. However, it was demonstrated that inhibition of both these pathways (syntheses of RNA and DNA) contributes to the antifungal activity of 5-fluorouracil [11].

Resistance to flucytosine is usually due to mutations in cytosine permease, which decreases the uptake of cytosine by fungal cells, and mutations in the enzyme cytosine deaminase, which is responsible for the conversion to 5-fluorouracil. Resistance can also result from increased synthesis of pyrimidines, which compete with the fluorinated antimetabolite, 5-fluorouracil, and thus decrease its antifungal activity [10].

8.1.5 Combination Therapy Against Fungal Infection

Mortality rates for fungal infections can be very high (approaching 80%) even after therapy with antifungal agents. It has been demonstrated that the use of a combination of the antifungal drug amphotericin B and various known antibacterial antibiotics such as azithromycin or rifamycins, shows fungicidal synergy in vitro [12]. Rifabutin (a rifamycin) is an antibacterial agent very similar to rifampin but with a broader spectrum of activity. Although rifabutin alone shows no antifungal activity, when used in combination with amphotericin B, the MIC of the latter drug is reduced to a level that is easily achievable in human tissues. The fungicidal synergies in the two examples mentioned above are mediated by inhibition of protein synthesis and RNA synthesis, respectively. The lack of antifungal activity of rifabutin is due to its inability to cross the fungal membrane. The reason for the synergy is that amphotericin B damages the fungal cell membrane sufficiently enough to let the rifabutin enter the cell and inhibit fungal RNA synthesis [12]. Such synergy between antifungal and antibacterial agents is another justification for not having separate definitions for the two terminologies but instead list them all under the term, "antibiotics" as has been discussed before (Sect. 1.1).

8.2 Antimalarial Drugs: Antibiotics That Inhibit Growth of Malarial Parasites

The Romans believed that malaria was caused by "mal aria" or "bad air" from the swamps. In 1880, Charles Louis Alphonse Laveran, a French army surgeon, was the first to discover that malaria was caused by a single-cell parasite that was later named plasmodium. He was awarded the Nobel Prize in 1907 for his discovery. Several species of plasmodium are known to infect humans to cause malaria, including *Plasmodium vivax, Plasmodium malariae, Plasmodium ovale,* and *Plasmodium falciparum.* The disease is transmitted by mosquitoes and affects hundreds of millions of people and kills at least one million people each year. Most of these deaths are caused by *P. falciparum.* Drugs used to treat or prevent malaria target either the parasite or the mosquitoes that function in the transmission of the parasite between humans.

8.2.1 DDT the Most Well-Known Insecticide

DDT (*d*ichloro-*d*iphenyl-*t*richloroethane) (Fig. 8.6) is an organic insecticide that was synthesized during World War II and was widely used to kill mosquitoes to prevent the spread of malaria and other insect-borne diseases. DDT was first synthesized by Othmar Ziedler in 1870, but the insecticidal property of DDT was not discovered

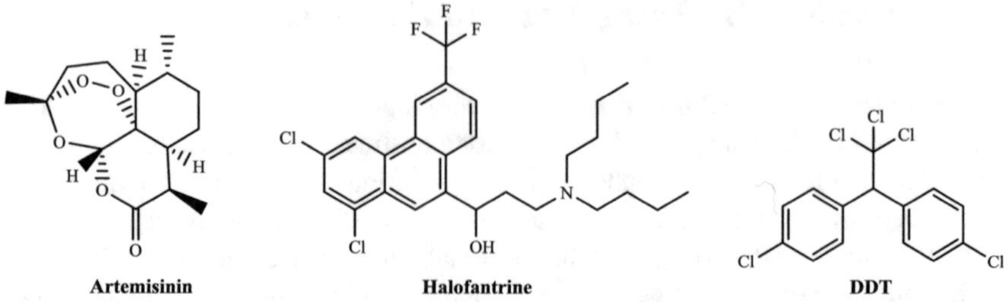

Fig. 8.6 Antimalarial drugs: artemisinin, halofantrine, and DDT

until 1939 by Paul Müller in Switzerland. Müller won the Nobel Prize for Medicine in 1948. DDT functions as an acute neurotoxic insecticide. DDT and other organochlorine class of insecticides have low water solubility and high lipid solubility. They are very resistant to degradation and so persist in the environment for a long time. The mechanism of toxicity of DDT is by hyperexcitation of the nervous system of insects. It functions by preventing the closing of the gates of sodium channels in the insect neurons after they are activated. This results in leakage of Na$^+$ ions through the nerve membrane. The hyperexcitability of the nerve results in repetitive discharges in the neuron after a single stimulus leading to uncontrolled spasms. Death of the insect is ascribed to respiratory failure after the disruption of nervous system function [13]. The author also reviewed the effects of various substituents in DDT and other alicyclic insecticides on activity. DDT was a widely popular insecticide and was primarily responsible for eradicating malaria from Europe and North America. However, currently it is banned in most of these places but continues to be used in most other countries mainly because it is highly effective in preventing the spread of malaria, it is very inexpensive and is not toxic for humans. Because of its lipid solubility, water insolubility, and extreme stability, DDT was found to accumulate in the food chain with predatory animals containing a high concentration of it, which results in toxicity. In the 1960s, there was huge public outcry against the use of DDT based on mainly a book, "Silent Spring" written by Rachel Carson. The book blamed the use of DDT for the decline in population of the bald eagle due to thinning of their egg shells. This led to the banning of the use of DDT in the developed world. Because of the continued use in other countries, resistance to DDT has now become widespread. One possible mechanism of resistance is by acquiring mutations in the sodium channel proteins, making them less sensitive to DDT.

8.2.2   Antimalarial Drugs: Quinine, Chloroquine, and Mefloquine

As mentioned before (Sect. 1.2), extract of cinchona bark was used in Peru, Bolivia and Ecuador to treat malaria as far back as the sixteenth century. The active

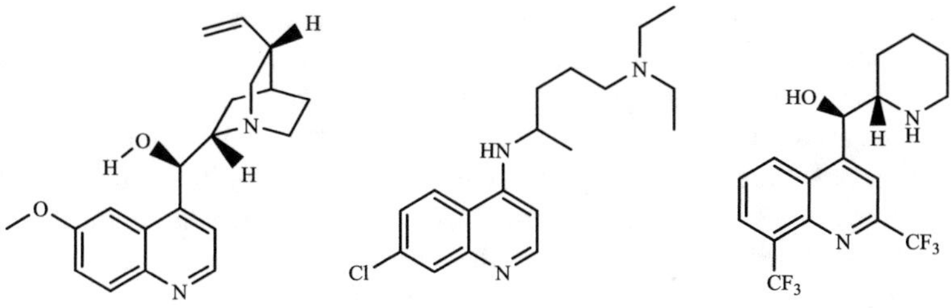

Fig. 8.7   Quinoline antimalarial drugs quinine, chloroquine, and mefloquine

component of the extract was later shown to be quinine (Fig. 8.7) which was the only available antimalarial drug until the 1940s when chloroquine, a 4-amino quinoline (not to be confused with the quinolones, Sect. 5.3) became a more popular drug of choice for both treatment and prophylaxis (use of antibiotics to prevent infections). Chloroquine was discovered by a German, Hans Andersag, in 1934 at Bayer I.G. Farbenindustrie A.G. laboratories in Eberfeld, Germany. He named his compound resochin. Mefloquine (brand name Larium) has now become popular since it needs to be taken only once a week. Another antimalarial drug, halofantrine is used for treatment but not for prophylaxis due to its toxicity.

Mechanism of Action   All four drugs have related structures and are expected to have similar mechanisms of action. Of these, chloroquine has been studied the most. In order to understand the mechanism, it is important to know the life cycle of the plasmodium parasite [14]. When an infectious mosquito bites, it injects sporozoites, which circulate and invade hepatocytes. After 1–2 weeks of asymptomatic hepatic infection, merozoites are released and invade erythrocytes. The asexual erythrocytic stage of infection is responsible for all clinical aspects of malaria. In erythrocytes, parasites develop into ring forms, mature trophozoites, and then multinucleated schizonts, which rupture and release more merozoites. Repeated cycles of erythrocyte invasion and rupture lead to symptoms of malaria. Some parasites develop into gametocytes, which may be taken up by mosquitoes, in which sexual reproduction and further development of the parasites lead to the generation of a new set of infectious sporozoites. One important property of *P. falciparum* is that erythrocytes infected with mature parasites adhere to vascular endothelium, and so are not taken to the spleen for clearance. So, it is only the ring form of *P. falciparum* cells that circulate in the blood. High numbers of circulating parasites lead to the manifestations of severe malaria.

Chloroquine is a 4-aminoquinoline with two basic centers with pKa values 9.8 and 8.6 for the quinolyl and tertiary nitrogens, respectively [15]. It is membrane soluble and accumulates in malaria-infected erythrocytes. Since the neutral pH of the cell is less than both pKa's of chloroquine, a large proportion of the chloroquine molecules will be protonated and thus have two positive charges. The few molecules that are not protonated are transported across the membrane into the digestive vacuole (lysosome) of the parasite. Once inside the lysosome, where the pH is 5.2 both the amino groups of most molecules will be protonated. Because of the two positive charges, the chloroquine is unable to cross the membrane, and thus the chloroquine accumulates in the lysosome as more and more molecules enter [16]. In the erythrocytes, the trophozoite digests the hemoglobin in order to utilize the amino acids. In the process, a toxic heme byproduct is formed. Normally, in the absence of chloroquine, the heme is detoxified by the enzyme heme polymerase, which polymerizes the heme to the nontoxic product hemozoin, which is insoluble and precipitates as a harmless black pigment. After the chloroquine is concentrated in the lysosome, it binds to heme and prevents its polymerization, leading to heme-mediated toxic effects and parasite death. Detailed mechanism of this process is not clear, but likely includes the binding of the drugs to either (or both) pre-crystalline soluble heme or (and) growing faces of the hemozoin crystal [15].

Resistance Development Chloroquine and the other quinolones are probably the most used antibiotics in human history. Of these, quinine and chloroquine are the least expensive and so have been used the most. Chloroquine has been widely used since the 1950s and has been very effective for both prophylaxis and treatment and is also safe in pregnancy. However, it is rather surprising that in spite of the widespread use in the past and in the present, not much resistance has developed to the drugs. The answer to this dilemma lies in the fact that the functions of chloroquine do not depend on binding to any protein. So the parasite cannot become resistant to the antibiotics simply by acquiring mutations in a gene. The target of chloroquine and related quinolines is the heme (called ferriprotoporphyrin IX) that is released after proteolytic digestion of hemoglobin within the digestive vacuole (lysosome) of the plasmodium parasite present in the erythrocytes. In order to decrease the binding of chloroquine to the heme, the structure of the heme needs to be changed, and that cannot be achieved by acquiring mutations in the parasite. Note that another antibiotic vancomycin functions by binding to a substrate and not to any protein. Thus, resistance to vancomycin is rare because the structure of the substrate cannot be changed by acquiring mutations in a gene. However, vancomycin resistance can still develop by the acquisition of a set of new genes that can be used to synthesize a different substrate (Sect. 3.3.3.7). Even such a rare mechanism is not possible for chloroquine resistance because the heme that chloroquine binds to is made by the host and not by the parasite. So the parasite cannot alter the structure of the heme to confer resistance to chloroquine. Therefore, chloroquine resistance is extremely rare, unique, complex, and has taken a long time to appear even in the face of massive use of the drug [15].

Because of the widespread use, some resistance has developed against the drug. However, the new drugs, mefloquine and halofantrine, which are supposed to work better than chloroquine, are too expensive to become popular in the developing countries, and so chloroquine continues to be used frequently. Resistance to chloroquine takes place by decreasing the accumulation of the drug in the digestive vacuole of the plasmodium. Chloroquine resistance involves greater genetic complexity than pyrimethamine resistance (Sect. 8.2.3), which can be conferred by a single mutation in the DHFR gene. Multiple mutations in the gene *pfcrt* (*plasmo-dium falciparum chloroquine resistance transporter*) are required to develop resistance to chloroquine. The gene codes for a transporter protein that is found in the vacuole membrane and may be involved in transport of the drug and/or in pH regulation. Eight mutations have been identified in chloroquine-resistant strains compared to sensitive strains. Seven of these eight mutations have been found to be common in all chloroquine-resistant strains obtained from diverse regions of Asia and Africa [17]. Method for quick detection of resistant strains of slow-growing pathogens will be discussed later in Sect. 8.2.5.

8.2.3 Antifolates as Antimalarial Antibiotics

Antifolates as antibacterial and anticancer drugs have been discussed in Chap. 4 (Sect. 4.3.6). Antifolate antimalarial drugs can be of two types: dihydrofolate reductase (DHFR) inhibitors such as pyrimethamine and cycloguanil and dihydropteroate synthetase inhibitors such as sulfonamides (Figs. 4.5, 4.6, and 8.8). Since these inhibitors act at two steps within the same pathway for the synthesis of tetrahydrofolate, the two types of inhibitors can be used in combination to obtain a synergistic effect. One such combination is Fansidar, which contains pyrimethamine and sulfadoxine (Fig. 8.8). Mechanisms of action of these drugs have been discussed in Sect. 4.3.6. Frequency of resistance development to the combination is much less than resistance to pyrimethamine when used alone. Note that even before starting to use the pyrimethamine-sulfadoxine combination there may already exist cross-resistance to it resulting from the resistance to the frequently used antibacterial antibiotic combination trimethoprim-sulfamethoxazole since the two combinations have similar modes of action (Sect. 4.3.6). High prevalence of markers for sulfadoxine and pyrimethamine resistance has been reported in *Plasmodium falciparum* even in the absence of drug pressure [18].

Another antimalarial drug is proguanil, which functions by inhibiting the DHFR enzyme. However, proguanil by itself does not have any effect on plasmodium DHFR. It is metabolized by the plasmodium to a cyclic triazine molecule, cycloguanil, which is the actual inhibitor of DHFR (Fig. 8.9). Thus, proguanil functions as a prodrug. However, this may not be the only mechanism of action for proguanil. It was observed that proguanil can act synergistically with another

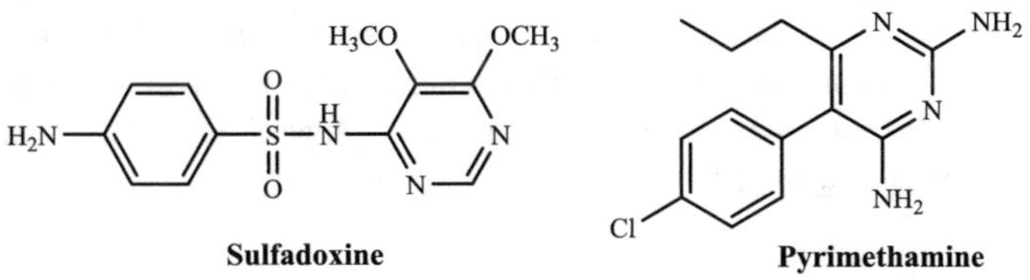

Sulfadoxine **Pyrimethamine**

Fig. 8.8 Antifolate antimalarials sulfadoxine and pyrimethamine

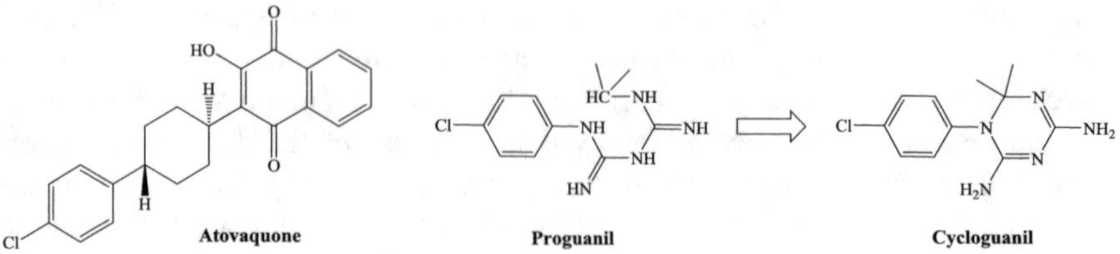

Fig. 8.9 Antimalarial drug atovaquone, prodrug proguanil and its conversion to cycloguanil

antimalarial drug, atovaquone but neither cycloguanil nor pyrimethamine had any synergistic effect when combined with atovaquone. This result suggests that proguanil in its prodrug form acts in synergy with atovaquone without being converted to cycloguanil [19]. Atovaquone is a naphthoquinone derivative and functions by collapsing the mitochondrial membrane potential in the plasmodium parasite. A combination of atovaquone and proguanil is sold under the trade name Malarone as an antimalarial drug.

Besides having antimalarial activity, the antifolates antibiotics also have antibacterial activity. Because of the widespread use of all antifolates for treatment of malarial as well as bacterial infections, significant resistance has developed to all of them, including the combination drugs. Mutations in the dihydrofolate reductase (DHFR) gene result in decreased binding to pyrimethamine or to cycloguanil, thus conferring resistance to the drugs. Some of these mutations have been identified. Method for quick detection of resistant strains of slow-growing pathogens will be discussed later in Sect. 8.2.5.

8.2.4 Artemisinins

The newest and most effective class of antimalarial drug is the artemisinin, a natural product developed in China in the 1960s. It is obtained from "quinghaosu" or sweet wormwood plant, *Artemisia annua*. The 2015 Nobel Prize in Medicine was awarded to Youyou Tu for her work on artemisinin. The mechanisms of action of artemisinins may include free-radical production in the parasite food vacuole and inhibition of a parasite calcium ATPase. One advantage of artemisinin is that it is effective against all stages in the life cycle of the parasite, including transmissible gametocytes. Thus it is highly effective in preventing the transmission of malaria. Artemisinin rapidly kills the malaria parasites at an early stage of their development, which explains its potency in the treatment of severe malaria. A number of artemisinin derivatives are also now available, including artesunate, artemether, artemotil, and dihydroartemisinin [14]. Recently it has been shown that artesunate inhibits glutathione S-transferase enzyme in *Plasmodium falciparum* [20].

8.2.5 Quick Detection of Resistant Strains

As mentioned above, widespread use of the various antimalarial drugs has resulted in resistance development. Some of these drugs, such as the antifolates, have also been used as antibacterial antibiotics causing higher frequency of resistance. Resistance to antifolates is usually due to point mutations. One such common mutation is at amino acid 108 in the active site of the enzyme DHFR. For example, pyrimethamine resistance is due to a Ser → Asn at position 108, while cycloguanil resistance is due to a Ser → Thr mutation at the same position. Note that these mutations are also abbreviated as Ser108Asn and Ser108Thr or as S108N and S108T respectively. As discussed above, resistance to chloroquine cannot be possible with just one mutation but involves eight mutations in the chloroquine transporter gene (Sect. 8.2.2). These eight mutations have been identified as M74I, N75E, K76T, A220S, Q271E, N326S, I356T, and R371I [17]. In most cases, these same mutations have been found in resistant strains from different patients in different geographic areas. These point mutations as well as whole genes that confer resistance to antibiotics are called resistance markers.

Early detection of resistance to antibiotics can help doctors determine which antibiotic to prescribe for a certain infection. For bacterial infection, this can be done by a broth dilution method or by a zone of inhibition method, as explained before (Sect. 2.2). However, these are less useful techniques for slow-growing pathogens, such as in case of malaria. In the USA, the Center for Disease Control (CDC) recommends that all cases of malaria diagnosed in the USA should be evaluated for drug resistance. Culturing the malaria parasite is slow, and moreover, the samples sent for testing may lose viability during transit. A much quicker method to detect antibiotic resistance in the infecting pathogen is by polymerase chain reaction, which is used to amplify the DNA in the region of the marker. This brings down the time required for the detection of antibiotic resistance from a few weeks to a few hours. The PCR method can work for all types of pathogens, such as bacteria, viruses, fungi, and parasites. As a method for the detection of Covid-19 virus that caused the 2019-pandemic, the technique has now become a household name. Also, viability of the pathogen in the sample is not necessary for doing PCR. Detection of the marker by PCR can help doctors select the best therapy for their patients. Another advantage of the PCR method is that multiple strains can be tested simultaneously for multiple markers.

The PCR method is briefly described here. The steps involved in PCR are shown in Fig. 8.10. In routine PCR, a template DNA is first denatured at high temperature to separate the two strands. It is then cooled so that the two strands hybridize to two short oligonucleotide primers which define the two ends of the region to be amplified. The binding of the primer to the template is known as annealing. The temperature is then raised to the optimal temperature for a thermophilic DNA polymerase so that the two template DNA strands can be copied by extension of each primer. The process is then repeated for multiple cycles. The first amplification of the defined

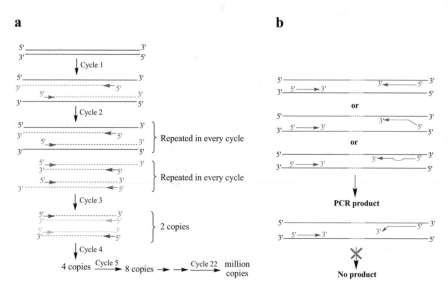

Fig. 8.10 Polymerase chain reaction (PCR). (**a**) Amplification of DNA during cycles of PCR. (**b**) Effectiveness of primers in PCR amplification. A few mismatches between primer and template are tolerated except even a single mismatch at the 3'-end

region of the template starts from the third cycle and then increases exponentially with each cycle (Fig. 8.10a).

PCR can also be used to detect antibiotic resistance not just for malarial parasite but for any infection. The presence of a known antibiotic resistance genes can be detected by amplifying part or whole of the gene whose sequence is known. Sequences of many antibiotic resistance genes are already known for many infecting species. Primers can be designed and are actually commercially available as part of kits for amplification of the genes. When multiple genes can be identified simultaneously by using multiple pairs of primers, the process is called multiplex PCR [21, 22].

PCR can also be used to identify specific point mutations that are known to confer resistance to various antibiotics. For efficient amplification, it is essential that the primer binds to the template, which takes place by hydrogen bonding between the complementary nucleotides. A few mismatches between the template and primer are tolerated and can still give a PCR product, although the annealing temperature may have to be lowered. Note that the mismatch will affect the polymerization only in the early cycles. After the third cycle, as the amount of PCR product increases exponentially, the effect of the mismatch will also decrease exponentially (Fig. 8.10a). The PCR product formed will have the primer sequence and not the template sequence at the two ends. However, if the mismatch is at the 3'-end, the polymerization reaction cannot initiate, and there will be no PCR product formed (Fig. 8.10b). This is true even if the mismatch at the 3'-end is for one base. This technique is known as amplification refractory mutation system (ARMS) [23]. This inability to give a PCR product is used to detect specific mutations in a DNA. Since most of the common mutations present in various resistant species are already known, primers can be designed and are actually commercially available to amplify part of the DNA containing this mutation. The 3'-ends of all these primers, each of which can serve as one of the two primers needed for each PCR amplification, are the same as the positions of the mutations, while the second primer of each set can be complementary to a common region at a distance that defines the length of the PCR product to be obtained. Thus the wild-type strain will give a PCR product only if the primer with the right 3'-end is used and will not give a product with the primer containing the 3'-end corresponding to any of the mutants. Similarly, the primer specific for a certain point mutation will give a product only if the infection is with that strain and not with any other strain. This way standard mutations in the infecting species can be identified quickly.

8.3 Antiviral Drugs: Agents That Inhibit Multiplication and Spread of Viruses

Although viral infections are as old as any other infections, it has taken the longest time to develop drugs for treating viral infections. However, since the mid 1980s more than 50 antiviral antibiotics (see Sect. 1.1 for use of the terminology) have been developed for clinical use. Some of these are broad-spectrum antivirals but a majority of them are specific for a certain viral species. Most of these antivirals are against the human immunodeficiency virus (HIV) while the remaining are against herpes, hepatitis B and C, and influenza A and B viruses [24]. Since viruses are not considered as living, killing viruses does not have a similar meaning as killing bacteria. Viruses consist of a genome (either DNA or RNA) that is present inside a protein envelope called a capsid. They may also contain a few enzymes that are not present in the host cells but are essential for the propagation of the virus. Viruses do not have all the enzymes and metabolites that are required for their replication but use those that are present in the host cells. Although there are disinfectants that are viricidal, meaning that they "kill" viruses usually by denaturing the protein envelope, all antiviral drugs function by specifically inhibiting the viral propagation.

8.3.1 Targets of Antiviral Antibiotics

Since viruses use the host cells' machinery for their replication, it is understandable why it took so long to develop antiviral drugs. The targets of an ideal antiviral agent cannot be any of the proteins that are also necessary for the host. The target should be other steps in the life cycle of the virus, which includes the following steps:

attachment to host cells, release of DNA and enzymes into the host cell, replication of viral DNA, synthesis of viral proteins, assembly of DNA and proteins to form the complete viral particles and finally, release of the particles from the host cell to infect other cells. Retroviruses, which contain RNA instead of DNA, have the additional step of reverse transcribing their RNA to DNA before replication and copying DNA to make RNA before packaging it to form the viral particles. Antiviral drugs have been developed targeting most of these steps in the life cycle of viruses. Vaccines stimulate the immune system to make antibodies that attack the virus before the start of its life cycle. Thus vaccines prevent infection but are not of much use in the treatment of infection and so will not be discussed further.

8.3.2   Antivirals That Inhibit Entry and Uncoating of Viruses

A peptide named Fuzeon was developed to inhibit the entry of virus into the host cell. Three other antiviral antibiotics, arildone, amantadine, and rimantadine (Fig. 8.11) function against influenza virus by inhibiting the uncoating of the protein capsid. The two molecules are derivatives of adamantane, which contains three fused cyclohexane rings. The effects of these drugs are maximum when they are present from the time of inoculation and decrease as the time interval from the inoculation of the virus to the addition of the drug increases [25]. Removal of the drug from infected cells by washing, reversed the effect, and viral replication resumed suggesting that the site of action of the drug is in the membrane prior to entry. Another antiviral drug pleconaril is being developed as an intranasal spray against rhinoviruses which causes common cold. It also functions by inhibiting uncoating of the virus.

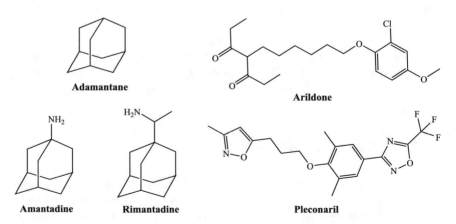

Fig. 8.11 Antivirals that inhibit entry and uncoating of viruses. Adamantane, which does not have antiviral activity, is included to show its similarity to amantadine and rimantadine

8.3.3   Antivirals That Inhibit Synthesis of DNA or RNA of Viruses

Maximum number of antivirals that have been developed belongs to this category. Most of these function as nucleoside or nucleotide analogs which resemble the natural substrate significantly enough such that they are incorporated into the DNA or RNA. However, this prevents further synthesis of the DNA or RNA because the next nucleotide cannot be added. These drugs work better against reverse transcriptase enzyme, which copies RNA to synthesize DNA. The first successful antiviral drug, acyclovir, approved by FDA in 1982, is a guanosine analog but lacks part of the ribose ring structure (Fig. 8.12) and is effective against herpes virus. Zidovudine (azidothymidine or AZT), approved by FDA in 1987, is a thymidine analog that is used against HIV. Another nucleoside analog developed as a reverse transcriptase inhibitor is lamivudine ($2'$,$3'$-dideoxy $3'$-thiacytidine, or 3TC), which was approved by FDA in 1995, is a cytidine analog. Note that the prefix thia- refers to a sulfur in place of a carbon, while thio- means a sulfur in place of an oxygen. Lamivudine is used against hepatitis virus and is used in combination with zidovudine for treatment against HIV. All three nucleoside analogs are actually prodrugs since they are phosphorylated first to the mono- then di- and finally to the triphosphates, which then function as substrates for the reverse transcriptase enzyme. However, the lack of $3'$-OH in all these three antiviral drugs prevents the further synthesis of DNA and so they act as chain terminators. Note that this same principle is used in DNA sequencing method done in the laboratory. Another nucleoside analog, 5-iododeoxyuridine, has the structure of thymidine with the methyl group replaced with iodine. Since methyl and iodo group have similar sizes, the drug functions as a substrate of DNA polymerase and is incorporated into DNA but later causes errors in replication and transcription. The molecule was initially tried as an anticancer drug but was not pursued further due to its toxicity. Instead, it became the first antiviral drug in 1962 and is used only topically to treat herpes infection. There are many other nucleoside analogs that function as reverse transcriptase inhibitors. Some of these, along with their dates of approval by FDA are listed here: ddI (dideoxyinosine, 1991), ddc (dideoxycytidine, 1992), d4T (analog of thymidine,

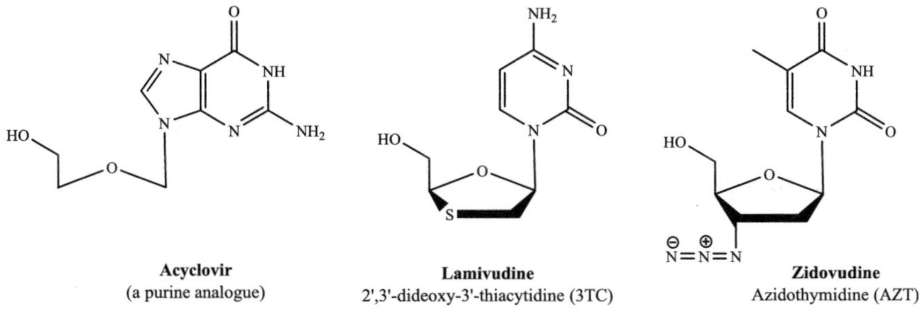

Acyclovir
(a purine analogue)

Lamivudine
2',3'-dideoxy-3'-thiacytidine (3TC)

Zidovudine
Azidothymidine (AZT)

Fig. 8.12 Nucleoside analogs as reverse transcriptase inhibitors

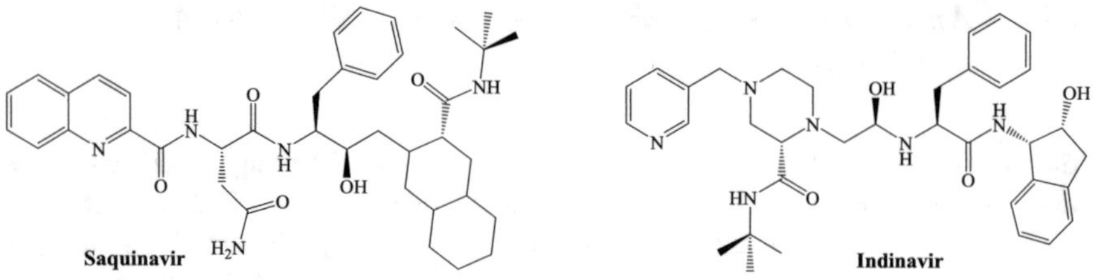

Fig. 8.13 HIV protease inhibitors saquinavir and indinavir

1995), ABC (analog of guanosine, 1998), FTC (analog of cytidine, 2003). Specificity of all these nucleoside analogs is based on the fact that these are more specific for viral reverse transcriptase and viral DNA polymerase compared to the host's DNA polymerase. However, the specificity is very limited, and there is significant toxicity associated with these drugs since they also inhibit the mammalian DNA polymerase. Besides the nucleoside analogs, several non-nucleosides have also been approved as antiviral drugs by the FDA. Also, the focus of antiviral drug development has now shifted to a much more effective class of drugs: the protease inhibitors (Fig. 8.13).

8.3.4   Protease Inhibitors as Antiviral Drugs

HIV makes the most efficient use of its small genome by synthesizing several proteins together as a long polypeptide which is then cleaved by a specific protease to obtain the various individual proteins that are needed for the assembly of the viral particle. HIV genome contains three major genes that code for the three long polypeptides: the Gag (group-specific antigen) precursor is cut to give structural proteins, the Pol (polymerase) precursor is cut to give the enzymes such as reverse transcriptase, integrase, and protease, and the Env (envelope) precursor is cut to give proteins that are further modified to form the glycoproteins of the envelope. The protease specifically cuts between a phenylalanine and a tyrosine or proline. The recognition sequence of the protease is unique for HIV since there is no protease present in humans with that specificity. Thus this represents a potential target for developing antiviral drugs (Fig. 8.13). Using computer modeling, several inhibitors of HIV protease have been developed, and many of them have been approved for clinical use [26]. Similar inhibitors have been developed for other viruses as well. Examples include saquinavir, ritonavir, indinavir, nelfinavir, and amprenavir against HIV and boceprevir, telaprevir, and simeprevir against Hepatitis C. Protease inhibitors are highly effective antiviral drugs and have been successful in bringing down the death rate due to HIV to less than half since they were first approved in the mid-1990s.

8.3.5   Neuraminidase Inhibitors as Antiviral Drugs

The final step in the viral life cycle is the release of the mature viruses from the host cell surface so that they can infect other cells. The process is the reverse of the viral entry step. During entry, the host cell membrane surrounds the virus to form endosomes. In the release step, the virus particles come out of the cell by the process of budding. The surface of the influenza virus contains two glycoproteins: hemagglutinin and the enzyme neuraminidase. During entry the viruses bind to the host cell by the interaction of the hemagglutinin on the surface of the virus with a sialic acid moiety in glycoproteins present on the surface of the host cell. During viral release, the same interaction would prevent release of the virus from the cell surface. The enzyme neuraminidase cleaves the sialic acid from the host glycoproteins and thus facilitates release of the viral particles. The neuraminidase also cleaves the sialic acid from the viral protein. Since this is an essential step in the life cycle of the virus, the enzyme neuraminidase is a target for development of antiviral drugs. Two such drugs, oseltamivir (Tamiflu) and zanamivir (Relenza) are effective antiviral drugs (Fig. 8.14) against flu viruses.

8.3.6   Antiviral Drugs with Unusual Mechanisms of Action

8.3.6.1   Antisense Oligonucleotides

Translation of viral mRNA to make viral proteins can be a potential target for development of antiviral drugs. Antisense oligonucleotide DNA that is complementary to some important part of the viral RNA genome is designed and synthesized. Binding of the oligonucleotide to the targeted part of the genome prevents translation of the RNA. One such antisense drug is fomivirsen that is used to treat opportunistic eye infection caused by cytomegalovirus in AIDS patients. Other strategies are under development for using antisense therapeutics. For example, antisense oligonucleotides can be used to inhibit protein synthesis by mediating the catalytic degradation of target mRNA. Double-stranded RNA oligonucleotides, known as short-interfering RNAs (siRNAs), can also be used to mediate the catalytic degradation of complementary mRNAs [27].

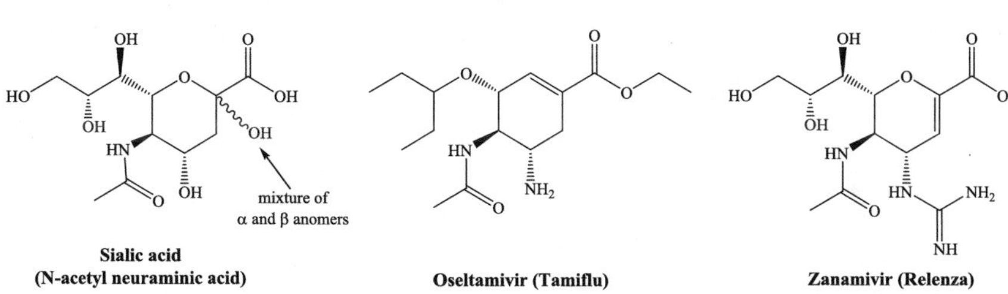

Fig. 8.14 Neuraminic acid and neuraminidase inhibitors: Tamiflu and Relenza

8.3.6.2   Interferons

Another unusual antiviral therapy is interferon therapy. Interferons are proteins secreted by the host cells in response to infection by pathogens, including viruses, bacteria, parasites, or cancer cells. Interferons interfere with viral replication (hence the name), trigger the immune system and also send signal to the surrounding uninfected host cells such that they can take action to resist infection by the pathogen. Cells respond to the interferon by reducing protein synthesis in the cell or by destroying the RNA in the cell which again, in turn reduces protein synthesis. They also induce apoptosis in virus-infected cells. Because of these effects, interferon or agents that stimulate the synthesis of interferon can be used as antiviral agents. One such drug, imidazoquinoline, is used topically against infections, including carcinoma. Interferon gene from humans can be cloned into bacteria to mass produce the protein, which is then used topically as antiviral or anticancer agents.

8.3.7   Resistance Development Against Antiviral Drugs

Although many of the antiviral drugs described above are very effective, there is always the problem of rapid development of resistance to the drugs [28]. Mechanism of resistance is usually by acquiring point mutations in the target. Method of acquisition of point mutations in bacteria has been discussed in Sect. 2.4. However, the process of acquiring point mutations is much faster in viral infections than in bacterial infections. This is because of the following reasons. Most viruses multiply at a much faster rate than bacteria, and so there is a higher probability of acquiring mutations. The virus load (number of viral particles) in infected patients is much higher than in case of bacterial infection. So there is a higher probability of developing mutations. The error rate of bacterial DNA replication is extremely low because of the proofreading activity of DNA polymerase and also because of DNA repair systems (Sect. 2.4). In contrast, reverse transcriptase does not have any proofreading activity, and there is no repair system for correcting mutations in RNA that are packaged into viral particles. Resistance development against antiviral drugs takes place within a few weeks in case of monotherapy and a few months in case of combination therapy. This time is much shorter than the total treatment time needed for curing the infection. So before the patient is cured, the virus may develop resistance to the drugs.

References

1. Dutcher JD (1968) The discovery and development of amphotericin B. Dis Chest 54:40–42
2. Brajtburg J, Powderly WG, Kobayashi GS, Medoff G (1990) Amphotericin B: current understanding of mechanisms of action. Antimicrob Agents Chemother 34:183–188
3. Ellis D (2002) Amphotericin B: spectrum and resistance. J Antimicrob Chemother 49(S1):7–10
4. Dismukes WE (2000) Introduction to fungal drugs. Clin Infect Dis 30:653–657
5. Vanden Bossche H, Marichal P, Gorrens J, Coene MC, Willemsens G, Bellens D, Roels I, Moereels H, Janssen PA (1989) Biochemical approaches to selective antifungal activity. Focus on azole antifungals. Mycoses 32(Suppl 1):35–52
6. Lupetti A, Danesi R, Campa M, Tacca MD, Kelly S (2002) Molecular basis of resistance to azole antifungals. Trends Mol Med 8:76–81
7. Ghannoum MA, Rice LB (1999) Antifungal agents: mode of action, mechanisms of resistance, and correlation of these mechanisms with bacterial resistance. Clin Microbiol Rev 12:501–517
8. Ferreira GF, Baltazar LM, Santos JRA, Monteiro AS, Fraga LAO, Resende-Stoianoff MA, Santos DA (2013) The role of oxidative and nitrosative bursts caused by azoles and amphotericin B against the fungal pathogen *Cryptococcus gattii*. J Antimicrob Chemother 68: 1801–1811
9. Mercer EI (1991) Morpholine antifungals and their mode of action. Biochem Soc Trans 19:788–793
10. Vermes A, Guchelaar H-J, Dankert J (2000) Flucytosine: a review of its pharmacology, clinical indications, pharmacokinetics, toxicity and drug interactions. J Antimicrob Chemother 46:171–179
11. Waldorf AR, Polak A (1983) Mechanisms of action of 5-fluorocytosine. Antimicrob Agents Chemother 23:79–85
12. Clancy CJ, Yu YC, Lewin A, Nguyen MH (1998) Inhibition of RNA synthesis as a therapeutic strategy against *Aspergillus* and *Fusarium*: demonstration of in vitro synergy between rifabutin and amphotericin B. Antimicrob Agents Chemother 42:509–513
13. Coats JR (1990) Mechanisms of toxic action and structure-activity relationships for organochlorine and synthetic pyrethroid insecticides. Environ Health Perspect 87:255–262
14. Rosenthal PJ (2008) Artesunate for the treatment of severe falciparum malaria. N Engl J Med 358:1829–1836
15. Natarajan JK, Alumasa J, Yearick K, Ekoue-Kovi KA, Casabianca LB, de Dios AC, Wolf C, Roepe PD (2008) 4-N, 4-S & 4-O Chloroquine analogues: influence of side chain length and quinolyl nitrogen pKa on activity vs. chloroquine resistant malaria. J Med Chem 51:3466–3479
16. Warhurst DC, Steele JCP, Adagu IS, Craig JC, Cullander C (2003) Hydroxychloroquine is much less active than chloroquine against chloroquine-resistant *Plasmodium falciparum*, in agreement with its physicochemical properties. J Antimicrob Chemother 52:188–193
17. Wellems TE, Plowe CV (2001) Chloroquine-resistant malaria. J Infect Dis 184:770–776
18. Marks F, Evans J, Meyer CG, Browne EN, Flessner C, von Kalckreuth V, Eggelte TA, Horstmann RD, May J (2005) High prevalence of markers for sulfadoxine and pyrimethamine resistance in *Plasmodium falciparum* in the absence of drug pressure in the Ashanti region of Ghana. Antimicrob Agents Chemother 49:1101–1105
19. Srivastava IK, Vaidya AB (1999) A mechanism for the synergistic antimalarial action of atovaquone and proguanil. Antimicrob Agents Chemother 43:1334–1339
20. Lisewski AM, Quiros JP, Ng CL, Adikesavan AK, Miura K, Putluri N, Eastman RT, Scanfeld D, Regenbogen SJ, Altenhofen L, Llinás M, Sreekumar A, Long C, Fidock DA, Lichtarge O (2014) Supergenomic network compression and the discovery of EXP1 as a glutathione transferase inhibited by Artesunate. Cell 158:916–928
21. Strommenger B, Kettlitz C, Werner G, Witte W (2003) Multiplex PCR assay for simultaneous detection of nine clinically relevant antibiotic resistance genes in *Staphylococcus aureus*. J Clin Microbiol 41(9):4089–4094

22. Bockelmann U, Dorries H-H, Ayuso-Gabella MN, de Marc MS, Tandoi V, Levantesi C, Masciopinto C, Van Houtte E, Szewzyk U, Wintgens T, Grohmann E (2009) Quantitative PCR monitoring of antibiotic resistance genes and bacterial pathogens in three European Artificial Groundwater Recharge Systems. Appl Environ Microbiol 75:154–163
23. Newton CR, Graham A, Heptinstall LE, Powell SJ, Summers C, Kalshekerl N, Smith JC, Markham AF (1989) Analysis of any point mutation in DNA. The amplification refractory mutation system (ARMS). Nucleic Acids Res 17:2503–2516
24. Razonable RR (2011) Antiviral drugs for viruses other than human immunodeficiency virus. Mayo Clin Proc 86:1009–1026
25. Kim KS, Sapienza VJ, Carp RI (1980) Antiviral activity of arildone on deoxyribonucleic acid and ribonucleic acid viruses. Antimicrob Agents Chemother 18:276–280
26. Wlodawer A, Vondrasek J (1998) Inhibitors of HIV-1 protease: a major success of structure assisted drug design. Annu Rev Biophys Biomol Struct 27:249–284
27. Spurgers KB, Sharkey CM, Warfield KL, Bavari S (2008) Oligonucleotide antiviral therapeutics: antisense and RNA interference for highly pathogenic RNA viruses. Antiviral Res 78:26–36
28. Strasfeld L, Chou S (2010) Antiviral drug resistance: mechanisms and clinical implications. Infect Dis Clin North Am 24:413–437

Chapter 9
Alternative Approaches for Antibiotic Discovery

Abstract Antimicrobial resistance is on the rise and very few new antibiotics are being discovered. Some alternative approaches for antibiotic discovery in order to combat the threat of antimicrobial resistance have been discussed. These include drug repurposing, anti-virulence approaches, inhibition of quorum sensing, inhibition of biofilms, bacterial toxin neutralization, metal ion chelation, antibiotics from plants, nanoparticles as antibiotics, and antibiotics against persister bacteria. Methods for improved drug delivery targeting infected sites and intracellular bacteria are also discussed.

9.1 Drug Repurposing

Antimicrobial resistance is on the rise, so it is of utmost importance to discover new antibiotics. However, very few new antibiotics have been discovered in the last two decades. There are several reasons for this [1]. It is not cost-effective for pharmaceutical companies to invest money for discovering new antibiotics. They can make more money on lifestyle medicines for the maintenance of chronic diseases such as diabetes or high cholesterol. Another obstacle is that new antibiotics have to go through the process of approval by regulatory authorities such as the FDA in the USA. For this, the pharmaceutical company has to demonstrate that the new drug is better than the existing ones and not just another one of similar kind.

Although the search for new antibiotics is desirable, it is a very slow process. An alternative approach that will save time and expense is to use existing drugs for other diseases and test them for antibiotic properties or chemically modify existing drugs that have already been approved for other purposes, such that they now function as antibiotics. Such a strategy is called "drug repurposing" or "drug repositioning" or "drug reprofiling" or "drug redirecting" or "drug rediscovery." Some scientists prefer to differentiate drug redirecting from the other terms in that it involves drug modification to change/redirect the cellular targets and the subsequent indications [2]. One big advantage of repurposing drugs is the bypassing of the need for spending enormous amount money, time, and human resources to bring the drug to the market. Moreover, since these drugs have already been approved for clinical use, a lot of necessary data already exists, especially on toxicity or lack thereof. Since these have not yet been used as antibiotics, it is unlikely that there is any pre-existing resistance to these drugs.

There are several disadvantages of repurposing. One drawback is that pharmaceutical companies cannot make a lot of money from such a drug with an expired patent. However, in the EU and the USA, there are ways to extend the market exclusivity period of protected drugs [3]. Under the current threat of the Covid-19 pandemic, there has been a flood of reports of repurposing drugs without sufficient scientific evidence. Attempted repurposing of hydroxychloroquine as a cure for Covid-19 should sound an alarm bell and result in more regulation of the process. One antibiotic that has frequently been prescribed as a cure for Covid-19 is azithromycin. Whether there is sufficient evidence for antiviral activity of an antibacterial antibiotic or not, one definite consequence of this practice is that it will contribute to increase of antimicrobial resistance.

Drug repurposing is not a new concept. A bibliometric survey concluded that >60% of all drugs have been tested against more than one disease [4]. One common repurposing goal is to use antibacterials as antiviral drugs. For example, the macrolide antibiotic leucomycin A3 was shown to have activity against influenza A virus [5]. There have been many reviews written on the numerous other examples of existing drugs that have been successfully repurposed for other diseases. Only a few of these drugs are discussed here as examples.

9.1.1   Repurposing the Anticancer Drug YM155

One example of drug redirecting is the anticancer drug YM155 which was identified by a high-throughput screening of chemical libraries [2]. The anticancer activity of YM155 is due to the fused imidazolium structure, which causes reactive oxygen species (ROS)-mediated enhancement of apoptotic functions. YM155 was also found to have a weak antibacterial activity against methicillin-resistant *Staphylococcus aureus* (MRSA) (MIC of 50 µg/ml). The group highlighted in red at the N3 position in Fig. 9.1 was replaced with various groups to screen for antibacterial activity. The analog in which it was replaced with a hexyl group gave the best activity against MRSA (MIC of 3.13 µg/ml), which is in a clinically achievable range. The mechanism of antibacterial activity involves the generation of reactive oxygen species, presumably during respiratory electron transport.

9.1.2   Chloroquine

Chloroquine was first synthesized in 1934 but was not used initially due to fear of toxicity. Later, starting from 1946, chloroquine has been successfully used to treat malaria. It also been repurposed for treatment of many other diseases. Since it is very

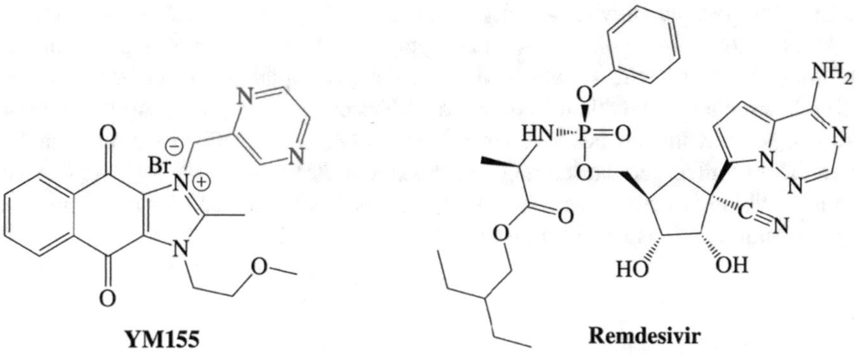

Fig. 9.1 Structures of YM155 and Remdesivir

effective against malarial parasite, the obvious repurposing approach was to test it against other parasitic diseases such amebic dysentery, giardiasis, trichomonas, etc. Chloroquine functions by invading the lysosome of malarial parasite. Besides its lysosomotropic mechanism of action, it also has other activities such as DNA intercalation, disruption of transcription, translation, and inhibition of tumor necrosis factor (TNF-α). Based on these activities, chloroquine has been investigated for use in cancer, viral and bacterial infections [4]. Anticancer activity of chloroquine is believed to be a result of its inhibitory effect on autophagy, which is the process by which damaged organelles are degraded in the lysosomes. However, such an interrelationship between autophagy, cancer, and chloroquine has been questioned [6]. The authors demonstrated that the antiproliferative effects of chloroquine are independent of its suppressive actions on autophagy. Since viruses also use autophagy to recover energy for survival, chloroquine also has been tested for antiviral property, but no positive results have been obtained in viral infection. Recently there was an attempt to promote hydroxychloroquine as a treatment for Covid-19, however, there is no scientific basis for such use of the drug.

9.1.3   Teicoplanin

The severity of the current Covid-19 pandemic has created an urgency to discover cures for the disease. A convenient but logical approach is to repurpose drugs that are already approved for other diseases. In a study, many approved drugs, mostly antibiotics were screened for inhibitory activity against SARS CoV-2 [7]. The virus contains a chymotrypsin-like (CL) protease called the main protease or 3CL^{Pro}, that is essential for the processing of polyproteins PP1A and PP1B that are translated from the RNA of the virus. Thus the protease plays an essential role in the initiation of replication of the virus. The authors tested in vitro binding to the 3CL^{Pro} and inhibition of proteolytic activity using an artificial substrate in order to screen for possible antiviral activity of the drugs. Several drugs such as lopinavir, hydroxychloroquine, chloroquine, azithromycin, atazanavir showed mild binding affinity for the protein. However, the antibiotic teicoplanin showed about 10- to 20-fold stronger activity than all other drugs. Teicoplanin is a glycopeptide antibiotic similar to vancomycin but less toxic (Sect. 3.3.3.6). It is effectively used in the treatment of infections caused by gram-positive bacteria, including MRSA. The use of an antibiotic to cure a viral infection again points to the need for redefining the term, "antibiotics" as has been discussed before (Sect. 1.1).

9.1.4 Remdesivir

Remdesivir is a new drug developed by Gilead. Research on it had started in 2009 when it was explored as treatment for hepatitis C (HCV) and respiratory syncytial virus (RSV) (usually causes mild cold-like symptoms). Later the antiviral profile of remdesivir was expanded to include emerging viruses, including Ebola, SARS, Marburg, and MERS. In 2018, NIH began a study of remdesivir and other drugs in patients with Ebola disease, but later the trial using remdesivir was discontinued. Because of the pressing need for new drugs in the early stages of the Covid-19 pandemic, clinical trial was started using remdesivir against coronavirus. In October 2020, FDA approved remdesivir for treatment of Covid-19.

Mechanism of Action Antiviral activity of nucleoside analogs can be of three types: mutagenic nucleosides, obligate chain terminators, and delayed chain terminators. Mutagenic nucleosides result in mutated RNA. Obligate chain terminators, such as azidothymidine (AZT), lack the OH group at $3'$ position and so stops RNA synthesis and delayed chain terminators allow a few more nucleotides to be incorporated and then causes the RNA synthesis to stop. Remdesivir belongs to the third type [8]. Remdesivir is a prodrug containing a nucleoside analog with the negative charges of phosphate blocked by two attached groups which facilitate permeability through the cell membrane. After entry into the cell, the groups are removed, forming the nucleoside monophosphate, which is then further converted to the diphosphate and then to the triphosphate. Since this is an analog of the natural substrate, it is incorporated into RNA. Moreover, since the molecule has a 3'-OH group, it can continue to add up to three more nucleotides. Beyond that, the RNA synthesis cannot continue because the cyano group at $1'$ position prevents translocation of the RNA chain during polymerization resulting in premature release of RNA.

9.2 Anti-virulence Approaches

Bacteria cause diseases by producing various virulence factors. The mere presence of bacterial cells does not cause disease. So, instead of developing antibiotics targeting bacterial growth, it can also be effective to develop drugs that can inhibit the virulence mechanism of the bacteria. The use of such anti-virulence drugs will decrease the use of conventional antibiotics and thus decrease the development of antibiotic resistance. Moreover, anti-virulence therapies do not directly affect bacterial viability and so can result in less resistance development because of reduced selective pressure on the bacteria. Since all bacteria do not have the same virulence mechanism, such antibiotics will be more specific and of narrow spectrum. Thus they will not kill all bacteria, including beneficial ones. Some possible examples are drugs that inhibit virulence processes such as quorum sensing [9] or biofilm formation [10], or chelation of essential metal ions such as iron [11].

9.2.1 Inhibition of Quorum Sensing

Many pathogens are known to utilize a quorum sensing system, which allows individual bacteria within a colony to communicate and coordinate with other cells in the colony in order to facilitate group behavior such as conjugation, competence for natural transformation, sporulation, biofilm formation, and virulence. Although bacteria are single-celled organisms, the system of quorum sensing allows them to regulate gene expression in response to cell-population density. Bacteria carry out quorum sensing by releasing signal molecules known as autoinducers. The concentration of autoinducer in the environment increases as the cell density increases. When a certain cell density is attained, the bacteria can sense the critical concentration of the autoinducer and activate the genes for various group behaviors. One mechanism by which bacteria use the quorum sensing system to evade the host's immune response is by delaying the release of virulence factors till a critical cell-population density is reached when it can overwhelm the host immune system. At this cell density, they activate genes for virulence and biofilm formation resulting in a more effective infection. There is also evidence of inter-species quorum sensing whereby two bacterial species can enhance each other's virulence.

Thus quorum sensing system and more specifically, the autoinducers can be promising targets for development of new antibiotics that can interfere with the cell-to-cell communication. *Pseudomonas aeruginosa* is an opportunistic pathogen that uses a quorum sensing system and biofilm formation to cause an effective infection. The quorum sensing system in *P. aeruginosa* utilizes acyl homoserine lactones as signal molecules. Mutants with a defective quorum sensing system have been found to be more susceptible to antibiotics. Several quorum sensing inhibitors have been identified, but none of them have reached a clinical trial stage. Recently,

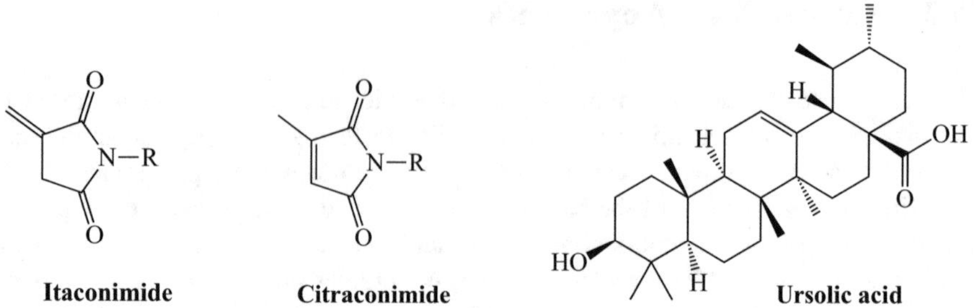

Fig. 9.2 Drugs with anti-virulence activities

itaconimide and citraconimide (Fig. 9.2) have been identified as novel quorum sensing inhibitors of *P. aeruginosa* [12].

Another strategy is to inhibit the activity of the autoinducer-producing enzymes without affecting bacterial growth. Several inhibitors of the enzyme have been developed [13]. One advantage of this strategy is that the autoinducer-producing enzymes are not present in mammalian cells but are present in several bacterial pathogens. Thus, these antibiotics will be highly specific against bacteria.

9.2.2   Inhibition of Biofilms

Antibiotics are not very effective in the presence of foreign bodies such as sutures and implants because bacteria form biofilms on foreign bodies. Biofilms are communities of microorganisms attached to their extracellular polymeric matrix. Bacteria in biofilms are protected from the effect of antibiotics for reasons that are not well understood. One possibility is that bacteria in biofilms are slow-growing or non-growing and so respond less to antibiotics. The extracellular polymeric substance (EPS) probably somehow chelates or sequesters the antibiotics making them less effective. Lyme disease, caused by *Borrelia burgdorferi*, is one example of diseases that are difficult to treat with antibiotics. According to the CDC data, more than 300,000 people get Lyme disease in the USA every year. Antibiotics are not only ineffective against the bacteria in biofilm, they actually promote formation of more biofilm. It is believed that the reason for this is that biofilm formation is triggered when the cells sense stress and presence of antibiotics causes that stress.

Alternative methods need to be developed for the treatment of bacteria in biofilms. One possibility is to hydrolyze the biofilm with an enzyme such as glycoside hydrolase, to allow the antibiotic to have a greater effect. It was shown that the periodontal pathogen, *Aggregatibacter actinomycetemcomitans* produces a glycoside hydrolase, dispersin B that can effectively hydrolyze its own biofilm [14]. It was demonstrated that this enzyme, dispersin B when combined with a sublethal concentration of colistin, reduced the viable counts of *Pseudomonas aeruginosa* by 2.5 orders of magnitude when used either prophylactically or on established 24-hour biofilms [15]. Other approaches for making antibiotics more effective against bacteria in biofilms is by disruption of the biofilm with human Dnase I [16] or by using diarylquinolines that can kill both planktonic cells as well as those growing in biofilms [17] or by adding exogenous metabolites to stimulate their central metabolic pathways [18] so that they can then be killed by antibiotics. Another inhibitor of biofilm is ursolic acid (Fig. 9.2). It is a natural product derived from many plants. It has antimicrobial and antitumor activities. Ursolic acid inhibits the growth of oral streptococci and inhibits their biofilm formation and suppressed expression of glucosyltransferase genes that are necessary for formation of plaque. It is effective in inhibiting multi-species biofilms formed by *S. mutans*, *S. sanguinis*, and *S. gordonii* [19]. Thus, ursolic acid has the potential to be used in dental caries treatment.

Probiotics have also been used to prevent the formation of biofilms. Most of the hospital-acquired urinary tract infections (UTI) are associated with the insertion of urinary tract devices, including catheters or stents. So it is of utmost importance to discover alternative treatment for UTIs. One such approach can be to use probiotics which are also known as non-pathogenic friendly bacteria because of their health-promoting properties. Pre-coating the surface of urinary tract devices with probiotics can prevent biofilm formation by pathogenic bacteria and thus make them more susceptible to antibiotics [20].

9.2.3 Bacterial Toxin Neutralization

Pathogenic bacteria produce various toxins as virulence factors. Examples are leukotoxin, hemolysin, and endotoxins. Bacterial toxin neutralization can be an effective anti-virulence therapy. *Clostridium difficile*, which causes diarrhea, produces two virulence factors, toxin A (TcdA) and toxin B (TcdB). Several small molecules have been identified that can inhibit the activity of toxin B in vitro. Another approach for neutralizing bacterial toxins is to use liposomes that mimic the cell membrane and so are able to sequester the toxins. Combining artificial liposomes with conventional antibiotics works better than mono-therapies [13]. Bacterial endotoxins (LPS) are known to cause septic shock. Antibiotics such as polymyxins can be used to neutralize LPS.

9.2.4 Metal Ion Chelation

Bacteria need metal ions such as iron, manganese, and zinc for their growth. The human body naturally sequesters these metal ions to prevent infection by bacterial pathogens. This process is called nutritional immunity. Most of the iron in the human body is bound to proteins such as transferrin, lactoferrin, hemoglobin, myoglobin, ferritin, and hemosiderin and so is not available for the infecting pathogen.

Moreover, due to aerobic environment, most of the iron in the body is present in the ferric form, which has extremely low solubility in water, which makes the availability of iron even less. In order to acquire this very little free iron, bacteria have several iron sequestering mechanisms. For example, Pseudomonas bacteria synthesize a siderophore called pyoverdines to acquire iron from the extracellular medium.

Molecules that can chelate this limited amount of available iron or other metal ions are expected to have antibiotic property. Lactoferrin, an iron chelator in humans, was shown to have bactericidal activity against the periodontal pathogen *A. actinomycetemcomitans* [21]. Growth of the bacteria is also severely inhibited by Maillard reaction products (MRP), also known as Amadori products, which are formed during autoclaving by a reaction between the aldehyde groups in reducing sugars and amino groups in proteins [22]. They are also formed during most cooking processes because proteins and carbohydrates are present in all food. It was shown that Maillard reaction products inhibit growth of bacteria by chelating iron [23]. Many known chelating agents have been studied for their binding to metal ions and for inhibiting bacterial growth. A few synergistic pairings of these chelators were also identified [24].

9.3 New Sources of Antibiotics

According to the original definition, antibiotics can only be obtained from other microorganisms. However, majority of the commercial antibiotics used today will not fall into that definition. So it will be wise to redefine the term "antibiotics" to be more inclusive rather than being restrictive as discussed before (Sect. 1.1). Finding new sources of antibiotics and developing improved methods for discovering new antibiotics are necessary to combat the rising crisis of antimicrobial resistance (AMR) development.

9.3.1 New Appropriate Methods for Antibiotic Discovery

Most of the new antibiotics were discovered within the first five decades after the discovery of the first one and so this period can be called the golden era for antibiotic development. This was followed by five more decades of very little success in antibiotic discovery, although the amount of research done and the number of papers published in this period have been much higher (Sect. 2.1). One reason for the lack of discovery of new antibiotics is that the same old scientific methods are repeatedly used to discover new antibiotics, and so it is not surprising that the same or similar antibiotics with minor modifications are being discovered and rediscovered. What is needed is to develop new methods for testing antimicrobial properties and new methods to culture microbes that cannot be grown by conventional methods.

As mentioned before (Sect. 1.4), 99% of bacterial species on this planet have not yet been cultured. So it is clear that developing methods that will allow us to grow these bacteria will result in discovery of many new antibiotics. A novel method to culture many of these organisms from soil has been developed and the process has led to the discovery of several new potential antibiotics [25]. One such antibiotic is teixobactin which has been discussed in detail in Sect. 3.3.4.

Another example of deficiency in methodology as an impediment, is the lack of discovery of antibiotics that can kill persisters, which are defined as a subset of a microbial population that has the ability to survive exposure to a bactericidal drug concentration, by virtue of the fact that they are non-growing [26]. Since persisters are non-growing bacteria, a modified method has been developed using non-growing bacteria for testing antibiotic activity against persisters. This is discussed in more detail in Sect. 9.4.

9.3.2 Antibiotics from Plants

Development of resistance is making all conventional antibiotics less effective. So it is imperative that new antibiotics be discovered to overcome this crisis. Scientists are increasingly looking at plants as a possible source of new antibiotics. Based on the original definition of antibiotics, plants do not even qualify to be a source of antibiotics, which were thought to be made only by microorganisms. However, all organisms are expected to make chemicals that will protect them from infection, and so they all can be possible sources for the discovery of new antibiotics. Since the plant kingdom is vast, we can be optimistic that in the future, many new antibiotics will be obtained from plants. Before the discovery of modern medicine, which was less than just a century ago, people had relied almost entirely on herbal medicine. Antibiotics are made by plants as secondary metabolites, which are compounds that are not directly involved in growth, development, or reproduction of the plant. Survivability of plant is not affected by the absence of secondary metabolites. Thousands of secondary metabolites have been identified that play an important role in competition and species interaction. Some classes of secondary metabolites are known to have antimicrobial activity. These include phenols, phenolic acids, quinones, flavones, flavonoids, flavonols, tannins, coumarins, lectins, polypeptides, alkaloids, terpenoids, and essential oils (Fig. 9.3). The term, "essential oil" can be misleading. It is not used in the same sense as essential amino acids or fatty acids, which are defined as nutrients that humans cannot make and so, should be obtained from diet. On the other hand, essential oils mean the hydrophobic extracts from plants that contain the scent or essence of the plant.

Numerous papers have been published on antimicrobial activity in extracts from plants. A comprehensive review of all these papers and of all antibiotics from plants has recently been published [27]. Many secondary metabolites from plants also show antitumor activity. A long list of purified compounds and mixture of compounds obtained from plants have been tested for antimicrobial activity. Many spices, such

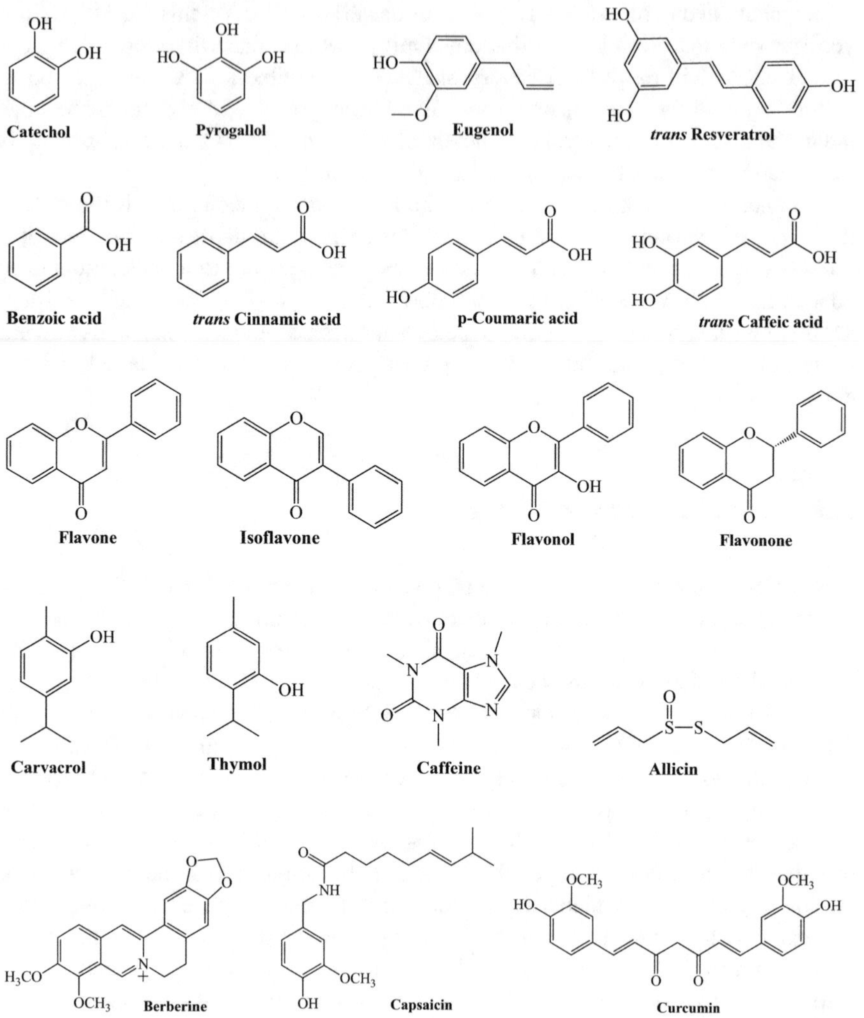

Fig. 9.3 Structures of some plant-derived compounds with antimicrobial activity

as clove, cinnamon, turmeric, garlic, neem, oregano, and thyme have been shown to have antibacterial and antifungal activity. Most of these are sold and used as dietary supplements and not as purified chemical compound with antibiotic activity. Although a majority of these show some antimicrobial activity, there actually has been very few true success stories because most of the antibiotic activities reported are weak. Those that show strong activity also have significant cytotoxic effect. Relatively strong antibiotic activity has been observed in garlic, clove, and cinnamon. Since all species have developed some defense mechanism against infections, it is not surprising that there have been so many reports of antimicrobial activity in plant extracts. Most of the studies have been done using essential oils, which are mixtures of many compounds. For developing new viable antibiotics from plants, we need better experimental designs using purified compounds.

Obtaining new purified compounds from plant sources have some inherent difficulties. Plant extracts are mixtures of numerous compounds. Identification and purification of the active compound in it is difficult, time consuming, and expensive. In most cases, the active component is present in very low proportion, and so a lot of starting material is needed to obtain a small amount of pure compound. Many of these active components function in synergy with other compounds in the extract and so, loss of activity may be observed during the purification process. On a positive note, once the two or more compounds working in synergy have been identified and the mixture of the purified compounds becomes commercially available, it will be a desirable antibiotic because it is more difficult for bacteria to develop resistance to a synergistic mechanism. So there is great prospect for success, but there is a long road ahead. A few more promising candidates from plant sources are discussed in the next section on persisters (Sect. 9.3.3).

9.3.3 Antibiotics Against Persisters and Slow-Growing Bacteria

One reason for the ineffectiveness of antibiotics is their inability to kill nongrowing cells. In a population of bacteria, there are always a few cells that are dormant or non-growing and so can withstand antibiotic treatment. These cells are not resistant mutants and will be susceptible to the same antibiotic when they start to grow later. Such temporarily nongrowing cells in a population of growing bacteria are called persisters since they have the ability to survive exposure to bactericidal concentration of an antibiotic. Persistence of bacteria is the reason why antibiotic regimens are for 10 days or more, even though more than 99% of the infecting bacteria are killed within the first day of antibiotic treatment [28]. One reason why antibiotics are ineffective against persisters is because cell wall synthesis and protein biosynthesis are downregulated in persisters, and thus these targets cannot be inhibited by antibiotics.

Several methods have been developed to kill persister cells. Mitomycin C and cisplatin can enter cells without using active transport and thus can kill persister cells. Another approach is to wake up persister cells by adding sugars or cis-2-decenoic acid so that they can then be killed by traditional antibiotics. In another approach, a 12 amino acid transporter sequence was added to tobramycin, which allowed it to permeate membranes of persister cells. Plant products have also been shown to kill persister cells. Essential oils from several spices were found to have activity against stationary phase culture of *Borrelia burgdorferi*, the causative agent of lyme disease [29].

Discovering antibiotics against persisters requires appropriate screening methods. All current commercial antibiotics were discovered using growing bacteria, so it is not surprising that most of them cannot kill nongrowing bacteria, which may explain the ineffectiveness of current commercial antibiotics. A modified method was developed in which bacteria were forced to be nongrowing by withholding nutrients. This led to the discovery that an extract from myrrh preferentially kills nongrowing bacteria [30].

Myrrh is a resin secreted by the aromatic plant, *Commiphora molmol*, belonging to the Burseraceae family, also known as the torchwood or incense family. Myrrh has been popularly used for many centuries in all parts of the world, especially for its antimicrobial activity, and has found their place in most religious and cultural practices. However, systematic scientific study of its antibiotic activity of myrrh is very limited. All published reports using modern scientific techniques showed its antimicrobial activity to be very weak. This is because the antibiotic in myrrh oil is not very effective against growing bacteria. If cells are incubated with myrrh in nutrient-free buffer, or on nutrient-free plates, they are killed at a very fast rate. Many of the currently used commercial antibiotics have been shown to have some activity against nongrowing cells but greater activity against growing cells [31]. Myrrh is the first known example of an antibiotic that preferentially kills nongrowing bacteria. One possible clinical significance of this observation is that myrrh oil can also kill bacteria in a nutrient-rich medium, provided growth of the bacteria is halted by the addition of a bacteriostatic antibiotic such as chloramphenicol.

A similarly modified method was developed to detect antibiotic activity against slow-growing bacteria. For this, zone of inhibition assay was done on agar plates containing low concentration of nutrients or low to zero concentration of salt, to demonstrate that an aqueous extract from rhubarb stalk preferentially kills slow-growing bacteria [32].

9.3.4   Nanoparticles as Antibiotics

Nanoparticles are particles whose size ranges from 1 to 100 nanometers. Because of their small sizes, nanoparticles have a very high surface area to volume ratio compared to bulk material. This is responsible for their unusual physical and chemical properties. One use of nanoparticles is to deliver antibiotics to the target tissues. This is discussed later in Sect. 9.4.2. However, some nanoparticles themselves can have direct antibiotic activity and are discussed here.

Most nanoparticles that have antimicrobial activity are usually inorganic nanoparticles, especially heavy metal-based nanoparticles. The most commonly used metal nanoparticles are silver-based systems. The cheaper metal copper also has antimicrobial activity. Bacteria require the transition metals in small amounts but in excess amounts they are all toxic. The nano sizes increase their surface area, which allows more points of contact with the bacteria and possibly increases the solubility of the metals. However, even with this increase, all these heavy metals have extremely low solubility in water. The mechanism by which copper metal kills bacteria, possibly includes the following steps. Bacteria bind to the surface of the metal. Copper ions on the surface of the metal are mis-recognized as nutrients and are taken in. The copper ions bind to proteins and inactivate them, affecting metabolic functions, especially the respiratory electron transport chain, and thus resulting in bacterial cell death. With this mechanism in mind, many buildings, especially hospital buildings are replacing touch surfaces such as door knobs and light switches with copper or copper alloys. Extrapolating this concept, it has now become fashionable to use copper water bottles. However, this practice will offer no benefit because the solubility of copper in water at standard pH is extremely low to have any beneficial effect. Using metal nanoparticles as disinfectants does have some merit, but using them as antibiotics in humans will be risky. Because the mechanism of action of the heavy metal ions is non-specific, they will have similar inhibitory activity against human cells, and so all these metals ions have associated toxicity. The toxicity aspect becomes more significant for targeting intracellular bacteria due to the longer exposure times required to reach the bacteria.

Metal oxide nanoparticles have also been used as antibiotics. Examples include ZnO, CuO, TiO_2, and Fe_3O_4. Recent examples of metal oxide nanoparticles with antibiotic activity are cerium oxide and the biodegradable material calcium sodium phosphosilicate, more commonly known as bioglass. Metal oxide nanoparticles have the advantage that they also have antimicrobial activity but have lower mammalian cell toxicity. Cerium oxide nanoparticles have the added advantage that they can enter mammalian cells and target intracellular bacteria. Cerium oxide nanoparticles have the same size and surface charge as bacteria and so can be transported into subcellular compartments as the bacteria and thus they can kill the intracellular bacteria [33]. Besides the toxicity due to the non-specific mechanisms of action of both metal and metal oxide nanoparticles, there is also an advantage of such non-specific actions. Since the inhibitory effect can be on multiple targets, it can be expected that resistance development against these mechanisms will be much less because mutations in one target protein will not be sufficient to provide resistance.

One new example of nanoparticles as antibiotics is the use of polymyxin-loaded cubosomes. Infections by gram-negative bacteria are difficult to treat because the outer membrane functions as an extra barrier against the antibiotics. Polymyxins B and E are considered as last-line therapies against antibiotic-resistant gram-negative bacteria. They are able to cause disorganization of the outer membrane by interacting with lipopolysaccharide (LPS) located in the membrane. Cubosomes, which are lyotropic liquid-crystalline nanoparticles, are able to solubilize the membrane as a result of their amphiphilic property. Cubasomes are nanoparticles that have been used to deliver antibiotics into cells. Since cubosomes and polymyxins disrupt the outer membrane of Gram-negative bacteria by different mechanisms, a new strategy has been described in which polymyxin-loaded cubosomes were used a polytherapy against gram-negative bacteria [34]. This successful polytherapy takes place by a two-step process. Electrostatic interaction between polymyxin and LPS destabilizes the outer membrane, while the cubosome causes further membrane disruption due to its surfactant-like property.

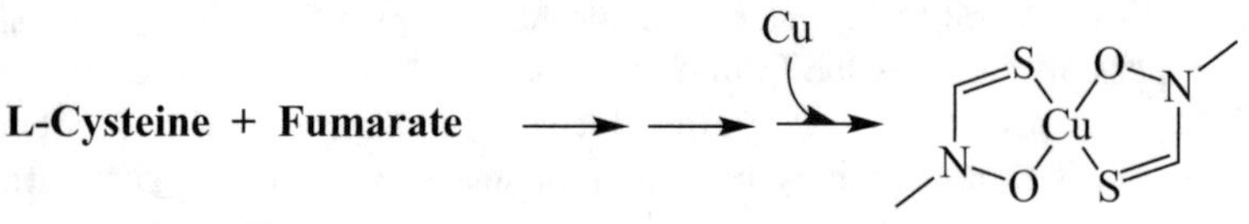

Fig. 9.4 Biosynthesis of Fluopsin C

One potential risk of using metal-based nanoparticles is that some bacteria can make use of it to their advantage. *Pseudomonas aeruginosa* actually has a unique ability to use excess copper ion to its advantage. It was shown recently that *P. aeruginosa* have the ability to synthesize a small-molecule copper complex, fluopsin C, in response to elevated copper concentrations [35]. Fluopsin C, which was discovered in 1970, is a broad-spectrum antibiotic that contains a copper ion chelated by two thiohydroxamates, which in turn are made from cysteine. Thus, this reaction serves two purposes: it protects *P. aeruginosa* from the toxic effect of excess copper and also converts the toxic copper to a broad-spectrum antibiotic that it can use to compete with other bacteria (Fig. 9.4).

9.4   Improving Delivery of Antibiotics

Compared to intravenous, transdermal, and transpulmonary routes of administering antibiotics, the most common, safe, and convenient route is by oral administration. However, even in this route, there are several barriers for absorption of the antibiotics through the intestinal lining and then transport to the blood vessels on the other side.

9.4.1   Steps to Increase Antibiotic Bioavailability

For an antibiotic to be effective, it must be soluble and stable in bodily fluids till it interacts with the bacterial cell. Most commercial antibiotics have undergone several modifications since their original discoveries. One main reason for these modifications is to improve their bioavailability. For example, the antibiotic erythromycin is prone to an intramolecular dehydration reaction. Azithromycin and clarithromycin (Fig. 6.8) were developed as analogs of erythromycin to decrease the rate of the dehydration reaction. As discussed before (Sect. 1.5), Penicillin G, the original antibiotic discovered was modified to make Penicillin V to make it stable in stomach acid so that it can be administered orally. Many antibiotics are modified to increase their water solubility, while for some the goal is to increase membrane permeability. Formulation of drugs is a broad subject that is beyond the scope of this book.

However, it is briefly mentioned here that many additives are included with all drugs to improve their bioavailability. For example, various bioenhancers are added to increase drug absorption.

9.4.2 Nanoparticles for Drug Delivery

The above-mentioned barriers can be overcome by using nanoparticle vehicles. A strategy for improving the effectiveness of antibiotics is by complexing them to micro or nano particles by chemical bonding (conjugation), physical binding (adsorption), or by encapsulation [36]. The delivery vehicle can be liposomes, dendrimers, peptides, or other polymers. Several such delivery vehicles, such as silver and titanium nanoparticles, have been approved by US FDA for antibacterial skin lotions and sunscreens. Antibiotics have also been conjugated on the surface of nanoparticles made of gold, iron oxide, or silica as delivery vehicle to target bacteria. Liposomes made of phospholipids can also deliver antibacterial drugs. Using antibiotics in nanoparticles can serve two purposes; (1) The nanoparticles can increase the bioavailability of the antibiotics and (2) The nanoparticles can be designed to deliver the antibiotic to the site of infection.

Nanoparticles Increase Bioavailability of Antibiotics The most significant barrier faced by antibiotics is the mucus layer lining the intestine. The mucus consists of mucin glycoproteins containing sialic acid, which is an α-keto acid sugar. Since the carboxylic acid is ionized at the pH of the intestine, the mucus has a negative charge which allows it to bind to positively charged molecules and subsequently remove them when the mucus layer sheds off. Another barrier to antibiotic transport through intestinal lining is efflux of absorbed molecules back into the intestine by P-glycoprotein, which is an ATP-dependent efflux pump (transporter) with broad substrate specificity and is expressed in tissues with excretory function, including the intestine. Other barriers to absorption are water-insolubility of some antibiotics, and degradation of some antibiotics by enzymes secreted by intestinal cells. For example, ciprofloxacin, vancomycin, rifampicin, and clarithromycin have low solubility and are poorly permeable through the intestinal cells. The large surface area of nanoparticles increases the solubility of the antibiotics and thus increases its bioavailability. Peptide-based antibiotics are unstable in the acidic environment of the stomach. Inside the nanoparticles, the antibiotics can be shielded from the harsh acidic conditions in the stomach. The nanoparticles can also shield the antibiotics from degradation by enzymes secreted by the intestinal cells [37]. Nanoparticles containing polyethylene glycol (PEG) coating have an electrically neutral surface and so do not bind to the mucus. Instead, the nanoparticle can pass through the mucus and directly interact with the intestinal cell membrane. The large surface area of the nanoparticles also ensures a greater contact area with the cells in the intestinal wall and thus results in better absorption. Nanoparticles made of chitosan, which are positively charged, can promote binding to the negatively charged mucous layer on the intestinal cell lining and thus increase the permeability of the antibiotic into the cells. Inhibitors of the P-glycoprotein efflux pump such as Tween or cetyltrimethylammonium bromide (CTAB) can also be included in the nanoparticles in order to prevent the antibiotic from being pumped from the intestinal cells back to the intestine.

Targeting Antibiotics to Infection Sites Cationic antimicrobial peptides and lipopeptides can kill the microbes by disrupting bacterial cell membranes. Because of their positive charges, these can target infection sites because most infecting bacteria have a negative surface charge at physiological pH (~7.2). On the other hand, many bacteria are found in varying acidic conditions within the body (stomach, pH 1.0–2.0, vagina, pH 4.0–5.0, and skin pH 4.0–5.5). Antibiotic-loaded chitosan microparticles and nanoparticles have been successfully used against *H. pylori*, which is associated with gastric ulcers [38]. Another method of targeting nanoparticles to the infection site is by taking advantage of the fact that bacteria secrete many virulence factors, including toxins, adhesins, and various enzymes such as phosphatases and lipases. Antibiotics can be targeted to the infection site by attaching a complementary substrate or antibody on the surface of the nanoparticle. Binding or reaction between the nanoparticle surface and the virulence factor will open up the nanoparticle to expose the antibiotic to the infecting bacteria [39].

9.4.3   Antibiotics Against Intracellular Pathogens

In some infections, the bacteria enter the mammalian cells and reside in the cytoplasm or in the organelles. Such intracellular pathogens may be able to continue with their normal activities inside the host cell. Bacteria in the cytoplasm are also capable of developing new antibiotic resistance by exchanging resistance genes by conjugation with other bacteria [40]. Some bacteria, such as Mycobacterium spp., are facultative intracellular, while some, such as Chlamydia spp., are obligate intracellular parasites, which cannot survive outside the cell. Intracellular bacteria cannot be neutralized by the host's immune system. Most antibiotics are unable to enter the host cells or their organelles, and thus intracellular bacteria also remain protected from the antibiotics. On the other hand, there are some pathogens such as *Mycobacterium tuberculosis* and *Listeria* monocytogenes that can resist action of antibiotics by remaining dormant inside macrophages and erythrocytes. [41]. Intracellular concentrations of antibiotics range from very low for hydrophilic antibiotics such as aminoglycosides and penicillins, to high, for lipophilic antibiotics such as macrolides, glycopeptides, and rifamycins. A large majority of antibiotics are unable to cross the host's cell membranes and so are ineffective against intracellular pathogens. So new methods are needed for the delivery of antibiotics into cells.

Encapsulation of Antibiotic Molecules One approach to targeting intracellular bacteria is by encapsulating antibiotic molecules in biodegradable polymers such as poly(lactic-co-glycolic acid) (PLGA) [42]. One example is targeting of *M.*

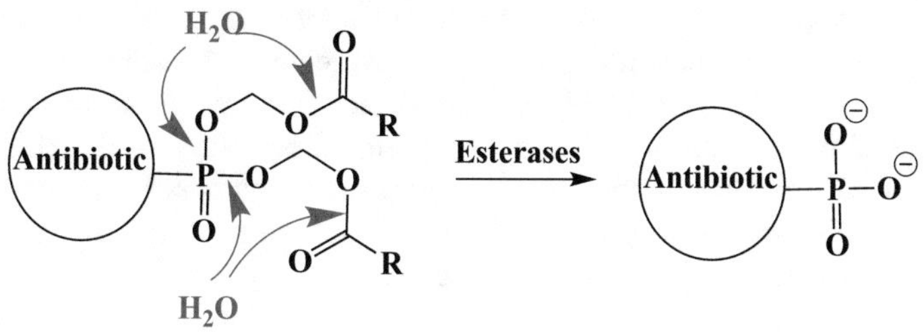

Fig. 9.5 Phosphonate antibiotic attached to pro-moieties. R is a lipophilic group. The red arrows show bonds that are hydrolyzed by esterases

tuberculosis, which can survive and remain dormant in alveolar macrophages and so is difficult to treat. Delivery of rifampicin in PLGA microparticles was shown to result in a 19-fold higher concentration of rifampicin in alveolar macrophages infected with TB compared to using the free drug in solution [43]. There are many other biodegradable polymers that can be used for antibiotic encapsulation, including poly (ε-caprolactone) (PCL), poly(lactic acid) (PLA), poly(γ-glutamic acid) (PGA), and poly(L-lysine). Hyaluronic acid, which is a nonsulfated glycosaminoglycan, is a component of the extracellular matrix, and is also often chemically modified and adapted for drug delivery.

Drugs can also be encapsulated into liposomes for delivery into cells. Liposomes are small spherical vesicles containing an aqueous region surrounded by a lipid bilayer (Sect. 7.1.2). The enclosed aqueous region can also contain drugs that can be delivered to cells. Recently these have been successfully used to deliver mRNA vaccines against Covid-19. PEGylated liposomes containing vancomycin have been used to target MRSA residing in human macrophages [44].

Intracellular Delivery as a Prodrug In a new approach, intracellular bacteria are targeted with an antibiotic that is masked by a prodrug. Since many metabolites are polar or charged, inhibitors of metabolic reactions, also known as antimetabolites, are often designed to have charges. One way of achieving that is by attaching a phosphonate functional group on the molecule. Note that, unlike a phosphate, a phosphonate has a stable carbon-phosphorus bond. A phosphate, in which the linkage to phosphorus is through an oxygen may be hydrolyzed by phosphatases. Although the phosphonate group increases interaction with the bacterial metabolic enzyme, the negatively charged molecules are unable to cross the host cell membrane and thus are ineffective against intracellular pathogens. In order to facilitate entry of the antibiotic through the host cell membranes, the negative charges of the phosphonate are masked by attaching lipophilic groups to the phosphonate via phosphodiester and carboxylester linkages. After entry into the cell, ideally, the bacterial esterases will hydrolyze the ester linkages to yield the original phosphonate antibiotic (Fig. 9.5). This strategy is called prodrugging, and the labile masking group is called a pro-moiety. An important requirement for the success of this strategy is that the lipophilic prodrug moiety should be resistant to host's serum esterases but sensitive to bacterial esterases such that they are hydrolyzed not in the bloodstream but after entry into the cell and interaction with the intracellular bacteria [45]. Appropriate pro-moieties can be designed to meet the substrate specificity requirements of the microbial esterases and, at the same time, differentiate from the specificity of the host enzymes. This way, it may be possible to design drugs for specific pathogens. Using transposon mutagenesis, the authors have identified two esterases GloB and FrmB in *S. aureus* that are able to hydrolytically remove the pro-moieties to yield active inhibitor molecule.

References

1. Kirienko NV, Rahme L, Cho Y-H (2019) Beyond antimicrobials: non-traditional approaches to combating multidrug-resistant bacteria. Front Cell Infect Microbiol 9:343. https://doi.org/10.3389/fcimb.2019.00343)
2. Jang H-J, Chung I-Y, Lim C, Chung S, Kim B-O, Kim ES, Kim S-H, Cho Y-H (2019) Redirecting an anticancer to an antibacterial hit against methicillin-resistant *Staphylococcus aureus*. Front Microbiol 10:350. https://doi.org/10.3389/fmicb.2019.00350
3. Simsek M, Meijer B, van Bodegraven AA, de Boer NKH, Mulder CJJ (2018) Finding hidden treasures in old drugs: the challenges and importance of licensing generics. Drug Discov Today 23:17–21
4. Baker NC, Elkins S, Williams AJ, Tropsha A (2018) A bibliometric review of drug repurposing. Drug Discov Today 23:661–672. https://doi.org/10.1016/j.drudis.2018.01.018
5. Sugamata R, Sugawara A, Nagao T, Suzuki K, Hirose T, Yamamota K-I, Oshina M, Kobayashi K, Sunazuka T, Akagawa KS, Omura S, Nakayama T, Suzuki K (2014) Leucomycin A3, a 16-membered macrolide antibiotic, inhibits influenza A virus infection and disease progression. J Antibiot 67:213–222
6. Eng CH, Wang Z, Tkach D, Toral-Barza L, Udwonali S, Liu S, Fitzgerald SL, George E, Frias E, Cochran N, De Jesus R, McAllister G, Hoffman GR, Bray K, Lemon L, Lucas J, Fantin VR, Abraham RT, Murphy LO, Nyfeler B (2016) Macroautophagy is dispensable for growth of KRAS mutant tumors and chloroquine efficacy. Proc Natl Acad Sci U S A 113:182–187. https://doi.org/10.1073/pnas.1515617113
7. Tripathi PK, Upadhyay S, Singh M, Raghavendhar S, Bhardwaj M, Sharma P, Patel AK (2020) Screening and evaluation of approved drugs as inhibitors of main protease of SARS-CoV-2. Int J Biol Macromol 164:2622–2631
8. Eastman RT, Roth JS, Brimacombe KR, Simeonov A, Shen M, Patnaik S, Hall MD (2020) Remdesivir: a review of its discovery and development leading to emergency use authorization for treatment of COVID-19. ACS Cent Sci 6:672–683
9. Defoirdt T (2018) Quorum-sensing systems as targets for antivirulence therapy. Trends Microbiol 26:313–328. https://doi.org/10.1016/j.tim.2017.10.005)
10. Maura D, Ballok AE, Rahme LG (2016) Considerations and caveats in anti-virulence drug development. Curr Opin Microbiol 33:41–46. https://doi.org/10.1016/j.mib.2016.06.001
11. Coraça-Huber DC, Dichtl S, Steixner S, Nogler M, Weiss G (2018) Iron chelation destabilizes bacterial biofilms and potentiates the antimicrobial activity of antibiotics against coagulase-negative *Staphylococci*. Pathog Dis 76:fty052. https://doi.org/10.1093/femspd/fty052
12. Fong J, Mortensen KT, Norskov A, Qvortrup K, Yang L, Tan CH, Nielsen TE, Givskov M (2019) Itaconimides as novel quorum sensing inhibitors of *Pseudomonas aeruginosa*. Front Cell Infect Microbiol 8:443. https://doi.org/10.3389/fcimb.2018.00443
13. Fleitas Martínez O, Cardoso MH, Ribeiro SM, Franco OL (2019) Recent advances in anti-virulence therapeutic strategies with a focus on dismantling bacterial membrane microdomains,

toxin neutralization, quorum-sensing interference and biofilm inhibition. Front Cell Infect Microbiol 9:74. https://doi.org/10.3389/fcimb.2019.00074

14. Ramasubbu N, Thomas LM, Ragunath C, Kaplan JB (2005) Structural analysis of dispersin B, a biofilm releasing glycoside hydrolase from the periodontopathogen *Actinobacillus actinomycetemcomitans*. J Mol Biol 349:475–486

15. Baker P, Hill PJ, Snarr BD, Alnabelseya N, Pestrak MJ, Lee MJ, Jennings LK, Tam J, Melnyk RA, Parsek MR, Sheppard DC, Wozniak DJ, Howell PL (2016) Exopolysaccharide biosynthetic glycoside hydrolases can be utilized to disrupt and prevent *Pseudomonas aeruginosa* biofilms. Sci Adv 2:e1501632

16. Kaplan JB, LoVetri K, Cardona ST, Madhyastha S, Sadovskaya I, Jabbouri S, Izano EA (2012) Recombinant human DNase I decreases biofilm and increases antimicrobial susceptibility in *Staphylococci*. J Antibiot 65:73–77. https://doi.org/10.1038/ja.2011.113

17. Balemans W, Vranckx L, Lounis N, Pop O, Guillemont J, Vergauwen K, Mol S et al (2012) Novel antibiotics targeting respiratory ATP synthesis in Gram-positive pathogenic bacteria. Antimicrob Agents Chemother 56:4131–4139. https://doi.org/10.1128/AAC.00273-12

18. Ali J, Rafiq QA, Ratcliffe E (2018) Antimicrobial resistance mechanisms and potential synthetic treatments. Future Sci OA 4:FSO290. https://doi.org/10.4155/fsoa-2017-0109

19. Lyu X, Wang L, Shui Y, Jiang Q, Chen L, Yang W, He X, Zeng J, Li Y (2021) Ursolic acid inhibits multi-species biofilms developed by *Streptococcus mutans*, *Streptococcus sanguinis*, and *Streptococcus gordonii*. Arch Oral Biol 125:105107. https://doi.org/10.1016/j.archoralbio.2021.105107

20. Carvalho FM, Teixeira-Santos R, Mergulhão FJM, Gomes LC (2021) Effect of *Lactobacillus plantarum* biofilms on the adhesion of *Escherichia coli* to urinary tract devices. Antibiotics 10:966. https://doi.org/10.3390/antibiotics10080966

21. Velliyagounder K, Kaplan JB, Furgang D, Legarda D, Diamond G, Parkin RE, Fine DH (2003) One of two human lactoferrin variants exhibits increased antibacterial and transcriptional activation activities and is associated with localized juvenile periodontitis. Infection. Immunity 71:6141–6147. https://doi.org/10.1128/iai.71.11.6141-6147.2003

22. Bhattacharjee MK, Sugawara K, Ayandeji OT (2009) Microwave sterilization of growth medium alleviates inhibition of *Aggregatibacter actinomycetemcomitans* by Maillard reaction products. J Microbiol Methods 78:227–230

23. Bhattacharjee MK, Mehta BS, Akukwe B (2021) Maillard reaction products inhibit the periodontal pathogen *Aggregatibacter actinomycetemcomitans* by chelating iron. Arch Oral Biol 122:104989–104995

24. Paterson JR, Beecroft MS, Mulla RS, Osman D, Reeder NL, Caserta JA, Young TR, Pettigrew CA, Davies GE, Williams JAG, Sharples GJ (2022) Insights into the antibacterial mechanism of action of chelating agents by selective deprivation of iron, manganese, and zinc. Appl Environ Microbiol 88(2). https://doi.org/10.1128/AEM.01641-21

25. Ling LL, Schneider T, Peoples AJ, Spoering AL, Engels I, Conlon BP, Mueller A, Till F, Schäberle TF, Hughes DE, Epstein S, Jones M, Lazarides L, Steadman VA, Cohen DR, Felix CR, Fetterman KA, Millett WP, Nitti AG, Zullo AM, Chen C, Lewis K (2015) A new antibiotic kills pathogens without detectable resistance. Nature. https://doi.org/10.1038/nature14098

26. Balaban NQ, Helaine S, Lewis K, Ackermann M, Aldridge B, Andersson DI, Brynildsen MP, Bumann D, Camilli A, Collins JJ, Dehio C, Fortune S, Ghigo JM, Hardt WD, Harms A, Heinemann M, Hung DT, Jenal U, Levin BR, Michiels J, Storz G, Tan MW, Tenson T, Van Melderen L, Zinkernagel A (2019) Definitions and guidelines for research on antibiotic persistence. Nat Rev Microbiol 17:441–448. https://doi.org/10.1038/s41579-019-0196-3

27. Chassagne F, Samarakoon T, Porras G, Lyles JT, Dettweiler M, Marquez L, Salam AM, Shabih S, Farrokhi DR, Quave CL (2021) A systematic review of plants with antibacterial activities: a taxonomic and phylogenetic perspective. Front Pharmacol 11:586548. https://doi.org/10.3389/fphar.2020.586548

28. Claudi B, Spröte P, Chirkova A et al (2014) Phenotypic variation of Salmonella in host tissues delays eradication by antimicrobial chemotherapy. Cell 158:722–733

29. Feng J, Zhang S, Shi W, Zubcevik N, Miklossy J, Zhang Y (2017) Selective essential oils from spice or culinary herbs have high activity against stationary phase and biofilm *Borrelia burgdorferi*. Front Med 4:169
30. Bhattacharjee MK, Alenezi T (2020) Antibiotic in myrrh from *Commiphora molmol* preferentially kills non-growing bacteria. Future Sci OA 6(4):FSO458. https://doi.org/10.2144/fsoa-2019-0121
31. McCall IC, Shah N, Govindan A, Baquero F, Levin BR (2019) Antibiotic killing of diversely generated populations of nonreplicating bacteria. Antimicrob Agents Chemother 63(7):e02360–e02318
32. Bhattacharjee MK, Bommareddy PK, DePass AL (2021) A water-soluble antibiotic in rhubarb stalk shows an unusual pattern of multiple zones of inhibition and preferentially kills slow-growing bacteria. Antibiotics 10:951. https://doi.org/10.3390/antibiotics10080951
33. Matter MT, Doppegieter M, Gogos A, Keevend K, Ren Q, Herrmann IK (2021) Inorganic nanohybrids combat antibiotic-resistant bacteria hiding within human macrophages. Nanoscale 13:8224–8234. https://doi.org/10.1039/D0NR08285F
34. Lai X, Han M-L, Ding Y, Chow SH, LeBrun AP, Wu C-M, Bergen PJ, Jiang J-H, Hsu H-Y, Muir BW, White J, Song J, Li J, Shen H-H (2022) A polytherapy based approach to combat antimicrobial resistance using cubosomes. Nat Commun 13:343. https://doi.org/10.1038/s41467-022-28012-5
35. Patteson JB, Putz AT, Tao L, Simke WC, Bryant H, Britt D, Li B (2021) Biosynthesis of fluopsin C, a copper-containing antibiotic from *Pseudomonas aeruginosa*. Science 374:1005–1009. https://doi.org/10.1126/science.abj6749
36. Van Giau V, An SSA, Hulme J (2019) Recent advances in the treatment of pathogenic infections using antibiotics and nano-drug delivery vehicles. Drug Des Devel Ther 13:327–343
37. Wu Z-L, Zhao J, Xu R (2020) Recent advances in oral nano-antibiotics for bacterial infection therapy. Int J Nanomedicine 15:9587–9610
38. Khutoryanskiy VV (2018) Beyond PEGylation: alternative surface modification of nanoparticles with mucus-inert biomaterials. Adv Drug Deliv Rev 124:140–149
39. Chen YL, Zhu S, Zhang L, Feng P-J, Yao X-K, Qian C-G, Zhang C, Jiang X-Q, Shen Q-D (2016) Smart conjugated polymer nanocarrier for healthy weight loss by negative feedback regulation of lipase activity. Nanoscale 8:3368–3375. https://doi.org/10.1039/C5NR06721A
40. Lim YM, deGroof AJ, Bhattacharjee MK, Figurski DH, Schon EA (2008) Bacterial conjugation in the cytoplasm of mouse cells. Infect Immun 76:5110–5119
41. Pinto D, São-José C, Santos MA, Chambel L (2013) Characterization of two resuscitation promoting factors of *Listeria monocytogenes*. Microbiology 159:1390–1401
42. Aksungur P, Demirbilek M, Denkbaş EB, Vandervoort J, Ludwig A, Unlü N (2011) Development and characterization of Cyclosporine A loaded nanoparticles for ocular drug delivery: cellular toxicity, uptake, and kinetic studies. J Control Release 151:286–294
43. Chew NY, Chan HK (2001) Use of solid corrugated particles to enhance powder aerosol performance. Pharm Res 18:1570–1577
44. Muppidi K, Wang J, Betageri G, Pumerantz AS (2011) PEGylated liposome encapsulation increases the lung tissue concentration of vancomycin. Antimicrob Agents Chemother 55:4537–4542
45. Miller JJ, Shah IT, Hatten J, Barekatain Y, Mueller EA, Moustafa AM, Edwards RL, Dowd CS, Planet PJ, Muller FL, Jez JM, John ARO (2021) Structure-guided microbial targeting of antistaphylococcal prodrugs. eLife 10:e66657. https://doi.org/10.7554/eLife.66657

Chapter 10
Global Action Plan and Antibiotic Stewardship

Abstract The World Health Organization has adopted a Global Action Plan to tackle antimicrobial resistance and to meet the goals of the plan, all member states have implemented their own National Action Plans. Such plans of some countries are discussed. Antibiotic Stewardship Programs for the responsible use of antibiotics have been adopted at local levels. The responsibilities of regulatory agencies, healthcare providers, patients, pharmaceutical companies, scientists, and the government for the proper use of antibiotics to prevent the development of antimicrobial resistance have been discussed. Subtherapeutic use of antibiotics for growth promotion of animals and banning of such practice have also been discussed in relation to antimicrobial resistance.

10.1 Global Action Plan

It is estimated that antimicrobial resistance (AMR) is responsible for about 700,000 deaths globally every year, including 230,000 people who die from multidrug-resistant tuberculosis. Recently there has been an increased emphasis on AMR since the start of the COVID-19 pandemic because of secondary bacterial infections or fear of it. COVID-19 patients are treated with antibiotic prophylaxis such as macrolides. Because of movement of people, animals, and goods throughout the world, antibiotic resistance can spread easily and quickly across the globe. So incidence of antimicrobial resistance in any country will not be confined there but can affect other countries in the world. An example of this was seen recently in the Covid-19 pandemic. So, steps to prevent the development of antibiotic resistance must be undertaken not just at the local or national level but at an international level. This has been discussed at length at the G7 and G20 meetings as well as at the World Health Organization and the United Nations. The World Health Organization (WHO) celebrates the World Antimicrobial Awareness Week every year from November 18 to 24 with the aim to increase awareness of global antimicrobial resistance and to encourage best practices to avoid further emergence and spread of drug-resistant infections.

In May 2015, the World Health Assembly adopted a Global Action Plan to tackle antimicrobial resistance [1]. The goal of the Global Action Plan is to ensure continuity of successful treatment and prevention of infectious diseases. The World Health Assembly also urged all Member States to develop by 2017 a National Action Plan that meets the objectives of the Global action plan. The five objectives of the plan are to improve awareness and understanding of antimicrobial resistance, to strengthen knowledge through surveillance and research, to reduce the incidence of infection, to optimize the use of antimicrobial agents, and to increase investment in new medicines, diagnostic tools, vaccines, and other interventions.

In its 5-year plan for 2019, the UK government called for a 15% decrease in human antibiotic use by 2024 and 25% decrease in the use of antibiotics in food-producing animals by 2020. In the USA, deaths from antibiotic-resistant infections in hospitals decreased by 28 percent from 2012 to 2017. Rapid detection and prevention strategies as well as vaccines have helped to reduce infections by drug-resistant *Streptococcus pneumoniae* and drug-resistant tuberculosis. However, the CDC is still concerned about other antibiotic-resistant infections that are increasing in the community such as drug-resistant *Neisseria gonorrhoeae, and* extended-spectrum beta-lactamase (ESBL)-producing Enterobacteriaceae and erythromycin-resistant group A *Streptococcus.*

10.2   National Action Plans

The WHO had urged all member countries to develop their own National Action Plans by 2017 to meet the objectives of the Global Action Plan. Many countries have implemented different versions of similar action plans. National Action Plans of some countries and regions that account for more than half of the world's population are specifically mentioned here.

10.2.1   National Action Plan of USA

According to data from the Center for Disease Control and Prevention (CDC), drug-resistant bacteria are responsible for more than 2.8 million infections and 35,000 deaths per year in the United States [2]. The CDC maintains a list of pathogens and based on the threat level, has classified them into three levels of seriousness: Urgent Threat Level Pathogens, Serious Threat Level Pathogens and Of Concern Threat Level Pathogens. Two notable trends described are that resistant infections and deaths from pathogens associated with hospitals are steadily declining, and resistance to essential antibiotics is increasing. Systemic misuse and overuse of antibiotics have led to the development of these drug-resistant bacteria. Any use of antibiotics, appropriate or inappropriate, increases the chance of resistance development. According to a recent CDC report [3], about 30% of antibiotics prescribed in the USA are inappropriate. They also reported that about 44 percent of outpatient antibiotic prescriptions are for treating patients with acute respiratory conditions, and of these prescriptions, about half are unnecessary. A large proportion of antibiotic prescriptions are written in these facilities, including those written by primary care physicians, nurse practitioners and physician assistants, and dentists [4].

The US federal government has established a National Action Plan for Combating Antibiotic-Resistant Bacteria (CARB) (2020–2025) to reduce infections occurring during health care delivery [5]. This is the Second National Action Plan, the first one was implemented in 2015. The plan proposes a coordinated set of five goals for addressing the problem of antibiotic resistance development. Goal 1: Slow the emergence of resistant bacteria and prevent the spread of resistant infections. Goal 2: Strengthen surveillance efforts to combat resistance. Goal 3: Develop diagnostic tests for identification and characterization of resistant bacteria. Goal 4: Support research for development of new antibiotics, other therapeutics, and vaccines. Goal 5: Improve international collaboration in all of the above four goals.

The US government has set several ambitious objectives for achieving the above goals. These include a significant reduction in nosocomial as well as community-acquired infections and providing support for development of new antibiotics and diagnostics. Positive effect of the National Action Plan can be seen in the statistics. In a 5-year period, there was a 18% decrease in the overall number of US deaths from antibiotic-resistant infections and a 30% decrease in the number of US deaths from resistant infections in hospitals.

10.2.2 National Action Plans of the European Commission

The European Commission is the executive branch of the European Union with 27 members. Each member state of the Commission has adopted an independent National Action Plan in alignment with the goals of the WHO Global Action Plan. They can all be accessed at the European Commission website [6]. The European Union countries have always been ahead of other countries in the prevention of antimicrobial resistance and took action long before WHO adopted the Global Action Plan. For example, as discussed in Sect. 2.10.2, as far back as the 1970s, the European Common Market had decided to phase out the use of tetracycline and later penicillin as growth promoters. Later in 1997, the European Union banned the subtherapeutic use of avoparcin, and in 1999 it banned the use of bacitracin, spiramycin, tylosin, and virginiamycin. In 2006, it banned the subtherapeutic use of all antibiotics for growth promotion (AGP) in animals. Following the ban by the European Union, it has been demonstrated that there has been a decline in the cases of antibiotic resistance. Compared to the European Union, the USA has been extremely slow in addressing the problem. It was not until 2017 that the USA banned the use of not all, but only the medically important antibiotics for growth promotion in animals.

10.2.3   National Action Plan of China

The prevalence of multidrug-resistant bacteria is high in China. Interestingly, although colistin is now not available in China, colistin-resistant bacteria have been detected in clinical samples and animals. As a populous, developing country with a large antimicrobial resistance burden, China has developed a National Action Plan to actively participate in global efforts against antimicrobial resistance [7]. The plan calls for coordination of various entities at a national level and embraces the one-health concept, integrating health, agriculture, and environmental protection departments. Measures to control environmental hazards (such as soil contamination) of antibacterial agents are also described. Need for basic research to control antimicrobial resistance is also emphasized. The Chinese government has also implemented a Work Program for the Reduction of the Use of Antimicrobials in Animals (2018–2021). Since China has a huge animal production industry, reducing or eliminating subtherapeutic use of antibiotics for growth promotion in animals is expected to be very challenging. However, as a result of this program, between 2014 and 2018, China has reduced subtherapeutic use of antibiotics in animals by 57% [8]. In China, the use of colistin for growth promotion in animals was banned in 2017, and finally in 2020 use of all antibiotics for growth promotion has been banned.

10.2.4   National Action Plan of India

India records the highest antibiotic consumption in the world [9]. Several factors contribute towards this high consumption rate. (1) Antibiotics are easily available without a prescription. Even if there is a prescription, there is no database that keeps track of prescriptions. (2) Many people usually self-medicate based on their prior experience and misplaced self-confidence about their medical knowledge. People also reconfirm their self-medication knowledge by getting free medical advice from the storekeeper selling the medications. Since none of these deciding parties are medical professionals, oftentimes people take antibacterial antibiotics for viral infections. (3) Manufacturers actively promote the sale of their antibiotics. Another important factor that contributes to antimicrobial resistance development is the fact that people often do not complete the required course of the antibiotics. This is again because of their lack of knowledge about the mechanism of antibiotic resistance development. People want to save money by keeping the remaining antibiotics for future infections. Because of the uncontrolled increase in antibiotic consumption, there is a high incidence of infection by antibiotic-resistant pathogens. This includes multi-drug resistant tuberculosis and "superbugs" such as the ESKAPE group of organisms (Sect. 2.8).

India's National Action Plan was released in 2017 and focused equally on human, environment, and food-animal sectors to achieve its goal [10]. Six strategic priorities outlined in the National Action Plan for antimicrobial resistance in India include improving awareness and understanding of AMR, improving surveillance, reduce infection, optimizing the use of antimicrobial agents, increasing investment in AMR activities, research, and innovations, and finally, increasing collaborations on AMR at national and international levels. One drawback that India faces is that, as in all other low and middle-income countries, lack of financial resources is the major challenge in the implementation of the plan. However, one advantage that India and other similar countries have is the concept of family physician. Unlike in the West, where a doctor-physician relationship is more like a business transaction, in India a family doctor has a more personal relationship with the whole family of the patient. This places them in the unique position of being able to educate the patient on the proper treatment strategy and emphasize that antibacterial antibiotics do not cure viral infections. They can also employ a "delayed prescribing" strategy in which the patient is asked to collect the prescription if symptoms worsen, thereby decreasing the use of antibiotics when they are not needed [10].

The use of antibiotics for growth promotion in animals is high in India, and rise of antimicrobial resistance due to this reason has been well documented. However, not much action has been undertaken till now to lower the subtherapeutic use of antibiotics [11].

10.3 Antibiotic Stewardship Programs

10.3.1 Definition

Antibiotic stewardship has been defined in many ways, mainly differing in which responsible party is being addressed. One commonly used definition is that antibiotic stewardship is the effort to measure and improve how antibiotics are prescribed by clinicians and used by patients. The main focus is on the health care providers, including physicians, nurses, pharmacists, and dentists. However, besides the doctors and the patients, there are many other players affecting the effectiveness of antibiotics, including national regulatory hospital administrators, drug manufacturers, researchers, agriculture and animal farming industry, and last but not the least, the politicians and lawmakers. To include all these factors, Antimicrobial Stewardship is best defined as "A coherent set of actions which promote using antimicrobials responsibly" [12].

10.3.2 Significance of Antibiotic Stewardship

PubMed citations of antimicrobial or antibiotic stewardship have increased sixfold in the last 10 years, thus pointing to the significance of the subject. Antibiotics are very effective in treating infections. However, not many new antibiotics are being discovered, and all the existing ones are gradually becoming ineffective due to resistance development resulting from misuse and overuse. Antibiotics are a precious resource. Judicious use of antibiotics is the only way to slowdown the development of resistance against them. There are several agencies that can make significant contribution towards this goal. These include regulatory agencies, healthcare providers, patients, farmers, pharmaceutical companies, scientists, and governments.

10.3.3   Regulatory Agencies

In 2014, the CDC recommended that all US hospitals should have an antibiotic stewardship program and released the Core Elements of Hospital Antibiotic Stewardship Programs to help hospitals achieve this goal. Starting from 2017, all hospitals were required to have an Antimicrobial Stewardship team consisting of infection preventionists, pharmacists, and a practitioner to write protocols and develop projects focused on the appropriate use of antibiotics. Effective from January 1, 2020, the antimicrobial stewardship requirements were expanded to also include outpatient health care organizations. Thus, the stewardship programs are applicable for hospitals, outpatient clinics, and long-term care facilities. The CDC has proposed some Core Elements of Antibiotic Stewardship to guide providers and facilities to improve antibiotic use and thus achieve the goals of the stewardship program.

10.3.4   Healthcare Providers

At local levels, many hospitals have developed Antibiotic Stewardship Programs to address the AMR problems. The program works by keeping physicians informed with regular updates on AMR and by giving guidance in improved antibiotic use. The stewardship program is not just for physicians but for all health-care workers and administrators as well as patients and their families. Antibiotic stewardship efforts should focus more on outpatient facilities to have a greater impact on reducing unnecessary antibiotic use. There is also a large variation between states in the number of prescriptions written. The reasons for such variation need to be analyzed to determine the frequency of inappropriate prescriptions and the ways to resolve the problem. Doctors write antibiotic prescriptions based on their experience and not by actually identifying an infection. In the best of worlds, ideally, all antibiotic prescriptions should be kept on hold after they are written till the pathogen is cultured, and in-vitro antibiotic susceptibility tests are conducted. However, that is not very practical to do mainly because of the time and expense involved. However, for the benefit of preventing AMR, we all should be working towards this goal. For example, the turnaround time for the laboratory tests can be brought down to less than 1 day if the samples for infectious diseases are given a higher priority, which is reasonable for the prevention of AMR and can be made possible by government mandate and subsidy. Dentists also play a big role in antibiotic stewardship and are discussed separately (Sect. 10.4). In less developed countries, doctors often prescribe antibiotics for viral infections in order to prevent secondary bacterial infection. Since this is a preventive measure, in most cases, these prescriptions are probably unnecessary and contribute to the antibiotic resistance pool. Healthcare providers often feel pressured by patients to prescribe antibiotics. However, that should never be an excuse to write prescriptions. Doctors often give in to patients' demands for a variety of reasons: the patient may be difficult to teach, many antibiotics don't do immediate direct harm to the patient (not considering AMR), the doctor may be uncertain about whether the patient has bacterial infection or not.

10.3.5 Farmers Use Antibiotics for Growth Promotion of Animals

More than 70% of all antibiotics produced in the USA is used in animals. It was observed that low doses of antibiotics in animal feed increases the growth of the animals, thus resulting in more profit for the farmers. This phenomenon is known as subtherapeutic use of antibiotics or antimicrobial growth promotion (AGP) and is discussed in detail in Sect. 2.10.2. Such use of antibiotics is unnecessary and contributes greatly to development of antibiotic resistance. Since subtherapeutic use constitutes an overwhelming majority of all antibiotic use in the country, it renders the National Action Plan almost useless. Since several antibiotics can have the same mechanism of action, bacteria that develop resistance to one specific antibiotic may also be resistant to the whole class of antibiotics that have similar mechanism of action. In order to prevent antimicrobial resistance development, it is imperative that the use of antibiotics for growth promotion be stopped. In fact many countries have already banned the use of all antibiotics for growth promotion. The European Union implemented the ban 17 years ago in 2006. A slight progress in this field was made in 2017 when the USA banned the use of not all but only the medically important antibiotics for growth promotion in animals. Over the years, there has been a slow but steady progress in decreasing the subtherapeutic use of antibiotics in chicken. The share of chicken grown in the USA without antibiotic for growth promotion increased from 3% in 2014 to 51% in 2018.

10.3.6 Patients Have a Great Responsibility

The first responsibility of the patient is to not pressurize the doctor to prescribe antibiotics. In much of the developing and under-developed world, antibiotics are available without a prescription and patients "confidently" self-medicate without even consulting a physician. Although such a practice needs to be stopped, it will not be easy to achieve that goal. This is because most people cannot financially afford to see a doctor, or there may be a shortage of doctors available. In countries where prescriptions are strictly required for antibiotics, people can get around the problem by taking antibiotics meant for pets, which are available without prescription. Scientifically, there is very little or no difference between antibiotics for humans and those for pets. Two common practices that do more harm than good are to not complete the course of antibiotics or to use less than the recommended dose every time. In both cases, the patient will feel good because the majority of the bacteria will actually be killed after just a few doses. The patients should be taught that there is an enormous difference between 99% bacteria killed and 99.99% bacteria killed and that 99% is not good enough, although it may temporarily make the patient feel good. Some patients have the wrong idea that by taking less than the prescribed dose, they can save the leftover medicine for the next time they have infection, which again, they will self-diagnose and self-medicate. This will result in double the problem because the same mistake will be made twice. Some patients also use the leftover antibiotics to treat other family members or friends who may get an infection later, again, putting them in danger. It is important that patients are taught the dangers of such practice. Patients should also be told not to get medical advice from the social media. Information found on the internet may appear to be genuine but most of the time they are not written by experts.

10.3.7   Uncontrolled and Unlicensed Drug Formulations

Another situation compounding the problem of AMR is the availability of adulterated, expired or poor-quality drugs. Presence of lower-than-needed concentration of the active ingredient in any drug formulation is a financial loss for the patient. However, that is only a minor concern in case of antibiotics. The sub-optimal concentration of the active antibiotic will result in emergence of resistant bacteria. So, one of the goals of stewardship program should be to stop sale of unlicensed drug formulations. This problem exists more in countries that do not have, or cannot afford the required resources for enforcement. A similar problem has been created, especially in richer countries, by the sudden explosive increase in the use of nutritional supplements, which are not regulated in the same way as modern medicine is. It is now accepted that plants can also be a source of antibiotics. So uncontrolled use of dietary supplements that can be a source of antibiotics will also contribute to the development of antimicrobial resistance. Thus, it may be a good idea to regulate some nutritional supplements, especially those in which the active ingredient has been significantly enriched. A similar concern exists for antibiotics that actually have been approved by regulatory agencies such as the FDA but are allowed to be sold without a prescription as active ingredient in ointments, mouthwashes etc. One such antibiotic is triclosan (Sect. 7.2.1), that is widely used in surgical scrubs and is allowed to be added in many health care products such as shampoos, deodorants, toothpastes, and mouthwashes. Triclosan functions by targeting fatty acid synthesis pathway. Widespread use of the antibiotic is expected to contribute to the development of resistance (Sect. 7.3.1).

10.3.8 Government Can Bring About the Fastest Changes

Irrespective of what form of government a country has, it can have the greatest and fastest impact on Antibiotic Stewardship or implementation of the Global or National Action Plan. This is because the government can make laws and provide funding for the various aspects of the plan.

10.3.8.1 Cost Factor

Very few new antibiotics have been discovered in the last few decades. One big reason for this is the extremely high cost of developing a new antibiotic. The process of bringing a new antibiotic to market is not cost-effective for the pharmaceutical industry. At the same time, the volume of sale of antibiotics is low because infection is not a chronic disease. Moreover, the cost of antibiotics is intentionally kept low because of intervention by the government in most countries. The patent laws are big impediment to discovering new antibiotics because when the patent expires, there will be even less income from the antibiotic. It is understandable that all patents should have an expiry date in order to enhance competition between companies. While this is true for all drugs, for antibiotics, there is one big difference. Antibiotics are the only drugs whose effectiveness decreases with time due to resistance development. For all other drugs, the effectiveness remains the same, although there may be better drugs that have been developed. So there is an apparent contradiction in the business model. To make money, companies have to sell more antibiotics, however, to avoid resistance development, the stewardship program advises to sell less of it. The best use of antibiotic is to not use it at all or use it rarely so that it remains effective longer and can be kept as the antibiotic of last resort. Since governments have already intervened in the business process by making these laws, they should intervene in the right way to make antibiotic discovery more sustainable. Many pharmaceutical companies have already given up on expanding research and development to discover new antibiotics.

10.3.8.2 Research Funding

The first step for developing new antibiotics is to provide funding for research in the field. There can be many possible sources of new antibiotics. Some of these are discussed in Chap. 9. However, detecting, identifying and purifying and/or synthesizing the active component will require research funding. Research is not a profitable business. For every experiment that leads to a viable drug candidate, there are hundreds that do not. Without government funding, not much progress can be made towards discovery of new antibiotics. The government funding agencies in the USA do spend a lot of money for research. To make the best use of that money, the method for determining winners of research grants needs to be modified. Currently only about 10% of grant applications are funded while the remaining 90% are not. This is like telling students in a class that no matter how well they perform in the exams, only 10% of students will get the highest grade, "A" and all the remaining students will get the failing grade, "F," with no intermediate grades possible.

10.3.9   Researchers Make the Greatest Contribution Towards Antibiotic Discovery

Scientific discoveries do not happen by chance or without effort (Sect. 1.5). A lot of time, effort, and money are required to discover each new antibiotic. However, the discovery process can be made more efficient by properly designing experiments, using better experimental methods for detecting antibiotic activity, searching for new sources of antibiotics and developing better methods for obtaining large amounts of the starting material, including better culturing methods for microorganisms that are otherwise not culturable. For example, it is known that 99% of bacterial species on this planet have not yet been cultured because of the lack of proper methods. A new method for culturing many of these organisms has been developed, and in the process, several new potential antibiotics have been discovered [13].

Plants can be a vast source of new antibiotics (Sect. 9.3). Recently there have been numerous papers published on antibiotic activity in extracts from plants. However, in most cases either the activity is weak or the active ingredient may be toxic for the human cells. To make better use of this vast resource for antibiotic discovery, better experimental methods are needed. If the same old scientific methods are repeatedly used to discover new antibiotics, it is not surprising that the same or similar antibiotics with minor modifications will be discovered and rediscovered. For example, since all current commercial antibiotics were discovered using growing bacteria for their detection, most of these antibiotics are unable to kill non-growing or slow-growing bacteria (Sect. 9.3.3). Modifying the detection method has led to the discovery of the first antibiotic that preferentially kills non-growing bacteria [14] and another antibiotic that kills slow-growing bacteria [15].

One possible solution to the problem of antimicrobial resistance is to repurpose old drugs as new antibiotics or vice versa (Sect. 9.1). However, sometimes this can lead to the wrong conclusion. One example is the use of antibiotics to cure lower back ache. Some researchers have reported that there is link between back pain and bacterial infection. They found that about half of patients with herniated discs have bacterial infection. In another study with 162 patients who had a slipped disc, half were treated with a 100-day course of antibiotic while the other half were treated with a placebo. The back pain was decreased in those who took the antibiotic. Repurposing of drugs is not a new concept (Sect. 9.1). However, this should always be thoroughly investigated before promoting or prescribing the drug. The studies that have been performed do not prove that improvement was due to antibiotics. Later, a double blind, randomized, placebo controlled, multicenter clinical trial involving 180 patients was done to determine if antibiotics can cure lower back ache [16]. They concluded that 3 months of treatment with amoxicillin did not provide a clinically important benefit compared with placebo. Therefore, the researchers do not support the use of antibiotic treatment for chronic low back pain and Modic changes.

10.3.10   Print and Online News Media Have a Role to Play

Scientific journals peer review all articles they publish. They should bear some responsibility to make sure that the articles are publishable. Articles that challenge well-established observations or scientific laws should be scrutinized more thoroughly by the process of peer review. One example is the concept of antibiotic resistance development. As explained earlier (Sects. 2.5 and 2.6), it is very important to complete the course of an antibiotic as prescribed by the doctor. It is a well-established fact that bacteria develop resistance to antibiotics when grown in the presence of sub-optimal concentration of the antibiotic or for a shorter duration than needed. In my opinion, it is irresponsible to publish any paper suggesting otherwise without providing much evidence. In a recent paper, it was suggested that we should stop advocating the "finish the course" message (citation not provided intentionally to prevent glorification of such publications). If it is left up to the patient to subjectively decide when to stop taking antibiotics, that will be a big blow to the stewardship program. There are numerous problems encountered in tackling the antibiotic resistance crisis. Official publications providing misinformation will only add to the problem.

The current Covid-19 pandemic has shown that some politicians as well as some news media blame the scientists because they did not have answers to all questions about the new virus or because their theories changed with time. Research is almost always done with good intentions. Researchers are one of the lowest paid professionals considering the amount of effort that is put in research and the educational background that is needed to effectively do the job and yet they do the job because of their passion for science. All scientific theories continuously evolve to accommodate new experimental observations. Although researchers do their best to design their experiments, they may be later proven to be flawed. That is the main reason why scientific theories evolve to become better. It is to be noted that this situation is different from non-scientists including politicians and other people with vested interests promoting theories or products that have not been scientifically proven to be true or effective. Politicians and news media should refrain from labeling scientists as villains simply because their theories change with time. It should be recognized that it is the very nature of science that theories will be modified as new data are obtained.

In these modern times, the social media have become an important part of everyone's lives. Information and misinformation published in social media do spread extremely rapidly and can bring about a sea-change in social behavior. So it is important that governments regulate social media to make them accountable for their contents. They should not be allowed to hide behind the argument that they are only platforms and not news media. After all, there is profit involved in social media, and any profitable business should come under some sort of government regulations.

10.4 Antibiotic Use in Dentistry

About 10% of all antibiotic prescriptions are written by dentists usually to manage pain and to avoid postsurgical complications. Dentists are also the major prescriber of clindamycin, which functions by inhibiting protein synthesis (Sect. 6.2). A default antibiotic prescription during dental procedures to prevent bacteremia and subsequent spread of bacteria to other organs was an old practice and is unfortunately still followed by some dentists. Another point of concern in dentistry is that antibiotics are always prescribed without first detecting, cultivating and identifying the pathogenic bacteria. The oral microbiome is very complex, with more than 200 bacterial species, which also interact with each other. With so many bacteria in close proximity, it may be difficult to diagnose bacterial infection within a tooth by doing laboratory cultures. So, antibiotics are used in dentistry on an empirical basis, and the antibiotics of first choice are usually the broad-spectrum ones, such as β-lactams and macrolides.

Most antibiotic stewardship programs usually focus on physician practices and overlook dental practices. However, since a large number of antibiotics are prescribed by dentists, this is an area that needs to be addressed. In order to reduce antibiotic resistance development, protocols for treatment with antibiotics need to be established.

10.4.1 Antibiotics for Prophylaxis

From 2011 to 2015, antibiotic prophylaxis prescribed for 168,420 dental visits for 91,438 patients in the USA was analyzed. It was concluded that 81% of antibiotics prescribed as prophylaxis before dental visits were unnecessary [17]. Clindamycin was more likely to be unnecessary relative to amoxicillin. Default antibiotic prescriptions are written for prophylaxis after dental procedures as a common practice but without any evidence of infection. Antibiotic prophylaxis is for preventing the risk of infection that is either local or has disseminated through the bloodstream to distant organs. For short-duration antibiotic prophylaxis, the dose is determined such that the plasmatic antibiotic concentration reaches three- to fourfold higher than the MIC. Amoxicillin is a preferred antibiotic for this purpose because of its moderate spectrum of activity and good bioavailability. Another antibiotic with similar properties usually prescribed for those who are allergic to β-lactams, such as amoxicillin, is clindamycin, which is a lincosamide with 50S subunit of the ribosome as target (Chap. 6). It is a broad-spectrum antibiotic that is effective against both aerobes and anaerobes and has high bioavailability in plasma, most tissues, and bones. This makes clindamycin a popular antibiotic with dentists in spite of its gastrointestinal side effects. Another antibiotic of choice for periodontal diseases is metronidazole, which is active against anaerobic pathogens (Chap. 5). Bacteria can infect immediately after dental surgery but are quickly cleared from the bloodstream by the host defense system in healthy patients within a few minutes to a few hours [18]. So, in most cases, an antibiotic is probably not needed. In order to reduce the development of AMR, it is necessary to establish protocols for prescribing antibiotic prophylaxis only for those patients who are more susceptible to such infections based on their health and the type of oral surgery [19].

Another common treatment in dentistry is the use of chlorhexidine, which functions as an antiseptic (Sect. 7.2.1). It is used as an ingredient in either a gel or a mouthwash. It is used as a gel for prophylaxis for dry sockets following tooth extraction. It can be more conveniently used as a mouthwash if there are multiple surgical sites [19]. Chlorhexidine in the various products should be used with caution because excessive use will lead to the development of resistance (Sect. 7.2.1). Resistance to chlorhexidine is usually due to the acquisition of a plasmid containing genes for efflux pumps [20, 21]. Resistance to chlorhexidine can also provide resistance to other drugs because efflux pumps can recognize multiple substrates. Moreover, plasmids usually contain resistance genes for several antibiotics resulting in simultaneous resistance development against multiple antibiotics. So it is recommended that chlorhexidine be used only for short periods when needed and not as routine mouthwashes.

10.4.2 Antibiotics for Toothache

Patients complaining of toothache are often prescribed antibiotics by physicians and dentists. However, according to the latest guidelines from The American Dental Association (ADA), in most cases, antibiotics are not recommended for toothaches. According to ADA, toothaches in healthy adults can best be treated with dental treatment and over-the-counter pain medications and not by antibiotics, which can actually do more harm than good. Of course, in some cases, antibiotics may be needed for example, when the patient has symptoms of fever, swollen lymph nodes, etc. One cause of toothache can be when a tooth decay progresses into the tooth. There will be an inflammatory response of the pulp to the bacteria. The pressure in the pulp will increase causing pain. Eventually the nerve and the pulp will die. The recommended treatment is removal of the nerve, pulp, and the bacteria. The inside of the tooth is then cleaned and sealed. Antibiotics should be prescribed only if the infection has spread into the surrounding tissues. However, some dentists routinely prescribe antibiotics irrespective of whether the infection has spread or not.

National dental and medical organizations from many countries have come together to support the WHO to increase awareness of global antimicrobial resistance. Guidelines for antimicrobial prescribing in dentistry are available on the respective websites of most countries. Other resources, including the Dental Antimicrobial Stewardship Toolkit, are also available from those websites. For example, the poster, "Antibiotics do not cure toothache" is available from the United Kingdom's (UK) Dental Antimicrobial Stewardship (AMS) toolkit. Obviously, such posters and leaflets are aimed at patients because the dentists and physicians are supposed to know this already. Problem arises when patients without this awareness tend to blame the doctor or dentist. Considering the intense pain that they feel and considering that they are paying a fee to see the doctor, there is a tendency to label the doctor as "not good" if they do not prescribe an antibiotic.

10.4.3   Antibiotics Following Tooth Extraction

After extractive oral surgery, antibiotics are often prescribed to prevent possible postsurgical infections. The most common complication after tooth extraction is the dry socket (alveolar osteitis), which is not an infection and rarely leads to an infection. It is caused by premature dislodging of blood clot, causing pain but rarely causes an infection. So currently, antibiotics are not recommended after tooth extraction or after implant insertion.

The Following key points of the Antibiotic Stewardship program were adopted by The American Dental Association.

1. Use of antibiotics prudently to minimize development of drug-resistant bacteria.
2. Reserve antibiotic prophylaxis for patients at high risk of post-treatment complications.
3. Prescribe antibiotics only when needed, select the right antibiotic at the right dose and for the right duration.

10.5   Vaccines Against Bacteria

One effective way to decrease antibiotic use is to develop vaccines against bacteria, and it is one of the priorities in the global as well as various national action plans to support research on development of antibacterial vaccines. WHO has recognized antimicrobial resistance as one of the top 10 threats to human health. Since detailed discussion of vaccines is not within the scope of this book on antibiotics, the topic is only briefly mentioned here. The subject has recently been reviewed in detail [22]. Some examples of currently available antibacterial vaccines are those against bacteria causing pertussis, tetanus, diphtheria, meningitis, pneumonia, cholera, typhoid, plaque, anthrax, tuberculosis, tularemia, typhus, and Q fever. Bacterial vaccines can be made of killed or live attenuated whole bacteria, bacterial capsular polysaccharides, toxoids, or purified proteins isolated from bacteria. Antibacterial vaccines will have an indirect effect on antimicrobial resistance levels. By decreasing the number of infections by the specific bacteria, they will contribute to reducing the consumption of antibiotics and thus lowering the AMR levels. While new vaccines will provide benefits to all age groups, it will be the senior population that will be most benefited because after, young children, they are the second-highest consumers of antimicrobials. One big advantage of vaccines is that they usually target an external structural part of the bacteria, and so the same vaccine will still work even if the bacteria acquire some new resistance genes. So vaccines are a valid approach to combat antimicrobial resistance.

A word of caution needs to be mentioned regarding promotion of bacterial vaccines. Since a vaccine is developed against a bacterial clone that is successfully spreading in the society, it is very effective in stopping that spread. However, at the same time, it will allow the expansion of clones that were previously minor and were not covered by the vaccine [23]. Thus any antibiotic resistance genes in these clones will now spread rapidly. Since resistance genes are often horizontally transferred, it can acquire the resistance genes present in the clones targeted by the vaccines. The resistance genes can be acquired either by conjugation or natural transformation (Sect. 2.6). One possible solution to the problem can be to develop vaccines against a resistance mechanism, for example, a vaccine against the enzyme that confers antibiotic resistance.

References

1. WHO (2015). https://www.who.int/antimicrobial-resistance/global-action-plan/en/
2. CDC (2019). https://www.cdc.gov/drugresistance/biggest-threats.html
3. CDC (2016). https://www.cdc.gov/media/releases/2016/p0503-unnecessary-prescriptions.html
4. Pewtrust (2020). https://www.pewtrusts.org/en/research-and-analysis/fact-sheets/2020/10/out patient-antibiotic-prescribing-varied-across-the-united-states-in-2018
5. HHS (2020). https://www.hhs.gov/sites/default/files/carb-national-action-plan-2020-2025.pdf
6. EC (2021). https://ec.europa.eu/food/plants/pesticides/sustainable-use-pesticides/national-action-plans_en
7. Xiao Y, Li L (2016) China's national plan to combat antimicrobial resistance. Lancet 16:1216–1218. https://doi.org/10.1016/S1473-3099(16)30388-7
8. Ma F, Xu S, Tang Z, Li Z, Zhang L (2021) Use of antimicrobials in food animals and impact of transmission of antimicrobial resistance on humans. Biosafety Health 3:32–38. https://doi.org/10.1016/j.bsheal.2020.09.004
9. Boeckel TPV, Gandra S, Ashok A, Caudron Q, Grenfell BT, Levin SA, Laxminarayan R (2014) Global antibiotic consumption 2000 to 2010: an analysis of national pharmaceutical sales data. Lancet Infect Dis 14:742–750. https://doi.org/10.1016/S1473-3099(14)70780-7

10. Ranjalkar J, Chandy SJ (2019) India's National Action Plan for antimicrobial resistance—an overview of the context, status, and way ahead. J Family Med Prim Care 8:1828–1834. https://doi.org/10.4103/jfmpc.jfmpc_275_19

11. Walia K, Sharma M, Vijay S, Shome BR (2019) Understanding policy dilemmas around antibiotic use in food animals & offering potential solutions. Indian J Med Res 149:107–118. https://doi.org/10.4103/ijmr.IJMR_2_18

12. Dyar OJ, Huttner B, Schouten J, Pulcini C (2017) What is antimicrobial stewardship? Clin Microbiol Infect 23:793e798. https://doi.org/10.1016/j.cmi.2017.08.026

13. Ling LL, Schneider T, Peoples AJ, Spoering AL, Engels I, Conlon BP, Mueller A, Till F, Schäberle TF, Hughes DE, Epstein S, Jones M, Lazarides L, Steadman VA, Cohen DR, Felix CR, Fetterman KA, Millett WP, Nitti AG, Zullo AM, Chen C, Lewis K (2015) A new antibiotic kills pathogens without detectable resistance. Nature. https://doi.org/10.1038/nature14098

14. Bhattacharjee MK, Alenezi T (2020) Antibiotic in myrrh from *Commiphora molmol* preferentially kills non-growing bacteria. Future Sci OA 6(4):FSO458. https://doi.org/10.2144/fsoa-2019-0121

15. Bhattacharjee MK, Bommareddy PK, DePass AL (2021) A water-soluble antibiotic in rhubarb stalk shows an unusual pattern of multiple zones of inhibition and preferentially kills slow-growing bacteria. Antibiotics 10:951. https://doi.org/10.3390/antibiotics10080951

16. BrÅten LCH, Rolfsen MP, Espeland A, Wigemyr M, Aßmus J, Froholdt A, Haugen AJ, Marchand GH, Kristoffersen PM, Lutro O, Randen S, Wilhelmsen M, Winsvold BS, Kadar TI, Holmgard TI, Vigeland MD, Vetti N, Nygaard ØP, Lie BA, Hellum C, Anke A, Grotle M, Schistad EI, Skouen JS, Grøvle L, Brox JI, Zwart J-A, Storheim K (2019) Efficacy of antibiotic treatment in patients with chronic low back pain and Modic changes (the AIM study): double blind, randomised, placebo controlled, multicentre trial. BMJ 367:l5654. https://doi.org/10.1136/bmj.l5654

17. Suda KJ, Calip GS, Zhou J, Rowan S, Gross AE, Hershow RC, Perez RI, McGregor JC, Evans CT (2019) Assessment of the appropriateness of antibiotic prescriptions for infection prophylaxis before dental procedures, 2011 to 2015. JAMA Netw Open 2(5):e193909. https://doi.org/10.1001/jamanetworkopen.2019.3909

18. Lockhart PB, Brennan MT, Sasser HC, Fox PC, Paster BJ, Bahrani-Mougeot FK (2008) Bacteremia associated with tooth-brushing and dental extraction. Circulation 117:3118–3125

19. Buonavoglia A, Leone P, Solimando AG, Fasano R, Malerba E, Prete M, Corrente M, Prati C, Vacca A, Racanelli V (2021) Antibiotics or no antibiotics, that is the question: an update on efficient and effective use of antibiotics in dental practice. Antibiotics 10:550. https://doi.org/10.3390/antibiotics10050550

20. Cieplik F, Jakubovics NS, Buchalla W, Maisch T, Hellwig E, Al-Ahmad A (2019) Resistance toward chlorhexidine in oral bacteria—is there cause for concern? Front Microbiol 10:587. https://doi.org/10.3389/fmicb.2019.00587

21. Cervinkova D, Babak V, Marosevic D, Kubikova I, Jaglic Z (2013) The role of the qacA gene in mediating resistance to quaternary ammonium compounds. Microb Drug Resist 19:160–167

22. Poolman JT (2020) Expanding the role of bacterial vaccines into life-course vaccination strategies and prevention of antimicrobial-resistant infections. npj Vaccines 5:84. https://doi.org/10.1038/s41541-020-00232-0

23. Henriques-Normark B, Normark S (2014) Bacterial vaccines and antibiotic resistance. Ups J Med Sci 119:205–208. https://doi.org/10.3109/03009734.2014.903324

Index

Printed in the United States
by Baker & Taylor Publisher Services